Third Generation Internet Revealed

Reinventing Computer Networks with IPv6

Lawrence E. Hughes

Foreword by Latif Ladid, President of the Global IPv6 Forum

Third Generation Internet Revealed: Reinventing Computer Networks with IPv6

Lawrence E. Hughes
Frisco, TX, USA

ISBN-13 (pbk): 978-1-4842-8602-9   ISBN-13 (electronic): 978-1-4842-8603-6
https://doi.org/10.1007/978-1-4842-8603-6

Managing Director, Apress Media LLC: Welmoed Spahr
Acquisitions Editor: Joan Murray
Development Editor: Laura Berendson
Coordinating Editor: Jill Balzano

Cover designed by eStudioCalamar

Cover image designed by Freepik (www.freepik.com)

Distributed to the book trade worldwide by Springer Science+Business Media New York, 1 New York Plaza, Suite 4600, New York, NY 10004-1562, USA. Phone 1-800-SPRINGER, fax (201) 348-4505, e-mail orders-ny@ springer-sbm.com, or visit www.springeronline.com. Apress Media, LLC is a California LLC and the sole member (owner) is Springer Science + Business Media Finance Inc (SSBM Finance Inc). SSBM Finance Inc is a **Delaware** corporation.

For information on translations, please e-mail booktranslations@springernature.com; for reprint, paperback, or audio rights, please e-mail bookpermissions@springernature.com.

Apress titles may be purchased in bulk for academic, corporate, or promotional use. eBook versions and licenses are also available for most titles. For more information, reference our Print and eBook Bulk Sales web page at http://www.apress.com/bulk-sales.

Any source code or other supplementary material referenced by the author in this book is available to readers on GitHub (https://github.com/Apress). For more detailed information, please visit http://www.apress.com/source-code.

Printed on acid-free paper

This book is dedicated to my beautiful wife of 28 years, Remy Hughes. She has been through thick and thin with me, and we have raised three children. She has been very supportive of me in my career, including spending tons of time on this book (as well as my book on AD Certificate Services, also from Apress). I have had to travel a great deal teaching PKI and IPv6, in addition to speaking at international IPv6 summits. We are both very proud of our grown kids, both of whom passed the IPv6 Certified Network Engineer course and are now working in IT in Texas. Here's hoping for many more years together as the world transitions to the Third Internet!

Table of Contents

About the Author

Lawrence Hughes received a BS in Mathematics from FSU in 1973 (there was no computer science major at the time). He worked all four years in college at the FSU Computing Center, working on a CDC 6400 mainframe. In 1975, he built one of the first home computers (a MITS Altair 8800) and later bought up CP/80 on it. A few years later, the Altair was upgraded to an IMSAI 8080, and then he replaced the Z80 CPU board with an 8086 board.

He founded Mycroft Labs in 1982 and created the MITE communication package for CP/M-80 and then later for MSDOS and Mac. In 1987 he cofounded Mika LP where he made a secure version of MITE called Whisper, for the US IRS.

In 1998 he took a position with SecureIT (a white-hat hacking and firewall company in Atlanta). This company was then bought by VeriSign, where Lawrence created their training in cryptography and PKI (Public Key Infrastructure) and delivered it internationally.

In 2000, Lawrence cofounded CipherTrust with Jay Chaudhry. This grew from six people literally in his basement to 250 people in 2006, when it was sold to Secure Computing for 273M.

In 2004 Lawrence founded InfoWeapons in Cebu City, Philippines, where he created the first secure DNS appliance that fully supported IPv6 (even in IPv6-only networks). This easily passed US DoD JITC certification.

In 2014, Lawrence cofounded Sixscape Communications Pte. Ltd. in Singapore. It specializes in digital certificate automation, cryptographic authentication, and secure messaging. All products are fully IPv6 compliant. Sixscape is still going well; see https://sixscape.com.

In 2019, Lawrence was inducted into the IPv6 Forum's "IPv6 Hall of Fame."

He now resides in Frisco, Texas, working in the PKI engineering group at the Bank of America.

About the Technical Reviewer

Fabio Claudio Ferracchiati is a senior consultant and a senior analyst/developer using Microsoft technologies. He works for Bluarancio (`www.bluarancio.com`). He is a Microsoft Certified Solution Developer for .NET, a Microsoft Certified Application Developer for .NET, a Microsoft Certified Professional, and a prolific author and technical reviewer. Over the past ten years, he's written articles for Italian and international magazines and coauthored more than ten books on a variety of computer topics.

Acknowledgments

I would like to thank Latif Ladid, president of the global IPv6 Forum, for his continued support and encouragement for my many experiences related to IPv6.

Many other IPv6 experts have provided isights and information. I would especially like to thank Eric Vyncke, author of *IPv6 Security* (Cisco Press), for deep insights into the security aspects of IPv6. I have also learned a lot from Silvia Hagen, author of *IPv6 Essentials* (O'Reilly), as well as Marc Blanchet, Ciprian Popuviciu, and many others. I have very much enjoyed meeting you at IPv6 summits and exchanging ideas.

I also would like to offer encouragement to the many engineers whom I have taught the IPv6 Forum's "Certified Network Engineer IPv6" training over the years. Keep it up! You are the cutting edge.

Finally, many thanks to Apress for accepting this book for publication.

Foreword

On February 5, 1999, I was chartered by the original IETF IPv6 Task Force and the IPv6 Deployment Working Group led by Jim Bound to promote the global rollout of IPv6, the successor to the foundation protocol of most networks, including the global Internet. I established the IPv6 Forum (see https://ipv6forum.com) and many national councils and chapters in over 90 countries. I have organized and helped present many international IPv6 summits over the years. This is one of the most important developments in IT in our time.

I have known Larry (the author of this book) for almost two decades. He has presented at many of these summits and conferences, run websites explaining IPv6, certified many telco and network engineers with the IPv6 Forum training, and created some breakthrough products (such as the first DNS (Domain Name System) appliance that supported IPv6) and now is pioneering *end-to-end direct* secure messaging over IPv6, something that is not possible (at least at the global scale) with the IPv4 + NAT Internet (which I call the "InterNAT").

We have identified a need for some 20 million network professionals to understand this new technology and created the curricula for several courses aimed at in-service professionals. Larry created courses based on these curricula and certified many telco, network, and government IT people with them in IPv6.

We also identified a need to certify network equipment and software as being compliant with the new IETF (Internet Engineering Task Force) standards for IPv6. We worked with the TAHI group of the WIDE project in Japan to create testing platforms to verify this compliance. Several testing centers have been deployed around the world and certified many products. Larry deployed and ran one of these in Cebu, Philippines, for many years. He also helped some gentlemen from the Indian government to help set up their own testing center. You can find out more about these programs at https://ipv6ready.org.

In 2007, Larry released a free PDF book called *The Second Internet*, which we published on the IPv6 Forum website. In 2010 he updated that with many changes that happened in the previous three years. Some 500K people worldwide have downloaded these free books and discovered the amazing new world of IPv6.

We recognized Larry's contributions to the adoption of IPv6 by inducting him into the "IPv6 Hall of Fame" in 2019.

Since then, Larry realized that ARPANET was really the "First Internet," making the IPv4 version the "Second Internet" and the IPv6 version the "Third Internet." He understood that this was not a minor change in one Application Layer protocol, such as HTTP 1.0 to HTTP 1.1, but a change that affected *all* application protocols. It is a true *generational* change, as sweeping as the change from ARPANET to the IPv4 Internet in 1983. He has now updated his free PDF book *The Second Internet* to the present day, now that IPv6 is being rapidly deployed globally (many countries now have over 50% of their traffic over IPv6). We are pleased that Apress/Springer have seen fit to publish this version and make it available globally.

Welcome to the Third Internet!

Latif Ladid, President of the global IPv6 Forum

https://ipv6forum.com

Introduction

This book is the end result of many years' effort. It covers a lot of territory, starting with the big picture of the evolution of the Internet, through three generations:

- *First generation*: ARPANET, 1969–1982, 8-bit addresses, host-to-host protocol

- *Second generation*: IPv4 era, 1983–2028?, 32-bit addresses, followed by NAT (Network Address Translation) and RFC (Request for Comments) 1918 private addresses

- *Third generation*: IPv6 era, 2014 (start of major deployment) to perhaps 2100?, 128-bit addresses, many major improvements from what we learned from IPv4

There are hundreds of RFCs (Internet standards) from the IETF (Internet Engineering Task Force) related to IPv6. This book serves as a guide to these RFCs, broken up into major subtopics including links to the actual standards (available free). The most important ones are highlighted. Hopefully this will help you find the ones most relevant to what you are trying to understand, without being overwhelmed by the total collection. I have tried to explain what the most important ones are about, but the real details are in the RFCs themselves.

To understand what is important about IPv6, I have presented a history of how IPv4 address depletion happened and where all those billions of addresses went. This approach also helps those who already know IPv4 to make the leap to IPv6 with "IPv6 is like IPv4 with the following improvements and extensions." I also explain the problems we are now facing with IPv4 since we are still using it 11 years after IANA (Internet Assigned Numbers Authority) ran out of public addresses. NAT was only ever supposed to be a temporary stopgap on the way to pure IPv6. It has made network architecture and software design far more complex than it should be. IPv6 allows a return to the simple monolithic IP address space of the early IPv4 Internet, with no need for NAT. IPv6 was based heavily on the hugely successful IPv4 design. It is *IPv4 on steroids*. IPv4 subnetting is very complicated, requiring many chapters or even an entire book to understand. IPv6 subnetting is simplicity itself: all subnets are /64. Period.

Many people think that the evolution of HTTP from 1.0 to 1.1 was a big deal. That was a minor change in *one* of the thousands of protocols in use on the Internet. A change in the Internet Layer (IPv4 to IPv6) affects *every* Application Layer protocol and will impact business models. Who will need an email account if we can all exchange email messages directly with each other, with no intermediary servers? DNS was okay for the IPv4 model with centralized servers, but we need something much better for billions of highly mobile nodes. The Web will continue to be around for some time, but IPv6 will make the evolution to true edge computing a reality – finally allowing us to take advantage of the incredible power of our computers and phones to run native applications that exchange data, instead of turning them into "dumb terminals" that only run web browsers.

I have tried to convey some of the amazing things that are now possible with the elimination of NAT, vast improvements in multicast, etc.:

- Any node (including phones) can now run servers, or do direct end-to-end connections, with no need for intermediary servers. Most snooping and hacking happen on intermediary nodes, which break TLS. S/MIME helps but is difficult for many users to deploy.

- Imagine being able to send emails and files and do chat and voice directly from one phone to another, with true end-to-end encryption and mutual strong authentication using client digital certificates – easy with IPv6 and PeerTLS.

- Since the first phones supported connections to the Internet, there were not enough public IP addresses for phones to get one, let alone a block (for hot spots). So they have *had* to use NAT, even double NAT (CGN). With IPv6 even phones can have public IP addresses and accept incoming connections (e.g., run servers).

- The United States, with 5% of the world's population, wound up with 41% of the public IPv4 addresses. Other countries were not amused. The IPv6 Internet is the first truly global Internet. Every country can have as many addresses as they can conceivably ever want. There are enough /48 blocks (each sufficient for the largest organization) for every human alive today to get over 5,000 of them. IP address scarcity is over. Even your phone can get 2 to the 64th addresses (an entire /64 block).

- With finally working multicast, we can do mass audio and video streaming on a practical basis – no more zillions of unicast connections (the YouTube model). This is as important as Ted Turner's insight that a satellite is basically a really tall broadcast antenna.

We are at a major inflection point in IT, possibly the largest one since the first computers were sold, or the first networks were created. IPv4 is at end of life. The IETF even had a working group called "Sunset IPv4" that planned how to shut down IPv4 on the global backbone without major disruptions.

This book could only have been written by a "silverback" like me, who lived through the whole time and experienced the changes firsthand. I lived through the mainframe era, the minicomputer era, early timesharing systems, the personal computer era, and the growth and evolution of the Internet.

If you want (or have) a job in computer networking, computer security, or network software design, if you stick with IPv4, you are already obsolete. Adapt or die. The future is here. I hope this book will encourage you not only to move ahead with technology but also to understand how to do that.

CHAPTER 1

Introduction

History of This Work and the Term "Third Internet"

This book is an update and expansion of my 2010 ebook, *The Second Internet*. That ebook has been available on the main website of the global IPv6 Forum (http://ipv6forum.com) since 2010 with some 500,000 downloads worldwide. *This* book is actually still about the new Internet based on IPv6, but since 2010 I have realized that the ARPANET[1] is *not* phase 1 of the First Internet; it IS the First Internet. That makes the Internet based on IPv4 (still what most people are using today) the real *Second Internet*, which makes the new Internet being created now, based on IPv6, the *Third Internet.*

One notable change since 2010 is that IPv6 is no longer just a *Draft Proposed Standard.* The Official IETF Standard for IPv6 has finally been released[2] – RFC 8200[3]: "Internet Protocol, Version 6 (IPv6) Specification," July 2017 (STD 86). This replaces RFC 2460 and several additions to it. So you need to get used to referring to RFC 8200 instead of RFC 2460!

The leap from the First Internet (ARPANET) to the Second Internet (IPv4 based) was clearly a generational change:

- The foundation protocol of the First Internet was usually referred to as NCP,[4] but officially was called the "host-host"[5] protocol. It was defined in a few RFCs. Many new RFCs (starting with RFC 791[6] in 1981) specified the new IPv4 and related protocols.

---

[1] https://en.wikipedia.org/wiki/ARPANET

[2] https://www.internetsociety.org/blog/2017/07/rfc-8200-ipv6-has-been-standardized/

[3] https://tools.ietf.org/html/rfc8200

[4] https://en.wikipedia.org/wiki/Network_Control_Program

[5] https://tools.ietf.org/html/rfc714

[6] https://tools.ietf.org/html/rfc791

© Lawrence E. Hughes 2022
L. E. Hughes, *Third Generation Internet Revealed*, https://doi.org/10.1007/978-1-4842-8603-6_1

1

- An IPv4-only node could not make a connection to, or exchange information with, an NCP-only node (and vice versa), without a complex gateway. The inability to interoperate often happens in generational changes. Translation between NCP and IPv4 was never accomplished.

- NCP had 8-bit addresses (max 2^8 or 256 addresses), while IPv4 has 32-bit addresses (max 2^{32} or 4.3 billion addresses). That is four times as many *bits* in each address as in NCP, but 2^{24} (16.7 million) times as many *addresses*. Each additional bit *doubles* the number of addresses.

- The First Internet lasted from 1969 until 1982 with significant growth and evolution during those years. The Second Internet began operation on January 1, 1983, grew extremely rapidly, and is still running, although it is developing more and more serious issues related to exhaustion of the IPv4 public address space.[7]

- While applications such as email, remote terminal emulation, and file transfer existed in the First Internet, all such apps had to be rewritten (significantly) to work over IPv4.

- Engineers familiar only with NCP had to go back to the books (and training classes) to master the new IPv4. All software and hardware devices that worked with NCP had to be rewritten to work with IPv4. There was no "dual-stack" period since that transition was done via a "flag day" (only NCP before January 1, 1983, only IPv4 from then on). There were many serious problems with doing such an abrupt transition, like worldwide email broke for several months. The IETF wisely decided to do a more gradual transition from IPv4 to IPv6.

- NCP node addresses were represented as a single one- to three-digit decimal number (e.g., "10"), while IPv4 addresses were represented using dotted decimal (e.g., "123.45.67.89"), which at the time looked very alien to NCP users.

---

[7]https://en.wikipedia.org/wiki/IPv4_address_exhaustion

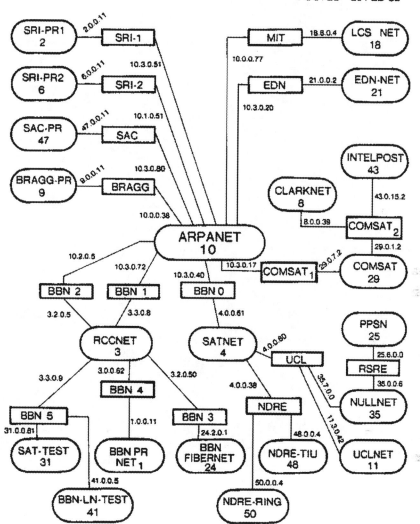

Figure 1-1. *Map of the entire First Internet circa 1982*

3

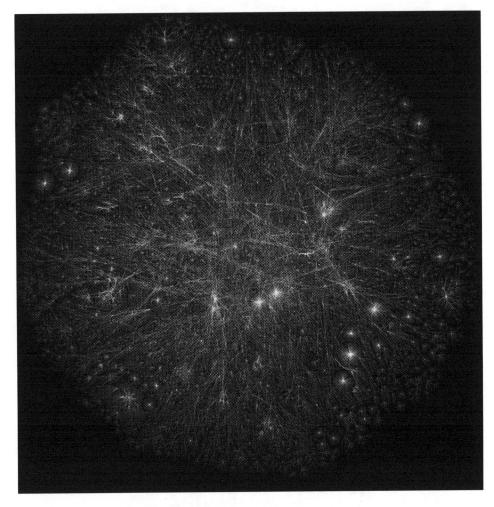

Figure 1-2. *One proposed map of the Second Internet, 2015*

The leap from the Second Internet to the Third Internet is of the same magnitude:

- The foundation protocol of the Second Internet (IPv4) was defined in several RFCs. Many new RFCs (starting with RFC 1881 in 1995) specified the new IPv6 and related protocols.

- An IPv6-only node cannot make a connection to, or exchange information with, an IPv4-only node (and vice versa), without a complex gateway. One solution is "dual stack," where every node has both IPv4 and IPv6 and hence can make connections to both the Second and Third Internets. Another is to run only IPv6 internally and provide access to external legacy (IPv4-only) nodes via a NAT64[8] gateway.

- IPv4 has 32-bit addresses (max 4.3 billion values), while IPv6 has 128-bit addresses (max 340 trillion, trillion, trillion values). This is again four times as many bits as in IPv4, but now 2^{96} times as many addresses as in IPv4. If you think of the IPv4 address space as the size of a basketball, the IPv6 address space is a sphere that would not only include the entire sun but go most of the way out to Venus. That's a *big* ball.

- The Second Internet began operation in 1983 and will probably be mostly phased out[9] by 2028 or so. There are no more public addresses for the Second Internet to grow with – all growth of the Second Internet today is in private Internets (networks that use RFC 1918 private addresses and are not directly connected to the *public* IPv4 Internet). Each of these private Internets is hidden behind an existing public IPv4 address with NAT[10] (or even behind multiple layers of NAT[11]).

- While applications such as email, remote terminal emulation, and file transfer existed in the Second Internet, most applications (aside from web scripts) must be rewritten to at least some extent to work over IPv6 (or, more commonly, over both IPv4 and IPv6). Since IPv6 has no NAT but ample global addresses, there are entirely new types of connections possible, such as servers on phones or end-to-end direct (e.g., connecting directly from my phone to yours, with no intermediary server).

---

[8] https://en.wikipedia.org/wiki/NAT64
[9] https://datatracker.ietf.org/wg/sunset4/about/
[10] https://en.wikipedia.org/wiki/Network_address_translation
[11] https://en.wikipedia.org/wiki/Carrier-grade_NAT

- Engineers familiar only with IPv4 are having to go back to the books and training classes to master the new IPv6. Engineers and developers who don't learn IPv6 will find it more and more difficult to remain employed, like NetWare engineers experienced once the transition to TCP/IP (Transmission Control Protocol/ Internet Protocol)–based networks took place. If you know IPv4 today, this book contains enough technical detail on IPv6 to get you well along your way to mastering IPv6. I have helped many senior network and telco engineers make the leap to IPv6 as a gold-certified IPv6 Forum trainer.

- IPv4 addresses were represented using *dotted decimal* notation (e.g., "123.45.67.89"), while IPv6 addresses are represented with what I call *coloned hex* notation (e.g., "2001:db8:ed3a:1000::2:1"), which looks very strange indeed to IPv4 engineers.

Why IPv6 Is Important

The Second Internet (aka the *Legacy* Internet) is now 36 years old. Think about what kind of CPU, amount of RAM, and which operating system you were using in 1983 – probably a Z80 8-bit CPU with 64 kilobytes of RAM and CPM/80 or, if you were a businessman, an 8088 "16-bit" CPU and MSDOS 1.0. If you were *really* lucky, you might have had an expensive hard disk drive with a massive 10 megabytes of storage. What, many of you reading this weren't even alive then? Ask your father what personal computing was like in 1983. I've been building, programming, and applying personal computers since my Altair 8800 in 1975. Hard to realize that is 44 years ago. Since 1983, network speeds have increased from 10 Mbps to 100 *Gbps* (10,000-fold increase). Access from home may have been 1200 baud (1.2 kbps) then, but 100 Mbps to 1 Gbps today. Amazingly we are still using essentially the same Internet Protocol. Think it's about time for an upgrade?

The Second Internet has impacted the lives of billions of people. It has led to unprecedented advances in computing, communications, collaboration, research, and entertainment (not to mention time-wasting, dating, gossiping, and even less savory activities). The Internet is now understood to be highly strategic in every modern country's economy. There are now people claiming that access to the Internet is a "human right."[12] It is difficult to conceive of a country that could exist without it. Many enormous companies (such as Google) would not have been possible (or even needed) without it. Staggering amounts of wealth have been created (and consumed) by it. It made "snail mail" (paper mail physically delivered) follow the Pony Express into oblivion (amazingly, governments everywhere are *still* trying to keep post offices going, even though most of them lose gigantic amounts of money every year and they mostly only deliver advertising circulars). The number of emails sent *daily* is at least four times the number of first-class mails sent *annually* (in the United States).

Estimates are that there are about 26 billion nodes[13] (computers, servers, or other network devices) connected to the Internet as of 2019. Neat trick for a protocol with only 4.3 billion theoretically possible unique addresses, eh?

But wait. There's more.

If you think *that's* impressive, wait until you see what its rapidly approaching successor, the Third Internet (made possible by IPv6), will be. One estimate (same link as earlier) predicts some *75* billion nodes by 2025. Entirely new and far more flexible communication and connectivity paradigms are coming that will make email and texting seem quaint (e.g., 5G[14]). Major areas of the economy – such as telephony, entertainment, and almost *all* consumer electronic devices (MP3 players, TVs, radios) – will be heavily impacted or even collapse into the Third Internet as just more network applications (like email and web did in the First and Second Internets). The Second Internet (the one you are likely using today, based on IPv4) that you think is so pervasive and so *cool* is tiny compared to the potential size of the Third Internet. One of the popular terms being used to describe it is *pervasive computing*.[15] That means it is going to be *everywhere*, even inside your body (embedded sensors communicating via a Personal Area Network (PAN) using your phone as a relay to the Third Internet).

Flash! The Second Internet is broken!

Most importantly, in the process of keeping IPv4 around *way too long*, we've already broken the Second Internet badly with something called NAT[16] (Network Address Translation – much more on this later). NAT has turned the Internet into a

---

[12] https://gizmodo.com/internet-access-is-now-a-basic-human-right-1783081865

[13] https://www.statista.com/statistics/471264/iot-number-of-connected-devices-worldwide/

[14] https://en.wikipedia.org/wiki/5G

[15] https://www.computersciencedegreehub.com/faq/what-is-pervasive-computing/

[16] https://en.wikipedia.org/wiki/Network_address_translation

one-way channel, introduced many major security issues, and is impeding progress on newer applications like Voice over Internet Protocol (VoIP) and Internet Protocol Television (IPTV).

NAT has fragmented the old monolithic pre-1995 Second Internet into millions of *private Internets, each* hiding behind one public IPv4 address. You can easily make outgoing connections from your node to servers like `www.facebook.com`, but it is difficult or impossible for other people to make connections to *your node*. NAT has divided the world into a few *producers* (like `www.facebook.com`) and millions of *consumers* (like you). You can post some content to their sites, but they own the sites and have complete control over what you can post and can withdraw your right to post at any time, for any reason.

In the Third Internet, anyone can be a *prosumer* (producer *and* consumer) of content. You will be able to run any server or even a global TV network from your node. You will be able to connect directly from your node to anyone else's node in the world (assuming no firewalls block that connection). There is no shortage of public addresses – we can all be first-class netizens. NAT was a necessary evil to keep the Internet on life support until the Third Internet was ready to be rolled out. The transition to IPv6 was supposed to be *finished* by 2010. NAT has now served its purpose and, like crutches when your broken leg has healed, should be cast aside. Its only purpose was to extend the life of the IPv4 address space while the engineers were getting IPv6 ready. IPv6 IS ready and in rapid adoption mode globally now. This book should be a "wake-up call" for everyone using the Internet.

Using a "horses and cars" metaphor, there is no reason to wait for the last horse to die (the last IPv4 node to be shut down) before we start driving cars (deploy IPv6). Another aspect of that is we no longer need horse doctors; we need car mechanics! Good news, everyone! IPv6 is ready for prime time today. My home is already fully migrated to dual stack (IPv4 + IPv6). It has been for over a decade.

Wait. How Can the Internet Grow to 75 Billion Nodes?

If there are only about 7.5 billion people alive, how can the Internet *possibly* grow to 75 billion nodes? The key here is to understand that the Third Internet (based on IPv6) is *the Internet of Things.*[17] A human sitting at a keyboard will be a relatively rare thing.

---

[17] `https://en.wikipedia.org/wiki/Internet_of_things`

However, IPv6 will make it far easier and cheaper to bring the next billion humans online using IPv6's advanced features and almost unlimited address space. Many Asian countries and companies (who routinely have 5- to 10-year horizons in their planning) already consider IPv6 to be one of the most strategic and important technologies anywhere and are investing heavily in deploying IPv6. 2018 was the *tipping point*[18] for IPv6. Adoption curves are starting to climb at steep rates reminiscent of the adoption of the World Wide Web back in the mid-1990s. By 2018, more than 50% of all global traffic was over IPv6 in many countries. IPv4 will be in decline, with worsening service and fewer and fewer public addresses, at any price. Also, today, many people have multiple devices connected to the Internet. In Singapore, 23% of the people have *five or more* nodes with Internet connectivity.

Why Was 2011 a Significant Year for the Second Internet?

There is an entire chapter in this book on the *depletion of the IPv4 address space*. What this means (in English) is that we are *running out* of public IPv4 addresses for the Second Internet. On February 3, 2011, there was a *very* important event in the history of the Internet. I woke my kids up to watch it live-streamed over the Internet, so they could tell their kids that they saw the beginning of the end of the IPv4 Internet. IANA allocated the final five unallocated blocks of IPv4 public addresses to the five RIRs (Regional Internet Registries).

In the mid-1990s, the folks in charge of the Internet realized we would soon run out of public IPv4 addresses and only managed to keep the Internet going through some clever tricks (NAT and private addresses), kind of like using private extension numbers in a company PBX phone system. However, even with this trick (which is now causing major problems), we have pretty much run out for good. All the groups that oversee the Internet – like the Internet Assigned Numbers Authority (IANA[19]), the Internet Corporation for Assigned Names and Numbers (ICANN[20]), the Internet Society (ISOC[21]), the Internet Engineering Task Force (IETF[22]), and the Regional Internet Registries (RIRs[23]) – have been saying for some time that the world *has* to migrate to IPv6 *now*.

---

[18] https://en.wikipedia.org/wiki/Tipping_point_(sociology)

[19] https://www.iana.org/

[20] https://en.wikipedia.org/wiki/ICANN

[21] https://en.wikipedia.org/wiki/Internet_Society

[22] https://en.wikipedia.org/wiki/Internet_Engineering_Task_Force

[23] https://en.wikipedia.org/wiki/Regional_Internet_registry

The five Regional Internet Registries are ARIN,[24] RIPE NCC,[25] APNIC,[26] LACNIC,[27] and AfriNIC.[28] They should know. They are the ones that manage and allocate public IP addresses to telcos, ISPs (Internet Service Providers), and cloud providers. They know that the IPv4 barrel is pretty much empty. We've *got* to provide tens of billions more globally unique Internet addresses, which has some far-reaching consequences. There is no additional source of IPv4 addresses, so these will have to be IPv6 addresses.

An Analogy: The Amazing Growing Telephone Number

When I was very young, my family's telephone had a five-digit phone number (let's say it was 5-4573). That covered only my small town (about 10,000 people at the time). As the number of phones (and hence unique phone numbers within my geographic region) grew, the telephone company had to increase the length of everyone's phone number. Our number became 385-4573 (seven digits), enough for 10^7 (10 million) phone numbers. This was enough to give everyone in my part of Florida a unique number, and we could ask the nice long-distance operator to connect us to people in other areas when we wanted to talk with them. When the telcos introduced the miracle of *Direct Distance Dialing*, our phone number grew to *ten* digits by adding an *area code*: for example, (904) 385-4573. In theory, this could provide unique numbers to 10^{10} (10 billion) customers. In practice some digit patterns cannot be used, so it is somewhat less than that, and today many people have multiple phone numbers (landline, cell phone, fax, modem, VoIP, etc.). Estimates are that the current supply of ten-digit numbers will last US subscribers at least 50 more years. Increases in the length of phone numbers may be an inconvenience to end users (and publishers of phone books), but the tricky problems are mostly in the big telephone company switches. Phone number lengths have been increased several times over the years, without leading to the collapse of civilization.

---

[24] https://www.arin.net/

[25] https://www.ripe.net/

[26] https://www.apnic.net/

[27] https://www.apnic.net/

[28] https://afrinic.net/

With 5G, *numeric* (aka E.164[29]) phone numbers are going away. In the future, your "phone number" will look like *sip:lhughes@sixscape.com*. There are an essentially unlimited number of SIP URIs (Uniform Resource Identifiers).[30] They are also conveniently organized into the same hierarchy used for email and web.

At the top (IANA) level, the final five unallocated blocks of IPv4 public addresses (16.7 million each) were given out to the five Regional Internet Registries[31] on February 3, 2011.[32] Since that date, if the RIRs asked for additional blocks of IPv4 addresses, IANA would tell them, "Sorry. The cupboard is bare." The RIRs had enough on hand to last a while, but *those* are gone now (except for Africa). I once bought some addresses from APNIC as a member and reserved a "/22" block of IPv4 addresses (a little over 1000 of the precious, and increasingly scarce, addresses for the Second Internet). These cost me about 1000 USD per year, but I could have used those for many things. You can think of this as staking out some of the last remaining lots in a virtual Oklahoma Land Rush. At the same time, I got my very own "/32" block of shiny new IPv6 addresses. You can think of this as getting an enormous spread of prime real estate in the virtual New World of the Third Internet. A few years ago, I got tired of paying the charges and returned those blocks to APNIC. I now have *one* public IPv4 at home, which I had to pay $50 for when I signed up with my current ISP and can keep so long as I have service with them (very effective marketing – if I gave this one up, it is unlikely I would ever get another).

There is a flourishing "gray market" for IPv4 addresses today. Going rate is about $16 per public IPv4. That price will go up until IPv6 is widely deployed, at which point that price will drop to zero quickly.

So Just What Is It That We Are Running Out Of?

There is a great deal of confusion and misunderstanding about this issue, as important as it is. Many people think that an "Internet address" is something like *www.ipv6.org*. That is not an Internet address; that is a *domain-qualified symbolic nodename*. That is an important part of a URI (Uniform Resource Identifier), which adds things such as a protocol designator (e.g., http:, mailto:, or sip:), possibly a nonstandard port number (e.g., ":8080"), and often a file path (e.g., "/files/index.html"). There are still a

---

[29] https://en.wikipedia.org/wiki/E.164

[30] https://en.wikipedia.org/wiki/SIP_URI_scheme

[31] https://en.wikipedia.org/wiki/Regional_Internet_registry

[32] https://www.nro.net/icann-nro-live-stream/

staggering number of possible domain-qualified nodenames that are easy to remember, more than could ever be used in the next hundred years. So just what is it that we are running out of?

The nodenames that you (and most humans) use to specify a particular node on the Internet, like *www.ipv6.org*, are made possible by something called the Domain Name System (DNS[33]). Those nodenames are not used in the actual packets as source and destination addresses (see the section on the IPv4 addressing model for the gory details). The addresses used in the packets in the Second Internet are 32-bit binary numbers. These are usually represented for us slow and stupid humans in *dotted decimal notation* like 123.45.67.89. With a 32-bit address, there are 2^{32} (about 4.3 billion) distinct values. When you use a *symbolic nodename* (known technically as a *fully qualified domain name*, or FQDN) in an application, that application sends it to a DNS server, which returns the numeric IP address associated with it. *That's* the address that is used in packets on the wire, for routing the packet to its destination.

The DNS nodenames are like the *names* of people you call; the IP addresses are like their *phone numbers*. DNS is like an online telephone book that looks up the "phone number" (IP address) for "people" (nodes) you want to "call" (connect to). Did you know that you can surf to an IP address? Try entering the URL *http://15.73.4.75*. That's a whole lot harder to remember than *www.hp.com*, which is why DNS was invented. It's these 32-bit numeric addresses (that most people never see) that we are running out of. The good news is that you can keep typing *www.hp.com* and DNS will return both the old-style 32-bit IPv4 address and a new-style 128-bit IPv6 address, which will be put into the network packets. Given the choice, your applications will prefer to use the new IPv6 address. You will hardly notice the difference unless you are a network engineer or a network software developer, except there's going to be a whole bunch of cool new stuff to do and new ways of doing old things. Plus, the Internet is going to work better than it ever has before.

Can you imagine trying to use telephones today with five-digit telephone numbers? In a few years, that's what IPv4 is going to feel like. I've been using IPv6 for over a decade, and IPv4 already looks antiquated to me. It's amazing we were able to build the current Second Internet with something so primitive and limited. I'm creating new apps for IPv6 already.

---

[33] https://en.wikipedia.org/wiki/Domain_Name_System

But You Said There Were 4.3 Billion IPv4 Addresses?

There are 26 billion nodes connected to the Second Internet, but only 4.3 billion IPv4 addresses? How does THAT work? Well, there are probably around 3 billion *usable* IPv4 public addresses (and essentially no new ones to allocate, except in very specific circumstances, like for IPv6 migration). The bulk of those nodes are not on the *public* IPv4 Internet, but in *private Internets* hiding behind NAT gateways. Pretty much all new nodes being added (like all those cell phones that can connect to the Internet) are in private Internets. There is no real shortage of addresses for private Internets. In theory, every public address could have as many as 16 million nodes (the number of possible nodes in the 10/8 private subnet) behind it. In practice a single NAT gateway can't handle anywhere near that many nodes, but there can still be hundreds or even *thousands* of nodes in each private Internet. The problem is that nodes in private Internets can't accept incoming connections, except via *NAT traversal*[34] (which introduces many security issues). NAT also breaks a lot of important protocols, like VoIP and IPsec. NAT was only ever meant as a temporary stopgap measure during the transition to IPv6. A lot of people today (including telcos and ISPs) seem to think we can just go on using IPv4 with NAT forever. **We can't.**

How did we get into this situation? Well, when the Second Internet was being launched, there were about 200 nodes on the First Internet, and 4.3 billion looked a lot like "infinity" to the people involved. So giant chunks of addresses were generously given out to early adopter organizations. For example, MIT and HP were given "class A" blocks of addresses (about 16.7 million addresses each, or 1/256 of the total address space). Smaller organizations were given "class B" blocks of addresses (each having about 65,535 addresses). Most of these organizations are not using anywhere *near* all those addresses, but they have only rarely been willing to turn them back in to be reallocated to newcomers. As detailed in the Organisation for Economic Co-operation and Development (OECD) study on IPv4 address space depletion and migration to IPv6, it is *very* difficult and time-consuming to recover these "lost" addresses. Also, some blocks of IPv4 addresses were used for things like multicast ("class D"), experimental use ("class E"), and other purposes like addresses for private Internets (RFC 1918[35]).

We are getting more efficient in our allocation of blocks of IPv4 addresses, but even with every trick we know, they are all gone now at the top (IANA) level and four of

---

[34] https://en.wikipedia.org/wiki/NAT_traversal
[35] https://tools.ietf.org/html/rfc1918

the five RIRs (ARIN, APNIC, RIPE, and LACNIC). There are something like 1.5 billion smartphones being sold each year (and this doesn't even count other devices that might need addresses). There may be tens of billions of IoT nodes. How do we connect all these? This can only be done by going to longer IP addresses (hence, a larger address space). This is one of the main things that IPv6 is about.

Is IPv6 Just an Asian Thing?

Some time ago, I heard some comments from US networking professionals and venture capitalists that IPv6 was an "Asian thing," something that is of little interest or concern to Americans. This shows an unusually provincial view of an extremely serious situation. This attitude was only partly due to the inequitable distribution of addresses for the Second Internet (there are over six IPv4 public addresses per American citizen, compared with only about 0.28 per person for the rest of the world). It has a lot more to do with a lack of knowledge of how certain parts of the Second Internet really work, compounded by a limited time horizon compared with Asian businessmen, who routinely plan 5–10 years ahead. American business schools teach that nothing is important beyond the next quarter's numbers. The depletion of IPv4 addresses is already here. Some American businessmen are now panicking ("Why didn't you *warn* us about this?").

Since 2010, US mobile telephone service providers have embraced IPv6 enthusiastically, more so than other regions or industries. They realized they could deploy *only* IPv6 for a far lower cost than trying to keep IPv4 alive one more year. Also, it was becoming a challenge to knit together multiple /8 subnets (the largest you can create with IPv4 private addresses), each of which is 16.7M addresses. Many telcos have far more than 16.7M customers. With IPv6 there is no such problem. Now that all Android phones include 464XLAT,[36] even legacy IPv4-only mobile apps work just fine. On iOS, Apple requires that apps work in an IPv6-only environment before they are approved for the App Store.

Any country or organization that (for whatever reason) doesn't migrate to IPv6 is going to still be "riding horses" while the rest of us are zipping around in these newfangled "cars." When I wrote the 2010 version of this book, I was having nightmares about the United States being just as reluctant to go to IPv6 as they were to adopt the

---

[36] https://sites.google.com/site/tmoipv6/464xlat

metric system (the United States is the only industrialized country *not* to have adopted the metric system, and I doubt they ever will). They *could* have decided to stay with IPv4. If they had, it would have become increasingly difficult for them to connect to non-US websites or for people in other countries to connect to US websites. It would have impacted all telephone calls between the United States and anywhere else in the world. It would have made IT products designed for the US market of little interest outside of the United States (kind of like automobiles that can't be maintained with metric tools). This would have isolated the United States even further and essentially leave leadership in Information Technology up for grabs. Japan, China, and South Korea are quite serious about grabbing that leadership, and they are well along their way to accomplishing this, by investing heavily in IPv6 since the late 1990s. Since then, America has finally "gotten religion" about IPv6, especially in mobile telephone service providers where IPv6 is approaching 100%.

Being good engineers, while the IETF has the "streets dug up" increasing the size of IP addresses, they fixed and enhanced many of the aspects of IPv4 (QoS, multicast, routing, etc.) that weren't done quite as well as they might have been (who could have envisioned streaming video 34 years ago?). IPv6 is not just bigger addresses. It's a whole new and remarkably robust platform on which to build the Third Internet.

So Exactly What Is This "Third Internet"?

Most things in computer technology evolve through various releases or generations, with significant new features and capabilities in the newer generations, for example, 2G, 3G, and 4G cell phones. The Internet is no exception. The remarkable thing, though, is that the Second Internet has lasted for 36 years already. The third generation has been quietly emerging for some time and is now well underway. 5G phones will be mostly based on IPv6. There are many technology trends going on right now, and some of them have been hyped heavily in the press. Some of them sound a lot like they might be the next generation of the Internet. Let's see if we can narrow down what I mean by "the Third Internet" by discussing some of the things that it is *not*.

Is It the *Next-Generation Network* (NGN) That Telcos Talk About?

Telcos around the world have been moving toward something they call NGN[37] for some time. Is that the same thing as the Third Internet? Well, there is certainly a lot of overlap, but, no, NGN is something quite different.

Historically, telephone networks have been based on a variety of technologies, mostly *circuit switched*, with call setup handled by SS7[38] (Signaling System 7). The core of the networks might be digital, but almost the entire *last mile* (the part of the telco system reaching from the local telco office into your homes and businesses) is *still* analog today. There was some effort at upgrading this last mile to digital with ISDN[39] (Integrated Services Digital Networks), but some terrible decisions regarding tariffs (the cost of services) pretty much killed ISDN in many countries, including the United States.

The ITU[40] (International Telecommunication Union), an agency of the United Nations that has historically overseen telephone systems worldwide, defines NGN as packet-switched networks able to provide services, including telecommunications, over broadband, with Quality of Service (QoS)–enabled transport technologies, and in which service-related functions are independent from underlying transport-related technologies. It offers unrestricted access by users to different telecommunication service providers. It supports generalized mobility, which will allow consistent and ubiquitous service to users.

In practice, telco NGN has three main aspects:

- In telco core networks, there is a consolidation (or *convergence*) of legacy transport networks based on X.25 and Frame Relay into the data networks based on TCP/IP (some still using IPv4, but more and more core networks are IPv6 today). It also involves moving from circuit-switched (mostly analog) voice technology (the Public Switched Telephone Network, or PSTN[41]) to Voice over Internet Protocol (VoIP[42]). So far, the move to VoIP is mostly internal to the

---

[37] https://en.wikipedia.org/wiki/Next-generation_network
[38] https://en.wikipedia.org/wiki/Signalling_System_No._7
[39] https://en.wikipedia.org/wiki/Integrated_Services_Digital_Network
[40] https://en.wikipedia.org/wiki/International_Telecommunication_Union
[41] https://en.wikipedia.org/wiki/Public_switched_telephone_network
[42] https://en.wikipedia.org/wiki/Voice_over_IP

telcos. What is in your house and company is good old POTS[43] (Plain Old Telephone Service).

- In the "last mile," NGN involves migration from legacy split voice and data networks to Digital Subscriber Line (DSL), making it possible to finally remove the legacy voice switching infrastructure. Today, more and more telcos are running FTTH[44] (Fiber to the Home), which is of course digital all the way.

- In cable access networks, NGN involves migration of constant bit rate voice to Packet Cable standards that provide VoIP and Session Initiation Protocol (SIP) services. These are provided over DOCSIS[45] (Data Over Cable Service Interface Specification) as the cable data layer standard. DOCSIS 3.0 does include good support for IPv6, though it requires major upgrades to existing infrastructure. There is also a "DOCSIS 2.0 + IPv6" standard, which supports IPv6 even over the older DOCSIS 2.0 framework, typically requiring only a firmware upgrade in equipment. That will likely get rolled out before DOCSIS 3.0 can be. Especially in the United States, DOCSIS 3.0 is finally being widely deployed, with speeds even above 1 Gbps.

A major part of NGN is **IMS**[46] (the *IP Multimedia Subsystem*). To understand IMS, I highly recommend the book *The 3G IP Multimedia Subsystem (IMS): Merging the Internet and the Cellular Worlds*, by Gonzalo Camarillo and Miguel A. Garcia-Martin. This was published by John Wiley & Sons, in 2004. This book says that IMS (which is the future of all telephony) was designed to work *only* over IPv6, using DHCPv6, DNS over IPv6, E.164 Number Mapping (ENUM), and Session Initiation Protocol/Real-Time Transport Protocol (SIP/RTP) over IPv6. IMS is *so* IPv6 specific that some of the primary concerns are how legacy IPv4-only SIP-based user agents (hardphones and softphones) will communicate with the IPv6 core. One approach is to use dual-stack SIP proxies that can in effect translate between SIP over IPv4 and SIP over IPv6. Translation of the media component (RTP) is a bit trickier and will be handled by Network Address Translation between IPv4 and IPv6. Newer IPv6-compliant user agents will be able to interoperate

---

[43] https://en.wikipedia.org/wiki/Plain_old_telephone_service
[44] https://en.wikipedia.org/wiki/Fiber_to_the_x
[45] https://en.wikipedia.org/wiki/DOCSIS
[46] https://en.wikipedia.org/wiki/IP_Multimedia_Subsystem

directly with the IMS core, without any gateways, and solve many problems. They are beginning to appear. One example is some dual-stack IP phones from the Korean company Moimstone.[47]

The first "Internet over telco wireless service" in early 2G networks was **WAP**[48] (Wireless Application Protocol). WAP 1.0 was released in April 1998. WAP 1.1 followed in 1999, followed by WAP 1.2 in June 2000. The Short Message System (SMS[49]) was introduced, but only IPv4 was supported. Speed and capabilities were somewhat underwhelming.

2.5G systems improved on WAP with **GPRS**[50] (General Packet Radio Service), with theoretical data rates of 56–114 Kbps. GPRS included "always on" Internet access, Multimedia Messaging Service (MMS[51]), and point-to-point service. It increased the speed of SMS to about 30 messages/second. Even Filipinos can't text *that* fast. As with WAP, only IPv4 was supported.

2.75G systems introduced **EDGE**[52] (Enhanced Data Rates for GSM Evolution), also known as EGPRS (Enhanced GPRS). EDGE service provided up to 2 Mbps to a stationary or walking user and 348 Kbps in a moving vehicle. IPv6 service has been demonstrated over EDGE but is not widely deployed.

3G systems introduced **HSPA**[53] (High-Speed Packet Access), which consisted of two protocols, **HSDPA** (High-Speed Downlink Packet Access) with theoretical speeds of up to 14 Mbps service and **HSUPA** (High-Speed Uplink Packet Access) with up to 5.8 Mbps service. Real performance was again somewhat lower, but better than with EDGE. HSPA had good support for IPv6.

The last gasp for 3G (sometimes called "3.9G") is **LTE**[54] (Long-Term Evolution). LTE is completely based on IP and was *supposed to be based on IPv6.* Early versions of the specification clearly described it with IPv6 mandatory and IPv4 support optional. It was later reworded to make most aspects "IPv4v6" (dual stack). The reality is mostly

---

[47] http://moimstone.com/eng/main.php

[48] https://en.wikipedia.org/wiki/Wireless_Application_Protocol

[49] https://en.wikipedia.org/wiki/SMS

[50] https://en.wikipedia.org/wiki/General_Packet_Radio_Service

[51] https://en.wikipedia.org/wiki/Multimedia_Messaging_Service

[52] https://en.wikipedia.org/wiki/Enhanced_Data_Rates_for_GSM_Evolution

[53] https://en.wikipedia.org/wiki/High_Speed_Packet_Access

[54] https://en.wikipedia.org/wiki/LTE_(telecommunication)

just IPv4. 3G was still based on two parallel infrastructures (circuit switched and packet switched). LTE is packet switched *only* ("all IP"). There are a few deployments of LTE (some of which are described incorrectly as "4G") around the world.

4G systems have been around for some time. These provide even higher-speed wireless transports. Originally 4G was supposed to be the big change to IP only, but IPv6 wasn't widely enough deployed, and vendors wanted to sell the higher speed as something really different.

So 5G is now being deployed. This will use an all-IP infrastructure for both wired and wireless. The specification for 5G claims peak downlink rates of as much as 1 Gbps and uplink rates of several hundred Mbps. 5G requires a "flat" IP infrastructure (no NAT), which can only be accomplished with IPv6. IPv4 address space depletion happened some time ago, so IPv4 is not even an option this time around. IPTV[55] is a key part of 5G, which requires fully functional multicast, scalable to very large customer bases. That also requires IPv6.

So clearly the telco's NGN is moving more and more *toward* IPv6. Some deployments are still mostly IPv4. However, NGN is just as clearly *not* the Third Internet described in this book. You might say that NGN (once it reaches 5G) will be just another one of the major subsystems hosted on the Third Internet, peer to email, the Web, IPTV, etc. 5G is also called "the Grand Convergence," referring to the long-awaited merging of "the Internet" and "telephony" into a single seamless network.

There will be *much more* to the Third Internet than just telephony, including most broadcast entertainment, exciting new possibilities for non-telephonic communication paradigms (fully decentralized instant messaging and peer-to-peer (P2P) collaboration), smart building sensor and control systems, and ubiquitous connectivity in essentially all consumer electronics, including MP3 players, electronic book readers, cameras, and personal health monitoring. It will also impact automotive design. See `www.car-to-car.org`[56] for some exciting new concepts in "cooperative Intelligent Transport Systems" that depend heavily on IPv6 concepts such as Networks in Motion (NEMO) defined in Request for Comments (RFC) 3963[57] and ad hoc networks. In fact, *only* IPv6 is being used in their designs, although it is a slightly modified version of IPv6 that is missing some

---

[55] `https://en.wikipedia.org/wiki/IPTV`

[56] `http://www.car-to-car.org/`

[57] `https://tools.ietf.org/html/rfc3963`

common functionality such as Duplicate Address Detection (DAD). Their modified IPv6 runs on top of a new, somewhat unusual Link Layer called the C2C Communication Network, which itself is built on top of IEEE 802.11p,[58] also known as Wireless Access in Vehicular Environments (WAVE).

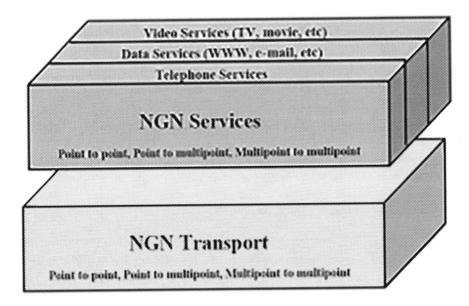

Figure 1-3. *NGN*

Is It Internet2 or National LambdaRail?

Internet2[59] is an advanced academic and industrial *consortium* led by the research and education community, including over 200 higher education institutions and the research departments of several large corporations. They have deployed a worldwide research network called *the Internet2 network*. While IPv6 is definitely being used on Internet2, they also use a lot of IPv4. Their focus is more on very high performance than which version of IP is used. The first part of the Internet2 network (called *Abilene*[60]) was built in 1998, running at 10 Gbps, even over Wide Area Network (WAN) links. It was associated with the National LambdaRail[61] (NLR) project for some time. Internet2 and NLR have

---

[58] https://tools.ietf.org/html/rfc3963

[59] https://en.wikipedia.org/wiki/Internet2

[60] https://en.wikipedia.org/wiki/Abilene_Network

[61] https://en.wikipedia.org/wiki/National_LambdaRail

since split and moved forward along two different paths. Today, most links in the global Internet2 network are running at 100 Gbps. This is 10–100 times faster than typical WAN links used by major corporations today.

Internet2 also features advanced research into secure identity and access management tools, on-demand creation and scheduling of high-bandwidth, high-performance circuits, layer 2 Virtual Private Networks (VPNs), and dynamic circuit networks (DCNs).

A recent survey of Internet2 sites showed that only a small percentage of them have even basic IPv6 functionality deployed, such as IPv6 DNS, email, or VoIP over IPv6. IPv6 is independent of their goals. Essentially, Internet2 is primarily concerned more with extreme high-end performance (100 Gbps and up) and very advanced networking concepts not likely to be used in real-world systems for decades. Although they do profess support for IPv6, they have not aggressively deployed it, and it is definitely not central to their efforts. They are doing little or no work on IPv6 itself or in new commercial applications based on IPv6. I guess those areas are not very exciting to academicians. They are *very* exciting to me – actually, more exciting than 100 Gbps links.

The real-world Third Internet I am writing about in this book will be built primarily with equipment that mostly has the same performance as the current Second Internet (no more than 1 Gbps on WAN links for some time to come and only that high in advanced countries). In much of the world today, 5–120 Mbps is considered good. Maybe 100 Gbps will be widely deployed by 2030–2040, but ultrahigh performance is not necessary to provide the revolutionary benefits described in this book. To give you an idea, Standard-Definition (SD) TV requires about 2 Mbps bandwidth per simultaneously viewed channel, and High-Definition (HD) TV requires about 10 Mbps bandwidth. That is about the most bandwidth-intensive application you will likely see for most users for some time to come. Voice only requires about 8–64 Kbps for good quality. In Japan and Korea today, home Internet accounts typically have about 50–100 Mbps performance. In my hotel room in Tokyo several years ago, I measured 42 Mbps throughput. That is enough for almost any use today. I now have 1 Gbps Internet service in my home in SG (for about S$49 a month). Most users, even in companies, would be really challenged to make effective use of 100 Gbps bandwidth. With that bandwidth you could download the entire Encyclopedia Britannica in just a few seconds (including images) or a typical Blu-ray movie (about 25 gigabytes) in about 2 seconds. With current caps on network traffic volume, you would go through your entire month's allowance in a matter of seconds. That is actually a serious concern even with 5G, with 1 Gbps potential speeds.

The necessary equipment and applications for the Third Internet can in many cases be created with software or firmware upgrades (except for older and low-end devices that don't have enough RAM or ROM to handle the more complex software and in high-end telco- and Internet Service Provider (ISP)–level products that include hardware acceleration).

The main technical advantages of the Third Internet will not be higher bandwidth, but the vastly larger address space, the restoration of the flat address space (elimination of NAT), and the general availability of working multicast. All these are made possible by migration to IPv6, which involves insignificant costs compared with supporting 100 Gbps WAN links. Perhaps generally available WAN bandwidth in that range will be what characterizes the *Fourth Internet*. I personally would just consider that "faster Third Internet."

So Internet2 is *not* the Third Internet I am writing about. Internet2 and NLR are primarily academic exercises that will not bear fruit for many decades. What they are doing is very important in the long run, but it does not address, and will not solve, the really major problems facing the Second Internet today. The Third Internet is being rolled out today and accounted for over 50% of global traffic in 2018. That is the beginning of the end for IPv4. Maybe 100 Gbps service will characterize the *Fourth Internet*.

Figure 1-4. *Logos for Internet2 and National LambdaRail*

Is It *Web 2.0*?

First, if you think that the terms "World Wide Web" and "Internet" are synonymous, let me expand your worldview a bit, in the same way that Copernicus did for people's view of our Solar System back in the mid-1500s. The "World Wide Web" is basically *one* service that runs on a much larger, more complex thing, which is called the *Internet*. The Web is a simple *client-server* system based on HTTP[62] (Hypertext Transfer Protocol) and HTML[63] (Hypertext Markup Language). Due to extremely serious limitations and inefficiencies of these standards, both have been enhanced and extended numerous times. The result is still not particularly elegant to real network software designers or engineers, but it has *clearly* had a major impact on the world. The technology of the Web was a refinement and convergence of several ideas and technologies that were in widespread use before HTML and HTTP were created by Tim Berners-Lee in the late 1980s, at CERN. But there is a lot to the Internet beyond the Web (email, instant messaging, video conferencing, VoIP, file transfer, peer-to-peer (P2P), VPNs, IPTV, etc.). There are thousands of Internet protocols, of which the Web uses *one* (HTTP).

Hypertext, WAIS/SGML, and Gopher

The terms *Hypertext* and *Hypermedia* were coined by Ted Nelson in 1965, at Brown University. These terms referred to online text documents (or rich media, including pictures, sound, and other media content) that contained *links* that allowed building paths from any word or phrase in the document to other parts of the same document or parts of other documents that were also online. In August 1987, Apple Computer released the first commercial Hypertext-based application, called HyperCard, for the Macintosh. There were already document storage and retrieval systems on the early Internet, such as WAIS[64] (Wide Area Information Server). WAIS was based on the ANSI Z39.50:1988 standard and was developed in the late 1980s by a group of companies including Thinking Machines, Apple Computer, Dow Jones, and KPMG Peat Marwick. As with the Web, there were both WAIS servers and clients. A later version of WAIS was based on ANSI Z39.50:1992, which included SGML (Standard Generalized Markup Language, ISO 8879:1986) for more professional-looking documents. There was another Internet application called Gopher[65] (University of Minnesota, circa 1991) that could distribute, search for, and retrieve documents. Gopher was also primarily text based and imposed a very strict hierarchical structure on information.

---

[62] https://en.wikipedia.org/wiki/Hypertext_Transfer_Protocol
[63] https://en.wikipedia.org/wiki/HTML
[64] https://en.wikipedia.org/wiki/Wide_area_information_server
[65] https://en.wikipedia.org/wiki/Gopher_(protocol)

HTML and HTTP

Tim Berners-Lee combined these three concepts (Hypertext, WAIS/SGML, and Gopher document retrieval) to create HTTP and HTML. HTML was a very watered-down and limited markup language compared with SGML. SGML is capable of creating highly sophisticated, professional-looking books. In comparison, HTML allows very limited control over the final appearance of the document on the client's screen. HTTP was a very simple protocol designed to serve HTML documents to HTTP client programs called *web browsers*. A basic HTTP server can be written in one afternoon and consists of about half a page of the C programming language (I've done it and retrieved documents from it with a standard browser). The first browser (Lynx,[66] 1992) was very limited (text only, but including Hypertext links). In 1993, at the National Center for Supercomputing Applications (NCSA) at the University of Illinois, the first Mosaic[67] web browser was created (running on X Windows in UNIX). Because it was created for use on X Windows (a platform with good support for computer graphics), many graphics capabilities were added. With the release of web browsers for PC and Macintosh, the number of servers went from 500 in 1993 to 10,000 in 1994. The *World Wide Web* has since grown to millions of servers and many versions of the web client (Internet Explorer, Mozilla Firefox, Safari, Opera, Chrome, etc.). It's been so successful that a lot of people today think that the "World Wide Web" *is* the Internet. It's really just one small part of it.

Web 2.0

The term *Web 2.0*[68] was first coined by Darcy DiNucci in 1999, in a magazine article. The current usage dates from an annual conference that began in 2004, called "Web 2.0," organized and run by Tim O'Reilly (owner of O'Reilly Media, publisher of many excellent books on computing).

Many of the promoters of the term *Web 2.0* characterize what came before (which they call *Web 1.0*) as being "Web as Information Source." Web 1.0 is based on technologies such as PHP, Ruby, ColdFusion, Perl, Python, and ASP (Active Server Pages). In comparison, Web 2.0 is "Network as Platform," or the "participatory Web." It uses the technologies of Web 1.0, plus new things such as Asynchronous JavaScript,

---

[66] https://en.wikipedia.org/wiki/Lynx_(web_browser)

[67] https://en.wikipedia.org/wiki/Mosaic_(web_browser)

[68] https://en.wikipedia.org/wiki/Web_2.0

XML, Ajax, Adobe Flash, and Adobe Flex. Typical Web 2.0 applications are the *Wiki*[69] (and the world's biggest wiki, the *Wikipedia*[70]), blogging sites, social networking sites like *Facebook*, video publishing sites like *YouTube*, photographic snapshot publishing sites like *Flickr*, Google Maps, etc.

Andrew Keen (British-American entrepreneur and author) claims that Web 2.0 has created a cult of digital narcissism and amateurism, which undermines the very notion of expertise. It allows *anyone anywhere* to share their own opinions and content, regardless of their talent, knowledge, credentials, or bias. It is "creating an endless digital forest of mediocrity: uninformed political commentary, unseemly home videos, embarrassingly amateurish music, unreadable poems, essays and novels." He also says that Wikipedia is full of "mistakes, half-truths and misunderstandings." Perhaps Web 2.0 has made it *too* easy for the mass public to participate. Tim Berners-Lee's take on Web 2.0 is that this is just a "piece of jargon." In the finest tradition of Web 2.0, these comments, which were found in the Wikipedia article on Web 2.0, probably include some mistakes, half-truths, and misunderstandings.

Basically, Web 2.0 does not introduce any revolutionary new technology or protocols; it is more a minor refinement of what was already being done on the Web, in combination with a new emphasis on end users becoming not just passive *consumers*, but also *producers* of web content. The Third Internet will actually help make Web 2.0 work better, as it removes the barriers that have existed in the Second Internet since the introduction of NAT to anyone becoming a producer of content. If anything, on the Third Internet, these trends will be taken even further by decentralizing things. There will be no need for centralized sites like YouTube or Flickr to publish your content, just more sophisticated search engines or directories that will allow people to locate content that will be scattered all over the world. Perhaps that will be the characterizing feature of *Web 3.0*? With IPv6 you can run any server (including a web server) on any computer you have, *including your phone*, and anyone in the world (who has IPv6) will be able to access it. Now *that's* a major change.

Web 2.0 is a really minor thing compared with the Third Internet. What isn't pure marketing hype is an evolutionary development of *one* of the major services (the *World Wide Web*) out of perhaps a dozen major subsystems that the Third Internet is capable of hosting. These include global telephony, newer forms of communication like decentralized instant messaging, major new peer-to-peer applications (not just

---

[69] https://en.wikipedia.org/wiki/Wiki
[70] https://en.wikipedia.org/wiki/Main_Page

file sharing), global broadcast entertainment via multicast IPTV, connectivity between essentially all consumer electronic products, personal healthcare sensor nets, smart building sensor nets, etc.

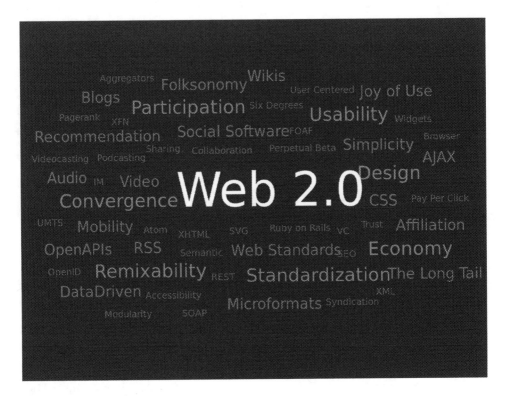

Figure 1-5. *Web 2.0 logo*

Whatever Happened to IPv5?

Two of the common questions people ask when they start learning about IPv6 are "If it's the next version after *IPv4*, why isn't it called *IPv5*?" and "What happened to the *first three* versions of IP?"

There is a 4-bit field in every IP packet header that contains the IP version number in binary. In IPv4, that field contains the binary value 0100 (4 in decimal) in every packet. An earlier protocol (defined in RFC 1190,[71] "Experimental Internet Stream Protocol, Version 2 (ST-II)," October 1990) used the binary pattern 0101 (5 in decimal)

---

[71] https://tools.ietf.org/html/rfc1190

in the IP Version field of the packet header. The Internet Stream Protocol was not really a replacement for IPv4 and isn't even used today, but unfortunately the binary pattern 0101 was allocated to it. The next available bit pattern was 0110 binary (6 in decimal). It would be even more embarrassing than explaining that there was no IPv5 to explain why the IP version number field for IPv5 contained the value 6. Now you know.

So what *did* happen to IPv1, IPv2, and IPv3? Those never made it out of the lab. The first version of IP that was released to the general public was IPv4.

ARPANET[72] (based on NCP) was the First Internet. It didn't use any version of the Internet Protocol – it used NCP. IPv4, the foundation protocol of the Second Internet, was the first public release of the Internet Protocol. IPv6, the foundation protocol of the Third Internet, is the second public release of the Internet Protocol. So we *could* have been talking about the transition from IPv1 to IPv2!

There have been rumors about an *IPv9*[73] protocol in China. A venture capital firm in Hong Kong actually asked me if China was already that far ahead of the rest of the world, and shouldn't we be supporting their version? It seems some researcher in a university there published a paper on an "IPv9," but it was never implemented and wasn't a replacement for IPv4 (let alone IPv6) anyway. It was a way to use ten-digit decimal phone numbers in a modified DNS implementation instead of alphanumeric domain names, for all nodes on the Internet. I guess if you speak only Chinese, a ten-digit numeric string may seem easier to use than an English domain name using Latin characters. Fortunately for Chinese speakers, we now have *Internationalized Domain Names*[74] in Chinese and other languages.

There are even internationalized top-level domains (TLDs) now. For an example, see `https://www.101domain.com/%E4%B8%AD%E5%9B%BD.htm`.[75]

Actually, there is a real RFC about IPv9, which you might enjoy reading. See RFC 1606,[76] "A Historical Perspective on the Usage of IP Version 9," *April 1*, 1994. This has nothing to do with the Chinese IPv9 and is much funnier. Please notice the release date of RFC 1606. There is a tradition of releasing gag RFCs on April 1. Some of them are hilarious (well, maybe you have to be a geek to see the humor).

---

[72] `https://en.wikipedia.org/wiki/ARPANET`

[73] `https://www.telecomasia.net/content/strange-case-chinas-ipv9-0`

[74] `https://en.wikipedia.org/wiki/Internationalized_domain_name`

[75] `https://www.101domain.com/%E4%B8%AD%E5%9B%BD.htm`

[76] `https://www.ietf.org/rfc/rfc1606.txt`

Let's Eliminate the Middleman

One of the things that the Third Internet does better than anything is *disintermediation*.[77] Just as email eliminated the need for a central post office and Amazon.com has mostly eliminated the need for physical bookstores, the features of the Third Internet will eliminate the need for many other existing centralized organizations and services. With a real decentralized end-to-end connectivity model, there is no need for two users to connect to a central server (such as Skype or Messenger) in order to chat with each other. They will simply connect directly to each other. That's hard to do today, because of NAT and an acute shortage of public IPv4 addresses.

The restoration of the original (pre-NAT) flat address space and the plethora of public addresses will allow anyone or anything to connect directly to anyone or anything on the Third Internet. It's going to be a *very* different online world. Many business models will go by the way, and many new ones will explode on the scene and make some new entrepreneurs very wealthy. Someone will need to provide centralized directory and presence servers that will let people locate each other, so that they *can* connect directly to each other. I am working on that very problem now.

Several years ago, a gentleman in my previous hometown of Atlanta, Georgia (home to Coca-Cola and UPS), had a small UHF TV station (WTBS, channel 17) that mostly broadcast old movies and Atlanta Braves baseball games, both of which he loved. He was one of the first people to realize that he could relay his TV station's signal through a transponder on a geostationary satellite ("that's just a *really* tall broadcast antenna"), and the rest is history. The man was Ted Turner,[78] and his insight created the Turner Broadcasting System (TBS), which along the way produced CNN, CNN Headline News, Cartoon Network, Turner Network Television (TNT), and many other things. His success allowed him to *buy* the Braves baseball team and the entire film library of MGM (not to mention a famous starlet wife, sometimes also called "Hanoi Jane"). When he began relaying his channel 17 signal, his viewership went from maybe 10,000 to 10,000,000 virtually overnight. That was a world-changing insight.

Some bright entrepreneur is going to realize that global multicast IPTV over IPv6 is the same kind of opportunity. Wonder what they will create with the wealth thereby generated? What country will they be from? I'm betting on India.

---

[77] https://en.wikipedia.org/wiki/Disintermediation
[78] https://en.wikipedia.org/wiki/Ted_Turner

I did warn you that this is revolutionary, highly disruptive technology. However, with great disruption comes great opportunity.

Why Am I the One Writing This Book? Just Who Do I Think I Am, Anyway?

I have been personally involved in helping create and deploy the Third Internet for many years. I've spoken at IPv6 summits around the world, including Beijing, Seoul, Kuala Lumpur, Manila, Taipei, Potsdam, and Washington, DC. I have so far invested 25 years of my life and about $9M of my own personal funds (which came from selling a previous Internet-based venture called CipherTrust where I was cofounder). I've built a new company in Singapore with lots of expertise in PKI and IPv6. It is called Sixscape Communications[79] – we are "The Netscape for the IPv6 Internet." The Third Internet is by far the biggest business and technology opportunity I've seen in my 45+ years in IT.

Now, *have I gotten your attention?* Great! Now let's explore just what the Third Internet is all about.

Summary

This chapter covered the three generations of the global Internet, from ARPANET to IPv4 to IPv6. IPv6 is not just another version of one of the many Application Layer protocols like HTTP v1.1. It is deep in the network stack (in the Internet Layer), so it affects *all* Application Layer protocols.

The First Internet (ARPANET) only served a few thousand people, mostly in the United States. Most of the users were in the US military, the US government, and a few research institutions (mostly universities). It had 8-bit addresses and was based on the host-host protocol. It had many of the applications we still use today, like email, FTP, chat, etc. It lived from roughly 1969 to 1982.

The Second Internet (IPv4 generation) took over from the ARPANET in 1983 and grew to serve billions of users, worldwide. It used 32-bit addresses, but we ran out of unique IPv4 public addresses in 2011 and "broke" the Internet with NAT and private addresses. Many people are still using this today, but it will eventually be phased out.

---

[79] `https://sixscape.com/`

The Third Internet (IPv6 generation) began major deployment in about 2014 and will probably still be in widespread use in 2100. It uses 128-bit addresses, which means we no longer need NAT or private addresses. Every node can have a globally unique IP address, even phones and temperature sensors.

We also presented some analogies to help you understand what IPv6 is and what it isn't.

CHAPTER 2

History of Computer Networks Up to IPv4

A long time ago (in a galaxy not too far away), regular people started connecting computers together. A few brave souls tried to do this with dial-up 1200-baud modems over phone lines. Pioneers brought up Bulletin Board Systems (BBSs; message boards that one person at a time could dial into and exchange short messages, and later small files, with each other). I brought up the eighth BBS in the world, in Atlanta, in about 1977, using code from the original CBBS in Chicago (created by Ward Christensen and Randy Suess). I used a modem donated by my friend Dennis Hayes (of Hayes Microcomputer Products). Later there were thousands of online Bulletin Board Systems, all over the world. Soon there followed commercial "information utilities" like CompuServe[1] and The Source,[2] which were like giant Bulletin Board Systems (BBSs) with many more features. Tens of thousands of users could connect to these simultaneously. It was like the first crude approximation to the Internet of today, based on circuit-switched connections over telephone lines. Everything was text oriented (non-graphical) and very slow. 1200 bits/second was typical at first, although later modems with speeds of 2400 bits/second, 9600 bits/second, 14.4 Kbps, 28.8 Kbps, and finally 56 Kbps were developed and came into widespread use. Later these modems were primarily used to dial into an ISP to connect to the Internet, and some people are still using them this way.

---

[1] https://en.wikipedia.org/wiki/CompuServe

[2] https://en.wikipedia.org/wiki/The_Source_(online_service)

© Lawrence E. Hughes 2022
L. E. Hughes, *Third Generation Internet Revealed*, https://doi.org/10.1007/978-1-4842-8603-6_2

Real Computer Networking

While home computer users were playing around with modems and Bulletin Board Systems, the big computer companies were working on ways to connect "real" computers at higher speeds and with much more complex software.

Ethernet and Token Ring

Much of this was based on Ethernet,[3] which was created by a team at Xerox Palo Alto Research Center (PARC[4]) led by Robert Metcalfe[5] between 1973 and 1975. The first released version (1976) ran at 3 Mbps. Metcalfe left PARC in 1979 to create 3Com[6] and create commercial products based on Ethernet. Working together with Digital Equipment Corporation (DEC), Intel, and Xerox (hence the "DIX"[7] standard), 3Com released the first commercial products running at 10 Mbps. Ethernet was standardized in 1980 by the IEEE (Institute of Electrical and Electronics Engineers) as 802.3. Early versions ran on 10base2 (a small-diameter coax cable) or 10base5 (a larger-diameter coax cable). These used a "multidrop" architecture, which was subject to many reliability problems. With the introduction of the simpler to deploy and manage "unshielded twisted pair" (UTP) (actually four-pair, or eight-wire) cables (known as 10baseT, mid-1980s) and star architectures using "hubs" and later "switches," Local Area Networks (LANs) really took off. Today, virtually all Ethernet networks use twisted pair copper wire (up to gigabit speed) or fiber-optic cable (for higher speed and longer runs). I helped deploy a 10base2 coax Ethernet network in Hong Kong in 1993. Trust me, twisted pair cabling is a *lot* easier to work with.

IBM for many years pushed a competing physical layer network standard called "Token Ring"[8] (later standardized as IEEE 802.5). Token Ring was available in 4 Mbps and 16 Mbps versions. Later, a 100 Mbps version was created, but by then Ethernet

---

[3] https://en.wikipedia.org/wiki/Ethernet

[4] https://en.wikipedia.org/wiki/PARC_(company)

[5] https://en.wikipedia.org/wiki/Robert_Metcalfe

[6] https://en.wikipedia.org/wiki/3Com

[7] http://electronicstechnician.tpub.com/14091/css/Ieee-802-3-Ethernet-Dix-193.htm

[8] https://en.wikipedia.org/wiki/Token_ring

dominated the market, and Token Ring quietly died. FDDI[9] (Fiber Distributed Data Interface) still in use today is based on the Token Ring concept.

Network Software

Network software quickly evolved once Ethernet and Token Ring hardware became available. One of the main goals was to "hide" the differences between various hardware-level technologies (Ethernet, Token Ring, Wi-Fi, etc.) and the higher-level software. This led to the multiple layers of the network stack. The bottom layer is very hardware specific, and the upper layers introduce more and more hardware independence, so that applications can be written once and run over any hardware media.

DEC[10] was one of the first companies to create networking software with DECnet[11] (1975). IBM had System Network Architecture (SNA,[12] 1974). Xerox created the PARC Universal Packet protocol[13] (PUP protocol, late 1970s), which eventually evolved into Xerox Network Services[14] (XNS, early 1980s) at PARC. XNS was the basis for the late Banyan VINES[15] network OS, based on "VINES IP" (similar to but incompatible with IPv4 from TCP/IP). Banyan VINES included the first network directory service, called "StreetTalk."[16] XNS also was the basis for Novell NetWare[17] (IPX/SPX, 1983), which eventually added its own NetWare Directory Services (NDS,[18] 1993).

Microsoft worked with 3Com to create their own network OS, called LAN Manager. It used the SMB[19] (Server Message Block) protocol on top of either the NBF (NetBIOS[20] Frames) protocol or modified XNS. In 1990, Microsoft added support for TCP/IP as an alternate protocol (LAN Manager 2.0). With the release of Windows NT Advanced Server

---

[9] https://en.wikipedia.org/wiki/Fiber_Distributed_Data_Interface

[10] https://en.wikipedia.org/wiki/Digital_Equipment_Corporation

[11] https://en.wikipedia.org/wiki/DECnet

[12] https://en.wikipedia.org/wiki/IBM_Systems_Network_Architecture

[13] https://en.wikipedia.org/wiki/PARC_Universal_Packet

[14] https://en.wikipedia.org/wiki/Xerox_Network_Systems

[15] https://en.wikipedia.org/wiki/Banyan_VINES

[16] http://banyan-vines.bamertal.com/

[17] https://en.wikipedia.org/wiki/NetWare

[18] https://en.wikipedia.org/wiki/NetIQ_eDirectory

[19] https://en.wikipedia.org/wiki/Server_Message_Block

[20] https://en.wikipedia.org/wiki/NetBIOS

in 1993, Microsoft finally phased out LAN Manager. By Windows NT v3.51[21] (May 1995), Microsoft encouraged users to deploy *only* TCP/IP (4 years ahead of Novell's support for TCP/IP). This lead time allowed Microsoft to take over leadership in personal computer networks from Novell. Microsoft introduced their version of network directory services in Windows Server 2000, now known as Active Directory.[22] The SMB protocol still survives as Microsoft's "File and Printer Sharing" protocol (now layered on TCP/IP, instead of NetBIOS or XNS). An open source implementation of this is available as SAMBA.[23]

The Beginnings of the Internet (ARPANET)

While all this commercial activity was going on, the US military (at their Defense Advanced Research Projects Agency, or DARPA[24]), with the help of Bolt, Beranek, and Newman (BBN[25]) and Mitre,[26] were designing a new, decentralized communication system based on packet switching. Existing communication systems (telephone, radio, etc.) were *centralized* and hence subject to being completely disabled due to the failure or loss of a few central nodes. Packet-switched networks were highly *decentralized* and had a fascinating new property, which is that you could lose large parts of a network and the remaining parts would still work (assuming at least some links connected the working parts).

The first network protocol developed as part of ARPANET was called the 1822[27] protocol (named after BBN Report 1822) and was implemented by a Network Control Program,[28] so the protocol was often referred to as NCP. The first email was sent over NCP in 1971, and the File Transfer Protocol followed in 1973. On January 1, 1983 ("flag day"), NCP was turned off officially, leaving only IPv4 on the Internet. I consider the NCP era to be the First Internet and the IPv4 era as the Second Internet. That makes the evolving IPv6 era the Third Internet. Fortunately, there is no need for a flag day to go from IPv4 to IPv6, as they can coexist (and probably will for perhaps 5–10 years).

---

[21] https://en.wikipedia.org/wiki/Windows_NT_3.51

[22] https://en.wikipedia.org/wiki/Active_Directory

[23] https://en.wikipedia.org/wiki/Samba_(software)

[24] https://en.wikipedia.org/wiki/DARPA

[25] https://en.wikipedia.org/wiki/BBN_Technologies

[26] https://en.wikipedia.org/wiki/Mitre_Corporation

[27] www.networksorcery.com/enp/rfc/rfc878.txt

[28] https://en.wikipedia.org/wiki/Network_Control_Program

We learned not to do that from the NCP to IPv4 transition. They broke global email for months in that process.

In May 1974, Vint Cerf[29] and Bob Kahn[30] released the paper "A Protocol for Packet Network Interconnection."[31] This described a monolithic protocol called TCP that combined the features of both modern TCP and IPv4. Later Jon Postel[32] was instrumental in splitting apart TCP and IP as we know them today. Vint Cerf is today considered the "father of TCP/IP" and is now an "Evangelist" at Google. He understands very well the problems with the current implementation of IPv4 (and why these things were done). He advocates for users to migrate to IPv6, which restores his original concept of a flat address space (no NAT), where any node can connect directly to any other node. One of my proudest possessions is a copy of RFC 791 (IPv4) autographed in person by Vint Cerf.

If you'd like to read about the creation of the Second Internet, I recommend the book *Where Wizards Stay Up Late: The Origins of the Internet*,[33] by Katie Hafner and Matthew Lyon. It is of considerable interest to those of us creating the Third Internet, as we are facing some of the same problems they did. Only this time around, we've got over a billion legacy users (and staggering investments in hardware and software) to worry about. On the other hand, we've got three decades of operational experience with IPv4 to draw upon.

Higher-level software protocols were built on top of the TCP and IP layers, called "application protocols," such as SMTP[34] and IMAP[35] (for email), FTP[36] (for file transfer), Telnet[37] (for terminal emulation), and more recently HTTP[38] (used in the World Wide Web) and SIP[39] and RTP[40] (used in VoIP). The resulting suite of protocols became known

---

[29] https://en.wikipedia.org/wiki/Vint_Cerf

[30] https://en.wikipedia.org/wiki/Bob_Kahn

[31] https://ieeexplore.ieee.org/document/1092259

[32] https://en.wikipedia.org/wiki/Jon_Postel

[33] www.amazon.com/Where-Wizards-Stay-Up-Late/dp/0684832674

[34] https://en.wikipedia.org/wiki/Simple_Mail_Transfer_Protocol

[35] https://en.wikipedia.org/wiki/Internet_Message_Access_Protocol

[36] https://en.wikipedia.org/wiki/File_Transfer_Protocol

[37] https://en.wikipedia.org/wiki/Telnet

[38] https://en.wikipedia.org/wiki/Hypertext_Transfer_Protocol

[39] https://en.wikipedia.org/wiki/Session_Initiation_Protocol

[40] https://en.wikipedia.org/wiki/Real-time_Transport_Protocol

for its two most important protocols, TCP and IP, or TCP/IP[41] (its formal name is "the Internet Protocol Suite").

UNIX

About this time (1973), Bell Labs (a research group within AT&T) created an interesting new operating system (called PWB-UNIX[42]) and a new language (in which UNIX was written) called "C."[43] Because of a 1958 consent decree, AT&T as a regulated monopoly was not allowed to market or sell UNIX commercially. They licensed it (complete with source code) to several universities. One of these was the University of California at Berkeley[44] (UCB; also famous for being the center of campus-based communist student activities at the time). The team at UCB extended UNIX in several very important ways such as adding Virtual Memory. They also integrated the new network protocol from ARPA as the first commercial implementation of TCP/IP. The "Berkeley System Distribution"[45] of UNIX became a main branch. Over time, they rewrote most of it and wanted to release it for free. AT&T sued them in court, and it seems most of the examples of "stolen code" AT&T cited had actually been written at UCB. The judge ruled that if UCB rewrote the remaining 10% or so (so there was zero original AT&T code), they could release that. That rewrite became 386BSD,[46] the starting point for FreeBSD[47] (the first open source operating system). Later FreeBSD was chosen by Japan's Kame[48] project to deploy the first version (the "reference" implementation) of a IPv6 network stack, in an eerie echo of BSD UNIX's choice for the first commercial IPv4 implementation.

UNIX and TCP/IP became very popular on college campuses, and with high-end workstation vendors, such as Sun, Silicon Graphics, and Intergraph. Personal computers were not powerful enough to run UNIX until the Intel 386, at which point UCB ported the BSD version to the 386. However, as documented above, most personal computer networking was already moving to TCP/IP.

---

[41] https://en.wikipedia.org/wiki/Internet_protocol_suite

[42] https://en.wikipedia.org/wiki/PWB/UNIX

[43] https://en.wikipedia.org/wiki/C_(programming_language)

[44] https://en.wikipedia.org/wiki/University_of_California,_Berkeley

[45] https://en.wikipedia.org/wiki/Berkeley_Software_Distribution

[46] https://en.wikipedia.org/wiki/386BSD

[47] https://en.wikipedia.org/wiki/FreeBSD

[48] https://en.wikipedia.org/wiki/Kame

Open System Interconnection (OSI)

While all this was going on, the ISO[49] (International Organization for Standardization) in Europe was creating a very thoroughly engineered suite of network protocols called Open System Interconnection[50] (OSI), or more formally X.200[51] (July 1994).

> Because "International Organization for Standardization" would have different acronyms in different languages (IOS in English, OIN in French for *Organisation internationale de normalisation*), our founders decided to give it the short form ISO. ISO is derived from the Greek isos, meaning equal.

This is where the famous "seven-layer" network model comes from (TCP/IP is really based on a "four-layer" model, which has caused no end of confusion among young network engineers). At one point the US government decided to officially adopt OSI for its networking (this was called GOSIP,[52] or Government Open Systems Interconnection Profile, defined in FIPS 146-1, 1990). Unfortunately, OSI was really more of an academic specification, not a real working network system, like TCP/IP was. After many years, GOSIP was finally abandoned, and IPv4 was deployed, but GOSIP's legacy has hindered the adoption of IPv6 in the United States ("Here we go again – GOSIP phase 2!"). X.400[53] email and X.500[54] directory systems were built on top of OSI and will not run on TCP/IP without substantial compatibility layers. One small part of X.500 (called X.509,[55] "The Authentication Framework") was the source of digital certificates[56] and Public Key Infrastructure,[57] still used today. Lightweight Directory Access Protocol (LDAP[58]) was an attempt to create an X.500-like directory system for TCP/IP-based networks. That's about all that is left of the mighty OSI effort today, outside of computer science textbooks and Cisco Press books.

---

[49] https://en.wikipedia.org/wiki/International_Organization_for_Standardization
[50] https://en.wikipedia.org/wiki/Open_Systems_Interconnection
[51] www.itu.int/rec/t-rec-x.200-199407-i
[52] https://en.wikipedia.org/wiki/Government_Open_Systems_Interconnection_Profile
[53] https://en.wikipedia.org/wiki/X.400
[54] https://en.wikipedia.org/wiki/X.500
[55] https://en.wikipedia.org/wiki/X.509
[56] https://en.wikipedia.org/wiki/Public_key_certificate
[57] https://en.wikipedia.org/wiki/Public_key_infrastructure
[58] https://en.wikipedia.org/wiki/Lightweight_Directory_Access_Protocol

IPv6 was based heavily on IPv4 and was defined by the same group that defined IPv4 (the IETF[59]). It is the "natural" and straightforward evolutionary step after IPv4. At this point everyone has decided that IPv6 is inevitable, although there were many battles and brave resistance for years.

Email Standardization

By this time, essentially all computer vendors had standardized on TCP/IP, but there were still a lot of competing standards for email, including Microsoft's MS-Mail, Lotus's cc:Mail, and MCI Mail. The Internet folks used a much simpler email standard called SMTP[60] (Simple Mail Transfer Protocol). It first became the connecting backbone between various email products (everyone had their email to SMTP gateways, so users could exchange messages across organizations). Soon, everyone started using SMTP (together with POP3 and later IMAP) all the way to the end user. Today virtually all email worldwide is based on SMTP and TCP/IP. However, Microsoft is forcing their own proprietary protocols (EWS and now Graph) into this space with Office 365. By the way, most of Office 365 works fine over IPv6. Teams and Azure are still IPv4-only (at least VMs in Azure are IPv4-only).

Evolution of the World Wide Web

Several other Internet applications evolved, including WAIS (Wide Area Information Server, for storing and retrieving documents) and Archie[61] (the very first search engine). In turn, these efforts were merged with the idea of Hypertext (documents with multilevel links) and evolved into HTML (Hypertext Markup Language) and HTTP (Hypertext Transfer Protocol). The World Wide Web was off and running. The first web browser and web server were created at the National Center for Supercomputing Applications[62] (NCSA) at the University of Illinois, Urbana-Champaign campus. The people who created those software projects (primarily Marc Andreessen[63] and Eric Bina[64]) were

---

[59] https://en.wikipedia.org/wiki/Internet_Engineering_Task_Force
[60] https://en.wikipedia.org/wiki/Simple_Mail_Transfer_Protocol
[61] https://en.wikipedia.org/wiki/Archie_search_engine
[62] https://en.wikipedia.org/wiki/National_Center_for_Supercomputing_Applications
[63] https://en.wikipedia.org/wiki/Marc_Andreessen
[64] https://en.wikipedia.org/wiki/Eric_Bina

soon hired by Jim Clark,[65] one of the founders of Silicon Graphics, to start Netscape,[66] one of the most successful and important companies in the Second Internet. They created a new and more powerful web server (Netscape Application Server[67]) and web browser (Netscape Navigator[68]). As an interesting aside, the original browser created by Andreessen at NCSA (Mosaic[69]) later became the starting point for Microsoft's Internet Explorer web browser.

And That Brings Us Up to Today

That pretty much brings us up to the present day where the entire world has standardized on IPv4 for both LANs (Local Area Networks) and WANs (Wide Area Networks). Multiprotocol Label Switching (MPLS) is not a competitor to TCP/IP; it is one more alternative at the Link Layer, peer to Ethernet and Wi-Fi. More and more companies and organizations built TCP/IP networks and connected them together to create the Internet. Major telcos provided "backbone" WAN connections and dial-up access service (soon known as ISPs or Internet Service Providers[70]). As the number of users (and the amount of traffic) on the Internet grew exponentially, Internet Exchange Points (IXPs)[71] were created around the world. These are places where ISPs connect to each other so that traffic from a user of any provider can reach users of any other provider, worldwide.

If you'd like to understand more about the physical implementation of the Internet, see *Tubes: A Journey to the Center of the Internet,*[72] by Andrew Blum. Highly recommended.

---

[65] https://en.wikipedia.org/wiki/Jim_Clark

[66] https://en.wikipedia.org/wiki/Netscape

[67] https://en.wikipedia.org/wiki/Netscape_Application_Server

[68] https://en.wikipedia.org/wiki/Netscape_Navigator

[69] https://en.wikipedia.org/wiki/Mosaic_(web_browser)

[70] https://en.wikipedia.org/wiki/Internet_service_provider

[71] https://en.wikipedia.org/wiki/Internet_exchange_point

[72] www.amazon.com/Tubes-Journey-Internet-Andrew-Blum/dp/0061994952

Summary and a Look Ahead

In this chapter, we covered the history of computer networking from the early days of modems and BBSs then to "real" early computer networking (DECnet, XNS, SNA, NetBIOS, NetWare, etc.) up to the consolidation of multiple network protocols into a single standard, which was TCP/IP (with version 4 of IP, or IPv4). We also covered the "precursor" to the IPv4-based Internet, which was called ARPANET, based on the host-host protocol (or NCP).

We covered some of the key participants (Bob Kahn, Vint Cerf) who created the technology used in the Second Internet and companies who pioneered the most widely used protocols, hardware, and applications (e.g., 3Com, Netscape).

Today, the most widely used application on the Second Internet is the World Wide Web. This evolved due to the widespread deployment of NAT and private addresses. These limited most users to only being able to make *outgoing* connections to a small number of centralized serves (that have public IP addresses), thereby centralizing functionality. This is comparable to dumb terminals (browsers) being used to access central mainframes (web servers) where the actual computing takes place.

In the Third Internet, there is no shortage of public IP addresses or need for NAT – even phones can now have public IP addresses and host servers or do end-to-end direct connections. This will lead to extensive decentralization. Now people can take advantage of the amazing computing power in their desktops, laptops, and even phones, rather than centralized computing on web servers. Rather than sending the entire GUI over the network (HTTP/HTML), we will run native GUI applications on our devices and exchange only data over the network. There will still be a need for shared databases, but web browsers and servers will decrease in importance. Zero-Trust Networks are a start in this direction. This is a *major* paradigm shift made possible by IPv6.

The original concept of the Internet was "complexity at the edge, simplicity at the core" – the World Wide Web has stood that on its head, primarily because of the limitation of IPv4 with NAT and private addresses.

CHAPTER 3

Review of IPv4

This chapter is a brief review of IPv4,[1] the foundation protocol of the Second Internet. I am covering it in this chapter to help you understand what is new and different in IPv6. It is not intended to be comprehensive. There are many great books listed in the bibliography if you wish to understand IPv4 at a deeper level. The reason IPv4 is relevant in this book is because the design of IPv6 is based heavily on that of IPv4. First, IPv4 can be considered one of the great achievements in IT history, based on its worldwide success, so it was a good model to copy from. Second, there were several attempts to do a new design "from the ground up" with IPv6 (a "complete rewrite"). These involved *really* painful migration and interoperability issues. You need to understand what the strengths and weaknesses of IPv4 are to see why IPv6 evolved the way it did. You can think of IPv6 as "IPv4 on steroids," which takes into account the radical differences in the way we do networking today and fixing problems that were encountered in the first three decades of the IP-based Internet, as network bandwidth and the number of nodes increased exponentially. We are doing things over networks today that *no one* could have foreseen a quarter of a century ago, no matter how visionary they were.

Network Hardware

There are many types of hardware devices used to construct an Ethernet network running TCP/IP. These include nodes, Network Interface Cards (NICs), cables, hubs, switches, routers, and firewalls.

A *node* is a device (usually a computer) that can do processing and has some kind of wired or wireless connection(s) to a network. Examples of nodes are desktop computers,

---

[1] https://en.wikipedia.org/wiki/IPv4

L. E. Hughes, *Third Generation Internet Revealed*, https://doi.org/10.1007/978-1-4842-8603-6_3

notebook computers, netbooks, smartphones, hubs,[2] switches,[3] routers,[4] wireless access points,[5] network printers, network-aware appliances, and so on. A node could be as simple as a temperature sensor, with no display and no keyboard, just a connection to a network. It could have a display and keyboard or be a "headless node" with a management interface accessed via the network with Telnet, Secure Shell[6] (SSH), or a web browser. All nodes connected to a TCP/IP network must have at least one valid IP address[7] (per interface). If a node has only *one* network interface, such as a workstation computer, it is called a *host*. If a node has *multiple* interfaces connected to different networks, and the ability to forward packets between them, it is called a *gateway* or a *router*. Routers and firewalls are special types of gateways that can forward packets between networks and/or control traffic in various ways as it is forwarded. Gateways make it possible to build *internetworks*.[8] They are described in more detail under the "IPv4 Routing" section in this chapter.

A *NIC*[9] (or *Network Interface Controller*) is the physical interface that connects a node to a network. It may also be called an *Ethernet adapter* if the network is based on Ethernet. It should have a female RJ-45[10] connector on it (or possibly a coax or fiber-optic connector). It could be an actual add-in Peripheral Computer Interconnect (PCI[11]) card. It could be integrated on the device's motherboard. It could also be something that makes a wireless connection to a network, using Wi-Fi, WiMAX, or similar standard. Typically, all NICs have a globally unique, hard-wired MAC address[12] (48 bits long, assigned by the manufacturer). A node can have one or more NICs (also called *interfaces*). Each interface can be assigned one or more IP addresses and various other relevant network configuration items, such as the address of the default gateway and the addresses of the DNS servers.

---

[2] https://en.wikipedia.org/wiki/Ethernet_hub

[3] https://en.wikipedia.org/wiki/Network_switch

[4] https://en.wikipedia.org/wiki/Router_(computing)

[5] https://en.wikipedia.org/wiki/Wireless_access_point

[6] https://en.wikipedia.org/wiki/Secure_Shell

[7] https://en.wikipedia.org/wiki/IP_address

[8] https://en.wikipedia.org/wiki/Internetworking

[9] https://en.wikipedia.org/wiki/Network_interface_controller

[10] https://en.wikipedia.org/wiki/Registered_jack#RJ45

[11] https://en.wikipedia.org/wiki/Conventional_PCI

[12] https://en.wikipedia.org/wiki/MAC_address

Network cables today are typically *unshielded twisted pair*[13] (UTP) cables that actually have *four* pairs of plastic-coated wires, with each pair forming a twisted coil. They have RJ-45 male connectors on each end. They could also be fiber-optic cables for very high-speed or long-run connections. Often today, professional contractors install UTP cables through the walls and bring them together at a central location (sometimes called the *wiring closet*) where they are connected together with a hub or a switch to form a star network.[14] Cables typically are limited to 100 meters or less in length, but the maximum acceptable length is a factor of several things, such as network speed and cable design. Modern cables rated as "CAT5"[15] or "CAT5E" are good up to 100 Mbps, while cables rated as "CAT6"[16] are good up to a gigabit per second (1 Gbps). Today, you can get CAT7 cables for speeds up to 10 Gbps. Above that speed, you should be using optical fiber[17] NICs and cables. It is also possible for twisted pair cables to be shielded if required to prevent interference from (or with) other devices.

An *Ethernet hub*[18] is a device that connects multiple Ethernet cables together so that any packet transmitted by any node connected to that hub is relayed to all the other nodes connected to the hub. It typically has a bunch of female RJ-45 connectors in parallel (called *ports*). In effect it ties together the network cables plugged into it into a star network. Hubs have a speed rating, based on what speed Ethernet they support. Older hubs might be only 10 Mbps. More recent ones might be "fast Ethernet," which means they support 100 Mbps. If you have five nodes (A, B, C, D, and E) connected together with a hub and node B sends a packet to node D, all nodes, including A, C, and E, will see the traffic. The nodes not involved in the transaction will typically just discard the traffic. This dropping of packets not addressed to a node is often done by the hardware in the NIC, so that it never interrupts the software driver. Many NICs have the ability to be configured in *promiscuous mode*.[19] When in this mode, they will accept packets (and make them available to any network application) whether those packets are addressed to this node or not. If this mode is selected, the dropping of packets not addressed to you must be done in software. However, sometimes you *want* to see all

---

[13] https://en.wikipedia.org/wiki/Twisted_pair

[14] https://en.wikipedia.org/wiki/Star_network

[15] https://en.wikipedia.org/wiki/Category_5_cable

[16] https://en.wikipedia.org/wiki/Category_6_cable

[17] https://en.wikipedia.org/wiki/Optical_fiber_cable

[18] https://en.wikipedia.org/wiki/Ethernet_hub

[19] https://en.wikipedia.org/wiki/Promiscuous_mode

traffic on the subnet. For instance, this would be useful with intrusion detection, for diagnostic troubleshooting, or for collecting network statistics. Hubs come in various sizes, from 4 ports up to 48 ports, and can even be coupled with other hubs to make large network "backbones." You can also have a hierarchy of hubs, where several hubs distributed around a company actually connect into a larger (and typically faster) central hub. Hubs do no processing of the packets; they are really just a cluster of Ethernet extenders[20] (repeaters) that clean up and relay any incoming signals from any port to all the other ports. Hubs are quite rare today. Most such devices today are now actually *switches*.

A *network switch*[21] is similar to a network hub but has some control logic that minimizes unnecessary traffic. It partitions a LAN into multiple *collision domains*[22] (one per switch port). Again, say you have a switch with cables connected to nodes A, B, C, D, and E. If B sends a packet to D, that packet will be sent out *only* to the port to which D is connected. Switches *learn* what nodes are connected to what ports by maintaining a table of MAC addresses vs. port number. When a switch is first powered on, this table is empty. As the nodes send packets through the switch, it learns what port each node is connected to.

If node A (connected to port 1) sends a packet to node B (connected to port 2), the switch adds the MAC address of A and the port it was seen on (1) to its table. In the future, when packets for A's MAC address come in any port, they will only be sent out port 1. Since the switch hasn't previously seen the MAC address of B (as a source address), it doesn't know where B is located, so it sends this first packet out to all ports. If B replies to A's packet, the switch adds B's MAC address and port (2) to the table. In the future, packets sent to B's MAC address will only be sent out port 2. Each addition to the table expires after a certain amount of time, to allow nodes to be moved to other ports. An incoming packet sent to a *broadcast* address will always be sent out to all ports. This behavior holds down excessive traffic that would normally just be dropped anyway by the unaddressed nodes (not to mention unnecessary packet collisions). It also provides a small degree of privacy, even if someone enables their NIC in promiscuous mode. If your LAN is built using switches instead of hubs, you can typically only sniff traffic originating from or terminating on the network segment connected to your port of the switch. Most switches are oblivious to IP addresses – they work only with MAC addresses. Because of

---

[20] https://en.wikipedia.org/wiki/Ethernet_extender·

[21] https://en.wikipedia.org/wiki/Network_switch

[22] https://en.wikipedia.org/wiki/Collision_domain

this, they are *IP version agnostic*. This means they will carry IPv4 or IPv6 traffic (or even other kinds of Ethernet traffic) so long as that traffic uses Ethernet frames with MAC addresses.

If you are using a switch, but one of your connected nodes really *does* want to see traffic from other network segments, some switches have a *mirror port* function that will allow all traffic from any combination of ports to be copied to one port, to which you connect the node that wants to monitor that traffic. This must be configured, which requires a management interface of some kind. Like hubs, switches come in various speeds, from 10 Mbps up to 1000 Mbps (1 Gbps). Unlike hubs, you can mix different speed nodes (10 Mbps, 100 Mbps, and even 1000 Mbps) on a single switch, so the speed rating is the *maximum* speed for nodes connected to it. Switches also come in sizes from 4 ports up to 48 ports, and better ones can be "stacked" (linked together) to effectively build a single giant switch. Lower-end (cheaper) switches may have few if any configuration options and may not even have a user interface. *Smart* (or *managed*) switches typically have a sophisticated GUI management interface (accessible via the network, usually over HTTP) or Command Line Interface (accessible either via a serial port, Telnet, or SSH) that allows you to configure various things and/or monitor traffic. Switches also typically include support for monitoring or control using SNMP (Simple Network Monitoring Protocol). Very advanced switches allow you to configure VLANs (Virtual Local Area Networks[23]), which allow you to effectively create multiple sub-switches that are not *logically* connected together, on a single physical switch. Some of these advanced functions process IP addresses (*layer 3* functionality) and hence are IP version *specific* (an IPv4-only smart switch cannot process IPv6 addresses, but the basic layer 2 switch functionality may work fine). Very recent smart switches do support both IPv4 and IPv6 (dual stack), for layer 3 functionality with both IP versions.

RFCs: The Internet Standards Process

Anyone studying the Internet, or developing applications for it, must understand the RFC[24] system. RFC stands for *Request for Comments*. These are the documents that define the Internet Protocol Suite (the official name for TCP/IP) and many related

---

[23] https://en.wikipedia.org/wiki/Virtual_LAN

[24] https://en.wikipedia.org/wiki/Request_for_Comments

topics. Anyone can submit an RFC. Ones that are part of the *Standards Track* are usually produced by the **IETF** (Internet Engineering Task Force) *working groups*. Anyone can start or participate in a working group. Submitted RFCs begin life as a series of *Internet Drafts*, each of which has a lifespan of 6 months or less. Most drafts go through considerable peer review, and possibly quite a few revisions, before they are either abandoned or approved and issued an official RFC number (e.g., 793) and become part of the official RFC collection. There are other kinds of documents in addition to the Standards Track, including information memos (FYI), humor (primarily ones issued on April 1), and even one obituary, for Jon Postel, the first RFC editor and initial allocator of IP addresses, RFC 2468,[25] "I Remember IANA," October 1998. There is even an RFC *about RFCs*, RFC 2026,[26] "The Internet Standards Process, Revision 3," October 1996. That is a good place to start if you really want to learn how to read RFCs.

The Internet Standards Process is quite different from the standards process of the **ISO** (International Organization for Standardization) that created the Open System Interconnection (OSI) network specification. The ISO typically develops large, complex standards with multiple four-year cycles, with hundreds of engineers and much political wrangling. This was adequate for creating the standards for the worldwide telephony system but is far too slow and hidebound for something as freewheeling and rapidly evolving as the Internet. The unique standards process of the IETF is one of the main reasons that TCP/IP is now the dominant networking standard worldwide. By the time OSI was specified, TCP/IP was already created, deployed, and being revised and expanded. OSI never knew what hit it.

Learning to read RFCs is an acquired skill, one that anyone serious about understanding the Internet, and most developers creating things for it, should master. There are certain "terms of art" (terms that have precise and very specific meanings), like the usage of MUST, SHOULD, MAY, and NOT RECOMMENDED. As an example, the IPv6-ready tests examine all the MUST (mandatory) and SHOULD (optional) items from relevant RFCs.

RFCs are readily available to anyone for free. Compare this with the ISO standards, which can cost over $1000 for a complete set of "fascicles" for something like X.500. Today you can obtain RFCs easily in various formats by use of a search engine such as Google or Yahoo. The "official" source is the URL:

---

[25] www.ietf.org/rfc/rfc2468.txt

[26] www.ietf.org/rfc/rfc2026.txt

`www.rfc-editor.org/rfc/rfcXXXX.txt`   (where XXXX is the
RFC number)

There is also an official RFC search page, where you can search for phrases (like
"TCP") in different tracks, such as RFC, STD, BCP, FYI, or all tracks. You can retrieve
ASCII or PDF versions. It is at

`www.rfc-editor.org/rfcsearch.html`

There are over 8000 RFCs today. I have included many references to the relevant
RFCs in this book. If you want to see all the gory details on any subject, go right to the
source and read it. You may find it somewhat tough going until you learn to read "RFC-
ese." A number of books on Internet technology are either just a collection of RFCs, or
RFCs make up a large part of the content. There is no reason today to do that – anyone
can download all the RFCs you want and have them in soft (searchable) form. I have
not included the text of even a single RFC in this book (warning: if you try to read this
book somewhere without Internet access like on a plane, you may want to look ahead
and download any relevant RFCs while you have Internet access). The casual reader
should not need to reference the actual RFCs. The complete set of RFCs is easily tens of
thousands of pages and growing daily.

Most of the topics covered in this book also have considerable coverage on the
Internet outside of the RFCs, such as in Wikipedia. Again, if you want to drill deeper in
any of these topics, crank up your favorite search engine and have at it. The information
is out there. What I've done is to try to collect together the essential information in a
logical sequence, with a lot of explanations and examples, plus all the references you
need to drill as deep as you like. I taught cryptography and Public Key Infrastructure for
VeriSign for two years, so I have a lot of experience trying to explain complex technical
concepts in ways that reasonably intelligent people can easily follow. Hopefully you will
find my efforts worthwhile.

IPv4

The software that made the Second Internet (and virtually all Local Area Networks)
possible has actually been around for quite some time. It is technically a *suite* (family) of
protocols. The core protocols of this suite are TCP (the *Transmission Control Protocol*)
and IP (*Internet Protocol*), which gave it its common name, TCP/IP. Its official name is
the *Internet Protocol Suite*.

TCP was first defined officially in RFC 675, "Specification of Internet Transmission Control Program," December 1974 (yes, 45 years ago). The protocol described in this document does not look much like the current TCP, and in fact, the Internet Protocol (IP) did not even exist at the time. Jon Postel was responsible for splitting the functionality described in RFC 675 into two separate protocols: (the new) TCP and IP. RFC 675 is largely of historical interest now. The modern version of TCP was defined in RFC 795, "Transmission Control Protocol – DARPA Internet Program Protocol Specification," September 1981 (7 years later). It was later updated by RFC 1122, "Requirements for Internet Hosts – Communication Layers," October 1989, which covers the Link Layer, IP Layer, and Transport Layer. It was also updated by RFC 3168, "The Addition of Explicit Congestion Notification (ECN) to IP," September 2001, which adds ECN to TCP and IP.

Both of these core protocols, and many others, will be covered in considerable detail in the rest of this chapter.

Four-Layer ("DoD") IPv4 Architectural Model

Unlike the OSI network stack, which really *does* have seven layers, the DoD network model has four layers, as shown in the following.

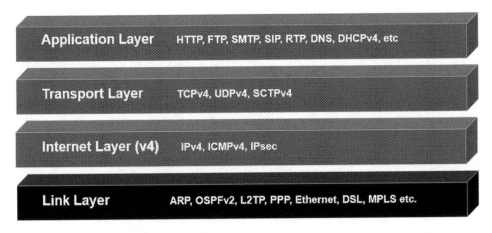

Figure 3-1. *Four-layer DoD model for IPv4*

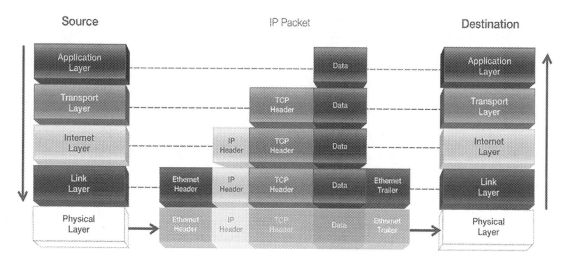

Figure 3-2. *Data flow in the four-layer model*

It just confuses the issue to try to figure out which of the seven OSI layers the various protocols of TCP/IP fit into. It is simply not applicable. It's like trying to figure out what color "sweet" is. The OSI seven-layer model did not even exist when TCP/IP was defined. Unfortunately, many people use terms like "layer 2" switches vs. "layer 3" switches. These refer to the OSI model. Books from Cisco Press and the Cisco certification exams are particularly adamant about using OSI terminology. I would be surprised if there is even a single actual OSI network running today. In this book we will try to consistently use the four-layer model terminology while referring to the OSI terminology when necessary for you to relate the topic to actual products or other books.

Note: Outgoing data begins in the application and is passed *down* the layers of the stack (adding headers at each layer) until it is written to the wire. Incoming data is read off the wire and travels *up* the layers of the stack (processing and removing headers at each layer) until it is accepted by the application. In the following discussion, for simplicity, I describe only the outgoing di**rection.**

The *Application Layer*[27] implements the protocols most people are familiar with (e.g., HTTP, SMTP, FTP). The software routines for these are typically contained in application programs such as browsers or web servers that make "system calls" to subroutines (or

---

[27] https://en.wikipedia.org/wiki/Application_layer

"functions" in C terminology) in the "socket API"[28] (an API is an Application Program Interface, or a collection of related subroutines, typically supplied with the operating system or C programming language compiler). The application code creates outgoing data streams and then calls routines in the socket API to actually send the data via TCP (Transmission Control Protocol) or UDP (User Datagram Protocol). Output to the *Transport Layer* is *[DATA]* using IP addresses.

The *Transport Layer*[29] implements TCP[30] (the Transmission Control Protocol) and UDP[31] (the User Datagram Protocol). These routines are internal to the socket API (hence live in Kernel Space[32]). In the case of TCP, packet sequencing, plus error detection and retransmission, is handled. The Transport Layer prepends a TCP or UDP packet header to the data passed down from the *Application Layer* and then passes the resulting packet down to the *Internet Layer* for further processing. Output to the *Internet Layer* is *[TCP HDR[DATA]]*, using IP addresses.

The *Internet Layer*[33] implements IP[34] (the Internet Protocol) and various other related protocols such as ICMP[35] (which includes the "ping" function among other things). The IP routine takes the data passed down from the *Transport Layer* routines, adds an IP packet header onto it, and then passes the now complete IPv4 packet down to routines in the *Link Layer*. Output to the Link Layer is *[IP HDR[TCP HDR[DATA]]]* using IP addresses.

The *Link Layer*[36] implements protocols such as ARP[37] (Address Resolution Protocol) that map IP addresses to MAC addresses for transmission between nodes in a single network link. It contains protocols such as Ethernet, Wi-Fi, and MPLS. It also contains routines that actually read and write data (as fed down to it by routines in the *Internet Layer*) onto the network wire, in compliance with Ethernet or other standards. Output to wire: Ethernet *frame* containing the IP packet, using MAC addresses (or other Link Layer addresses for non-Ethernet networks).

---

[28] www.ibm.com/support/knowledgecenter/en/SSLTBW_2.3.0/com.ibm.zos.v2r3.hali001/thesocketapi.htm

[29] https://en.wikipedia.org/wiki/Transport_layer

[30] https://en.wikipedia.org/wiki/Transmission_Control_Protocol

[31] https://en.wikipedia.org/wiki/User_Datagram_Protocol

[32] www.linfo.org/kernel_space.html

[33] https://en.wikipedia.org/wiki/Internet_layer

[34] https://en.wikipedia.org/wiki/Internet_Protocol

[35] https://en.wikipedia.org/wiki/Internet_Control_Message_Protocol

[36] https://en.wikipedia.org/wiki/Link_layer

[37] https://en.wikipedia.org/wiki/Address_Resolution_Protocol

Each layer "hides" the details (and/or hardware dependencies) from the higher layers. This is called "levels of abstraction." An architect thinks in terms of abstractions such as roofs, walls, windows, etc. The next layer down (the builder) thinks in terms of abstractions such as bricks, glass, mortar, etc. Below the level of the builder, an industrial chemist thinks in terms of formulations of clay or silicon dioxide to create bricks and glass. If the architect tried to think at the chemical or atomic level, it would be very difficult to design a house. Their job is made possible by using levels of abstraction. Network programming is analogous. If application programmers had to think in terms of writing bits to the actual hardware, applications such as web browsers would be almost impossible. Each Network Layer is created by specialists who understand the details at their level, and lower layers can be treated as "black boxes" by people working at the higher layers.

Another important thing about Network Layers is that you can make major changes to one layer, without impacting the other layers much at all. The connections between layers are well defined and don't change (much). This provides a great deal of separation between the layers. In the case of IPv6, the Internet Layer is almost completely redesigned internally, while the Link Layer and Transport Layer are not affected much at all (other than providing more bytes to store the larger IPv6 addresses). If your product is "IPv6-only," that's about the only change you would need to make to your application software (unless you display or allow entry of IP addresses). If your application is "dual stack" (can send and receive data over IPv4 or IPv6), then a few more changes are required in the Application Layer (e.g., to accept multiple IPv4 and IPv6 addresses from DNS and try connecting to one or more of them based on various factors or to accept incoming connections over both IPv4 and IPv6). This makes it possible to migrate (or "port") network software (created for IPv4) to IPv6 or even dual stack with a fairly minor effort. In comparison, changing network code written for TCP/IP to use OSI instead would probably involve a complete redesign and major recoding effort.

IPv4: The Internet Protocol, Version 4

IPv4 is the foundation protocol of the Second Internet and accounts for many of its distinguishing characteristics, such as its 32-bit address size, its addressing model, and its packet header structure and routing. IPv4 was first defined in RFC 791 "Internet Protocol," September 1981.

Relevant Standards for IPv4

**RFC 791, "Internet Protocol," September 1981
(Standards Track)**

**RFC 792, "Internet Control Message Protocol," September 1981
(Standards Track)**

**RFC 826, "An Ethernet Address Resolution Protocol,"
November 1982 (Standards Track)**

RFC 1256, "ICMP Router Discovery Messages," September 1991
(Standards Track)

RFC 2390, "Inverse Address Resolution Protocol," September 1998
(Standards Track)

**RFC 2474, "Definition of the Differentiated Services Field
(DS Field) in the IPv4 and IPv6 Headers," December 1998
(Standards Track)**

RFC 4650, "HMAC-Authenticated Diffie-Hellman for Multimedia
Internet KEYing (MIKEY)," September 2006 (Standards Track)

RFC 4884, "Extended ICMP to Support Multi-Part Messages," April
2007 (Standards Track)

RFC 4950, "ICMP Extensions for Multiprotocol Label Switching,"
August 2007 (Standards Track)

RFC 5494, "IANA Allocation Guidelines for the Address Resolution
Protocol (ARP)," April 1009 (Standards Track)

**RFC 5735, "Special Use IPv4 Addresses," January 2010 (Best
Current Practices)**

IPv4 Packet Header Structure

So what are these packet headers mentioned previously? In IPv4 packets, there is an IPv4 packet header,[38] then a TCP (or UDP) packet header, and then the packet data. Each header is a structured collection of data, including things such as the IPv4 address of the sending node and the IPv4 address of the destination node. Why are we getting down to this level of detail? Because some of the big changes from IPv4 to IPv6 have to do with the new and improved IP packet header architecture in IPv6. In this chapter, we'll cover the IPv4 packet header. Here it is.

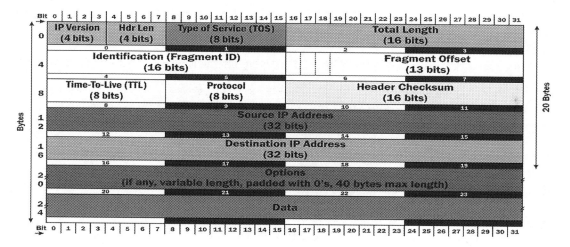

Figure 3-3. *IPv4 packet header*

The *IP Version* field (4 bits) contains the value 4, which in binary is "0100" (you'll never guess what goes in the first 4 bits of an IPv6 packet header!).

The *Header Length* field (4 bits) indicates how long the header is, in 32-bit "words." The minimum value is "5," which would be 160 bits, or 20 bytes. The maximum length is 15, which would be 480 bits, or 60 bytes. If you skip that number of words from the start of the packet, that is where the data starts (this is called the "offset" to the data). This will only ever be greater than 5 if there are options before the data part (which is not common).

---

[38] https://en.wikipedia.org/wiki/IPv4#Header

The *Type of Service* field (8 bits) is defined in RFC 2474,[39] "Definition of the Differentiated Services Field (DS Field) in the IPv4 and IPv6 headers," December 1998. This is used to implement a fairly simple QoS (Quality of Service). QoS involves management of bandwidth by protocol, by sender, or by recipient. For example, you might want to give your VoIP connections a higher priority than your video downloads or the traffic from your boss higher priority than your co-worker's traffic. Without QoS, bandwidth is on a first come–first served basis. 8 bits are not really enough to do a good job on QoS, and DiffServ is not widely implemented in current IPv4 networks. QoS is greatly improved in IPv6.

The *Total Length* field (16 bits) contains the total length of the packet (including the packet header) in bytes. The minimum length is 20 (20 bytes of header plus 0 bytes of data), and the maximum is 65,535 bytes (since only 16 bits are available to specify this). All network systems must handle packets of at least 576 bytes, but a more typical packet size is 1508 bytes. With IPv4, it is possible for some devices (like routers) to *fragment* packets[40] (break them apart into multiple smaller packets) if required to get them through a part of the network that can't handle packets that big. Packets that are fragmented must be *reassembled* at the other end. Fragmentation and reassembly is one of the messy parts of IPv4 that got cleaned up a lot in IPv6. A lot of hacking attacks exploit the messy scheme in IPv4.

The *Identification (Fragment ID)* field (16 bits) identifies which fragment of a once larger packet this one is, to help in reassembling the fragmented packet later. In IPv6 packet fragmentation is not done by intermediate nodes, so all the header fields related to fragmentation are no longer needed.

The next three bits are flags related to fragmentation. The first is reserved and must be zero (an April Fool's RFC[41] once defined this as the "evil" bit, which the sender should set if they are doing something malicious). The next bit is the **DF** (Don't Fragment) flag. If DF is set, the packet cannot be fragmented (so if such a packet reaches a part of the network that can't handle one that big, that packet is dropped). The third bit is the **MF** (More Fragments) flag. If MF is set, there are more fragments to come. Unfragmented packets of course have the MF flag set to zero.

---

[39] https://tools.ietf.org/html/rfc2474

[40] https://en.wikipedia.org/wiki/IP_fragmentation

[41] www.ietf.org/rfc/rfc3514.txt

The *Fragment Offset* field (13 bits) is used in reassembly of fragmented packets. It is measured in 8-byte blocks. The first fragment of a set has an offset of 0. If you had a 2500-byte packet, and were fragmenting it into chunks of 1020 bytes, you would have three fragments as follows:

Fragment ID	MF Flag	Total Length	Data Size	Offset
1	1	1020	1000	0
2	1	1020	1000	125
3	0	520	500	250

The *Time-To-Live (TTL)* field (8 bits) is to prevent packets from being shuttled around indefinitely on a network. It was originally intended to be lifetime in seconds (hence the name), but it has come to be implemented as "hop count." This means that every time a packet crosses a switch or router, the hop count is decremented by one. If that count reaches zero, the packet is dropped. Typically, if this happens, an ICMPv4 message ("Time Exceeded") is returned to the packet sender. This mechanism is how the *traceroute* command works. The primary purpose of TTL is to prevent *looping* (packets running around in circles).

The *Protocol* field (8 bits) defines the type of data found in the data portion of the packet. Protocol numbers are not to be confused with *ports*. Some common protocol numbers are

1	ICMP	Internet Control Message Protocol (RFC 792)
6	TCP	Transmission Control Protocol (RFC 793)
17	UDP	User Datagram Protocol (RFC 768)
41	IPv6	IPv6 tunneled over IPv4 (RFC 2473)
83	VINES	Banyan Vines IP
89	OSPF	Open Shortest Path First (RFC 1583)
132	SCTP	Streams Control Transmission Protocol (RFC 4960)

The *Header Checksum* field (16 bits) is the 16-bit one's complement of the one's complement sum of all 16-bit words in the header. When computing, the checksum field itself is taken as zero. To validate the checksum, add all 16-bit words in the header together including the transmitted checksum. The result should be 0. If you get any other value, then at least 1 bit in the packet was corrupted. There are certain multiple bit errors that can cancel out, and hence bad packets can go undetected. Note that since the hop count (TTL) is decremented by one on each hop, the IP header checksum must be recalculated at each hop. The IP header Checksum was eliminated in IPv6.

The *Source IP Address* field (32 bits) contains the IPv4 address of the sender (may be modified by NAT).

The *Destination IP Address* field (32 bits) contains the IPv4 address of the recipient (may be modified by NAT in a reply packet).

Options (0–40 bytes) is not often used. These are not relevant to this book. If you want the details, read the RFCs.

Data (variable number of bytes) is the data part of the packet – not really part of the header. This is not included in the IP header checksum. The number of bytes in the Data field is the value of "Total Length" minus the value of "Header Length."

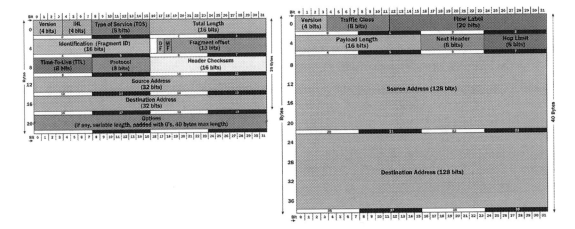

Figure 3-4. *IPv4 and IPv6 packet headers side by side*

Changes from IPv4 Header to IPv6 Header

- IPv4 *IHL* field discarded (IPv6 header is fixed length, 40 bytes long)

- IPv4 *Type of Service* field became the IPv6 *Traffic Class* field

- IPv4 *Total Length* field became the IPv6 *Payload Length* field (and no longer includes length of basic header)

- IPv4 *Fragmentation* fields moved to the IPv6 *Fragment Extension Header*

- IPv4 *Time To Live* field became the IPv6 *Hop Limit* field

- IPv4 *Protocol* field became the IPv6 *Next Header* field

- IPv4 *Header Checksum* field discarded (not needed in IPv6 header)

- 32-bit IPv4 *Source Address* field grew to 128-bits in IPv6 header

- 32-bit IPv4 *Destination Address* field grew to 128-bits in IPv6 header

- New 20-bit *Flow Label* field added to the IPv6 header

IPv4 Addressing Model

In IPv4, addresses are 32 bits in length. They are really just integer numbers from 0 to 4,294,967,295. For the convenience of humans, these numbers are typically represented in *dotted decimal notation*. This splits the 32-bit addresses into four 8-bit fields and then represents each 8-bit field with a decimal number from 0 to 255. These decimal numbers cover all possible 8-bit binary patterns from 0000 0000 to 1111 1111. The decimal numbers are separated by "dots" (periods). Leading zeros can be eliminated. The following are all valid IPv4 addresses represented in dotted decimal:

123.45.67.89	A globally routable address (valid anywhere on the Internet)
10.3.1.51	A private address (valid only within a single LAN, as per RFC 1918)
255.255.255.255	The broadcast address for IPv4
127.0.0.1	The loopback address for IPv4

Originally there were five *classes* of IPv4 addresses, as defined in RFC 791,[42] "Internet Protocol," September 1981:

Class A: First bit 0 (0.0.0.0–127.255.255.255), 8-bit network number, 24-bit node within network number, subnet mask 255.0.0.0. There are 128 class A networks, each containing 16.7M addresses.

Class B: First 2 bits "10" (128.0.0.0–191.255.255.255), 16-bit network number, 16-bit node within network number, subnet mask 255.255.0.0. There are 16,384 class B networks, each containing 65,536 addresses.

Class C: First 3 bits "110" (192.0.0.0–223.255.255.255), 24-bit network number, 8-bit node within network number, subnet mask 255.255.255.0. There are 2M class C networks, each containing 256 addresses.

Class D: First 4 bits "1110" (224.0.0.0–239.255.255.255), used for multicast.

Class E: First 4 bits "1111" (240.0.0.0–255.255.255.255), experimental/reserved (not forwarded by most routers).

Network Ports

Each IP address on a network node has 65,536 ports associated with it (the port number is a 16-bit value, and 2 to the 16th is 65,536). Any of those ports can either be used to make an outgoing connection or to accept incoming connections. There is a list of

---

[42] https://tools.ietf.org/html/rfc791

well-known ports[43] that associates particular ports with certain protocols. For example, port 25 is associated with SMTP. There is nothing magical (or email-ish) about port 25. SMTP will work just as well on any other port, for example, 10025. Use of port 25 for SMTP is simply a convention that many people adopt. Such conventions make it easier to locate the SMTP server on a node you might not be familiar with. To be specific, ports are a Transport Layer thing, and there are really 65,536 TCP ports and another 65,536 UDP ports for each address. IP and ICMP, which are Internet Layer things, do not have any port(s) associated with them.

Anyone can reserve port numbers[44] with IANA. I happen to have been awarded two port numbers, 4604 for my Identity Registration Protocol (IRP) and 4605 for my SixChat protocol. IANA reviewed both requests and determined that they were innovative (did not duplicate any other protocols) and viable (did something useful) and met all requirements for a modern protocol (e.g., support for Explicit TLS).

Service Name ⊠	Port Number ⊠	Transport Protocol ⊠	Description ⊠	Assignee ⊠	Contact ⊠	Registration Date ⊠	Modification Date ⊠
irp	4604	tcp	Identity Registration Protocol	[Sixscape_Communications_Pte_Ltd]	[Lawrence_E._Hughes]	2014-03-17	2014-08-26
	4604	udp	Reserved				

Figure 3-5. *IRP port number registration*

When you deploy an Internet server (e.g., an SMTP server for sending and receiving email), the software opens a *socket* (a programming abstraction) in *listen mode* on a particular port (in the case of SMTP, port 25). An email client that wants to connect to it creates its own socket in *connect mode* and tells it to connect to a particular IP address (that of the SMTP server) using a particular port (in this case 25). When the connection attempt reaches the server, the server detects the attempt and accepts the connection (actually the port on the server that the connection is accepted on will be any available port, typically higher than 1024). A well-written server would then make a clone of itself (this is called *forking* in UNIX speak) and then go back to listening for further connections, while its clone went ahead and processed the connection. When the processing is complete on a given connection, the sockets used would be closed (on both server and client), and the clone of the server will quietly commit suicide. In

---

[43] https://en.wikipedia.org/wiki/List_of_TCP_and_UDP_port_numbers

[44] www.iana.org/assignments/service-names-port-numbers/service-names-port-numbers.xhtml

theory you could have thousands of clones of the server all simultaneously handling email connections on a single server (given sufficient memory and other resources). Busy web servers (like those at Google) often have *many* thousands of connections being processed at any given time (but never more than 65,000 on a given interface – each connection uses up one port).

If *threads* are used instead of *processes*, the scheme is similar but has far less overhead.

In UNIX, ports with numbers under 1024 are *special*, and only software that has root privilege can use them. Most common Internet services use ports in that range. There are many *well-known ports*, but here are a few of the more common ones:

```
22 – SSH (Secure Shell)
25 – SMTP (client to server and server to server E-mail protocol)
53 – DNS (Domain Name System)
80 – HTTP (world wide web)
110 – POP (server to client E-mail retrieval)
143 – IMAP (more modern server to client E-mail retrieval)
389 – LDAP (directory service)
443 – HTTPS (world wide web over SSL)
```

IPv4 Subnetting

This leads us naturally into the topic of IPv4 subnetting.[45] This is one of the more difficult areas of networking for people learning to work with IPv4. All addresses have two "parts," the first part being the address of the network (e.g., 192.168.0.0) and the second being the node within that network (e.g., 0.0.2.5). These two parts can be split apart at some "bit boundary." In this case, the address of the network is in the first 16 bits, and the node within the network is in the last 16 bits. The addresses of all nodes in such a network share the same first 16 bits, but each has a unique last 16 bits. So such a network might have nodes with addresses 192.168.2.5, 192.168.3.7, and 192.168.200.12, but *not* one with the address 192.169.2.1 (that address is in network *192.169.0.0*, not network 192.168.0.0).

A *subnet*[46] *mask* is a 32-bit value in which the first n bits (n=1–32) have the value 1 and the remaining 32-n bits have the value 0. It is used to split an IPv4 address into its two parts (the first n bits and the last 32-n bits). In the network just described, the subnet

---

[45] www.cisco.com/c/en/us/support/docs/ip/routing-information-protocol-rip/13788-3.html

[46] https://en.wikipedia.org/wiki/Subnetwork

mask is *255.255.0.0* (the first 16 bits have the value 1; the last 16 bits have the value 0). You do a Boolean "AND" function of the address with the subnet mask to get the address of the network and a Boolean "AND" of the address with the *one's complement* of the subnet mask (in this case 0.0.255.255) to get the node within subnet. This is difficult to visualize in dotted decimal. It is rather more obvious in binary. The Boolean "AND" function produces a 1 if both inputs are 1; else, it produces a 0. The "one's complement" (Boolean "NOT") function changes each 0 to a 1 and each 1 to a 0. With the "AND" function, where there is a 1 in the mask, the corresponding bit of the address "flows through" to the result. Where there is a 0 in the mask, the corresponding bit of the address is blocked (forced to the value 0). The following example (with addresses and mask shown in both dotted decimal and binary) should make this clear.

```
Address                192.168.2.5      1100 0000 1010 1000 0000 0010 0000 0101
Subnet mask            255.255.0.0      1111 1111 1111 1111 0000 0000 0000 0000
AND                                     ---- ---- ---- ---- ---- ---- ---- ----
Network address        192.168.0.0      1100 0000 1010 1000 0000 0000 0000 0000

Address                192.168.2.5      1100 0000 1010 1000 0000 0010 0000 0101
NOT mask               0.0.255.255      0000 0000 0000 0000 1111 1111 1111 1111
AND                                     ---- ---- ---- ---- ---- ---- ---- ----
Node within network    0.0.2.5          0000 0000 0000 0000 0000 0010 0000 0101
```

Figure 3-6. *IP addresses and subnet masks*

For subnet mask 255.0.0.0 (class A), the first 8 bits are the network address, and the last 24 are the node within subnet.

For subnet mask 255.255.0.0 (class B), the first 16 bits are the network address, and the last 16 are the node within subnet.

For subnet mask 255.255.255.0 (class C), the first 24 bits are the network address, and the last 8 are the node within subnet.

Subnetting was easy when the three IP address classes (A, B, and C) were used. The first few bits of the address determined the subnet mask. If the first bit of the 32-bit address was "0," then the address was class A, and the subnet mask was 255.0.0.0. If the first 2 bits of the address were "10," then the address was class B, and the subnet mask was 255.255.0.0. If the first 3 bits of the address were "110," then the address was class C, and the subnet mask was 255.255.255.0. This could actually be done automatically, so no one worried about subnet masks.

One of the changes made in the IPv4 addressing model in the mid-1990s was to introduce Classless Inter-domain Routing,[47] in RFC 1519,[48] "Classless Inter-domain Routing (CIDR)," September 1993. It was later replaced by RFC 4632[49], "Classless Inter-domain Routing (CIDR)," August 2006.

When CIDR was introduced, there were two consequences. First, the split between the two parts of the address could come at any bit boundary, not just after 8, 16, and 24 bits. Second, several small blocks (e.g., /28 blocks) could be carved out of a bigger block anywhere in the address space (perhaps from an old class A block, such as 7.x.x.x), so you could no longer determine the correct subnet mask by looking at the first few bits of an address. For example, a "/8" (class A) block might be carved up into 65,536 "/24" (class C) subnets, which could be allocated to different organizations.

Let's say your ISP, instead of giving you a class C block, only gives you a "/28" block of real (routable) IPv4 addresses, which would be 16 real IPv4 addresses, for example, 123.45.67.0 through 123.45.67.15. First, two of these addresses are not usable (may not be assigned to nodes). 123.45.67.0 is the "network address," and 123.45.67.15 is the "broadcast address." That leaves 14 usable addresses (123.45.67.1 through 12.45.67.14). So what is your subnet mask? If you check the preceding table of useful CIDR block sizes, a /28 subnet has a subnet mask of 255.255.255.240. In binary that is 1111 1111 1111 1111 1111 1111 1111 0000 (first 28 bits are 1; last 4 bits are 0). However, by the old rules (first bit is a 0), these are really from a "class A" block, so the automatically generated subnet mask would have been 255.0.0.0, which is not correct.

Now, what if your organization really has 100 nodes that need IP addresses? How do you give each of them unique addresses if you only have 14 usable addresses to work with? That's where Network Address Translation (NAT) comes in (covered in the next chapter). If you think CIDR made your life more "interesting," wait until you see what NAT does to it! Getting rid of CIDR and NAT is one of the big wins in IPv6. In fact, you will find that the entire subject of subnets has become totally trivial.

---

[47] https://en.wikipedia.org/wiki/Classless_Inter-Domain_Routing
[48] https://tools.ietf.org/html/rfc1519
[49] https://tools.ietf.org/html/rfc4632

MAC Addresses

IPv4 addresses are not actually used at the lowest layer of the IPv4 network stack (the *Link Layer*). Each network hardware interface actually has a 48-bit "MAC address" burned into it by the manufacturer. The first 24 bits of this (called the "Organizationally Unique Identifier[50]" or OUI) specify the manufacturer and are purchased by vendors from the IEEE[51] (Institute of Electrical and Electronics Engineers). A given vendor may have multiple OUIs, but a given OUI is associated only with one vendor. The last 24 bits of this (called the "Network Interface Controller–Specific" part) are assigned by each manufacturer, to be unique within a given OUI. This means that the entire 48-bit value is globally unique. For example, Dell Computer has a number of OUIs assigned to them by the IEEE, including 00-06-5B, 00-08-74, and 00-18-8B. If you encounter a NIC with a MAC address that has one of those sets of 24 bits, it was made by Dell Computer. When you use the command "ipconfig /all" in Windows, you get a list of network configuration information for all your interfaces (some of which are "virtual"). If you look for "Local Area Connection," that is information about your main (or only) network connection to your LAN. Under that, you will see an item labeled "Physical Connection," followed by six pairs of hex digits, separated by dashes. That is the MAC address of your Network Interface Controller (NIC). Mine is 00-18-8B-78-DA-1A. This means my NIC was made by Dell (my whole computer was, but the MAC address doesn't tell you that). Actually, since the NIC I'm using is on the motherboard (not an add-on PCI card), this does tell me the motherboard was made by Dell.

You can look up the vendor of any device based on its OUI (or MAC-48 address). See `www.whatsmyip.org/mac-address-lookup/`

This site tells me that the Ethernet adapter in my desktop computer (MAC address 9C-5C-8E-8F-2F-B0) was created by ASUSTek Computer Inc.

Network switches come in two varieties. "Layer 2" switches (which I would call *Link Layer* switches) only work with MAC addresses. They don't even "see" IP addresses. Hence, "Layer 2" switches are IP version agnostic; they work equally well with IPv4 or IPv6 or a mixture of the two (dual stack). "Layer 3" switches (sometimes called "smart" switches) work with MAC addresses, but they *also* understand and can see IP addresses (these work at *both* the *Link Layer* and the *Internet Layer*, in terms of the four-layer Model). They can do things like create VLANs (Virtual LANs) to segregate traffic based

---

[50] https://en.wikipedia.org/wiki/Organizationally_unique_identifier
[51] https://en.wikipedia.org/wiki/Institute_of_Electrical_and_Electronics_Engineers

on IP addresses. An IPv4-only "layer 3 switch" cannot work with IPv6 traffic (or at least none of its "higher-level" functions will affect IPv6 traffic). There are now a few dual-stack "layer 3" switches on the market, such as the SMC 8848M, which I happen to be running in my home network. I can even manage it over IPv6 (via Web and SNMP) and create VLANs[52] based on IPv6 addresses.

Mapping from IPv4 Addresses to Link Layer Addresses

The software in the *Application Layer*, the *Transport Layer*, and the *Internet Layer* of the IPv4 stack work with IP addresses. But the *Link Layer* (and the hardware) works with MAC addresses (or other Link Layer addresses). How do IPv4 addresses get mapped onto Link Layer addresses?

Address Resolution Protocol (ARP)

There are two protocols in IPv4 (that don't even exist in IPv6) called ARP[53] (Address Resolution Protocol) and InARP[54] (Inverse Address Resolution Protocol). These protocols live in the Link Layer. ARP maps IP addresses onto Link Layer addresses. This is kind of like the mapping between FQDNs and IP addresses done in the Application Layer by DNS, but down in the Link Layer. InARP maps Link Layer addresses onto IP addresses (kind of like a reverse DNS lookup).

ARP is defined in RFC 826,[55] "An Ethernet Address Resolution Protocol," November 1982. ARP operates only within the network segment (routing domain) that a host is connected to. It does not cross routers. It is used to determine the necessary Link Layer addresses to get a packet from one node in a subnet to another node in the same subnet (which could be a "default gateway" node that knows how to relay it on further). But for the hop from the sender to the default gateway, it is the same problem as getting the packet to any other local node. When the sender goes to send a packet, if the recipient's address is on the local link, an ARP request is done for the recipient's address, and the packet is sent to the recipient. If the recipient's address is *not* on the local link, an ARP

---

[52] https://en.wikipedia.org/wiki/Virtual_LAN

[53] https://en.wikipedia.org/wiki/Address_Resolution_Protocol

[54] www.oreilly.com/library/view/internet-core-protocols/1565925726/ch03s01s03s01.html

[55] https://tools.ietf.org/html/rfc826

request is done instead for the sender's default gateway address, and the packet is sent to the default gateway node, which will then worry about forwarding it on toward the real recipient.

Say Alice (one IPv4 node) wants to send a packet to Bob (another IPv4 node, on the same Ethernet network segment). Assume Alice does not currently know Bob's MAC address. Each machine has a table of IP addresses and MAC addresses (called an *ARP table*). At this time, there is no entry in Alice's ARP table with Bob's IP address and MAC address. So Alice *first* sends an Ethernet ARP request to all machines on the network segment (using the Ethernet broadcast address), with the following info:

```
opcode = REQUEST
hardware type = Ethernet,
protocol = IPv4
sender's IP address = Alice's IPv4 address (she knows her own IP address)
sender's MAC address = Alice's MAC address (she knows her own MAC address)
recipient's IP address = Bob's IPv4 address (she knows Bob's IP address)
recipient's MAC address = all zeros (she doesn't know Bob's MAC address yet)
```

All machines on the network segment will receive the packet, but everyone other than Bob will ignore it ("Not for me – IGNORE!"). Bob understands Ethernet protocol and IPv4. He recognizes his own IPv4 address ("It's for ME!"). He adds Alice's IPv4 address and Alice's MAC address into his ARP table (for future reference) and then sends a response Ethernet ARP packet back to Alice, using her MAC address (which he now knows) instead of the broadcast address, with the following info:

```
opcode = RESPONSE
hardware type = Ethernet
protocol = IPv4
sender's IP address = Bob's IPv4 address (he knows his own IP address)
sender's MAC address = Bob's MAC address (he knows his own MAC address)
recipient's IP address = Alice's IPv4 address (from the request)
recipient's MAC address = Alice's MAC address (from the request)
```

Only Alice gets the response (this was not a broadcast). Alice sees that this is a RESPONSE, and the sender's address tells her whom the response was from. Alice then adds Bob's IP address and MAC address into her ARP table. Now that she knows how to send things to Bob, she goes ahead with sending the packet that she originally was trying to send. This process is called *address resolution*, hence the name *Address Resolution Protocol*.

The ARP table has expiration times (TTL), and when an entry becomes "stale," it will be discarded, and the next time a packet is sent to that address, a new fresh entry will be added to the ARP table.

In Windows, you can view your ARP table at any time, in a DOS window, with the command "arp –a". The results might look something like the following.

```
C:\Users\lhughes.HUGHESNET>arp -a

Interface: 172.20.2.1 --- 0xb
  Internet Address        Physical Address        Type
  172.20.0.1              00-1b-21-1d-c1-59        dynamic
  172.20.0.11             00-17-a4-ec-11-9c        dynamic
  172.20.0.12             00-e0-81-47-fa-ce        dynamic
  172.20.0.13             00-15-f2-2e-b4-1c        dynamic
  172.20.0.21             00-18-f3-2e-32-87        dynamic
  172.20.0.88             00-14-fd-12-fa-5a        dynamic
  172.20.1.6              00-1e-90-1e-5b-4f        dynamic
  172.20.1.8             00-1e-65-97-de-e0        dynamic
  172.20.1.9              00-15-f2-2e-b4-1c        dynamic
  172.20.255.255         ff-ff-ff-ff-ff-ff        static
  224.0.0.22             01-00-5e-00-00-16        static
  224.0.0.252            01-00-5e-00-00-fc        static
  224.111.140.122        01-00-5e-6f-8c-7a        static
  226.196.145.70         01-00-5e-44-91-46        static
  237.62.223.84          01-00-5e-3e-df-54        static
  239.255.255.250        01-00-5e-7f-ff-fa        static
```

Figure 3-7. *Reverse ARP listing*

Inverse ARP (InARP)

There is another protocol called Inverse ARP (InARP) that maps Link Layer addresses onto IP addresses. InARP is defined in RFC 2390,[56] "Inverse Address Resolution Protocol," September 1998.

InARP is needed only by a few network hardware devices (like ATM). It works almost exactly like ARP, except different opcodes are used and the sender sends the recipient's MAC address (which it knows), but zero fills the recipient's IP address (which it wants to know). The recipient recognizes its own MAC address and responds with the same information that it does to an ARP. The older RARP (Reverse ARP) protocol is now deprecated.

---

[56] https://tools.ietf.org/html/rfc2390

Types of IPv4 Packet Transmissions

The most common type of packet transmission is *unicast*.[57] This is when one node (A) sends a packet to just one other node (B). A and B can be in the same local link or halfway around the world. So long as routable IP addresses are used and a routing path is available between A and B, it is still called unicast.

Another kind of transmission is *broadcast*[58] (covered in more detail in the following). Here a node can transmit a packet to *all* nodes in the local link. Any node not interested in a broadcast packet will just drop it. If the packet was an ICMP Echo Request (ping), all nodes on the local link might reply to it, which could cause a lot of excess traffic.

There is another kind of transmission called *anycast*.[59] Here a node can transmit a packet to a single node out of a set of some collection of nodes (e.g., the "nearest" DNS server). Usually only a single node will accept the transmission and reply to the sender. This mechanism is somewhat limited in IPv4 but works really well in IPv6. DNS anycast is used with the root DNS servers to allow multiple copies of each root server, to handle the load and minimize turnaround on root server requests. DNS anycast is usually done at the BGP routing level.

There is one more kind of transmission called *multicast*.[60] Here one node can send a single stream of packets, such as a digitized radio program, and any number of recipient nodes can *subscribe* to that multicast and receive it. Usually listening is a passive act; no responses are sent to the sender. The sender has no knowledge of which or even how many nodes are receiving the transmission. It is efficient because other nodes further along the network handle replication of the traffic to nodes beyond them. This is analogous to many radios receiving a transmission from a single radio transmitting station. This is covered in more detail in the following. This is supported in IPv4 but works far better in IPv6.

IPv4 Broadcast

Any node can send a packet to a special IPv4 address (255.255.255.255), and all nodes on the local link will receive it. Any destination address that has all ones in the "node within

---

[57] https://en.wikipedia.org/wiki/Unicast

[58] https://en.wikipedia.org/wiki/Broadcasting_(networking)

[59] https://en.wikipedia.org/wiki/Anycast

[60] https://en.wikipedia.org/wiki/Multicast

subnet" field is broadcast (e.g., 172.16.255.255 in 172.16/16). Usually, there is some kind of information in the packet that allows most nodes to realize that packet does not concern them (e.g., if a broadcast packet contains a DHCPv4 request, all nodes that don't have a DHCPv4 server will ignore it). This mechanism can help locate servers or solve other problems (like not yet having a valid IP address), but it can put unnecessary loads on all nodes that aren't involved. It can also lead to broadcast storms, which involve massive amounts of useless traffic clogging or totally shutting down an IPv4 network. As an example, a "smurf attack"[61] sends zillions of pings to the broadcast address with the source address containing the *spoofed* address of the node under attack (not the address of the actual sender). All nodes on the local link "respond" to the poor node under attack, which *amplifies* the attack. There are certain kinds of misconfigurations or hardware failures in network switches that can cause broadcast storms as well.

Packets sent to the broadcast address do not cross routers (or VLAN boundaries), so appropriate use of these can limit the extent of disruption due to excessive broadcasts or storms. The set of nodes that a broadcast will reach is called a *broadcast domain*. Switches do not block broadcasts – they relay packets with a broadcast destination address out all ports (unlike packets with a unicast destination address).

Broadcast is used in the DHCPv4, to allow a node to find and communicate with the DHCPv4 server before it even gets an address.

Broadcast does not exist in IPv6, because it can be so trouble-prone. Other mechanisms (e.g., multicast or solicited node multicast) are used to locate DHCPv6 servers or solve other problems for which broadcast may be used in IPv4.

IPv4 Multicast

Multicast allows a node to transmit a stream of data to one of a number of special "multicast" addresses. Multicast supports only UDP, not TCP. Any number of other nodes can subscribe to that address and receive the datagrams. As one example, this could be used to send "broadcast" (in the media sense) radio or television programs. Multicast packet transmission differs from broadcast packet transmission in that only nodes that have *subscribed* to that multicast address receive the packets.

Sites like YouTube, and services like "on-demand" television, use traditional unicast (one sender connecting to one recipient) transmissions to each user. This requires a

---

[61] https://en.wikipedia.org/wiki/Smurf_attack

great deal of bandwidth and a powerful network infrastructure at the transmission site, especially if there are a large number of recipients (potentially millions). Multicast is necessary to bring costs and network bandwidth requirements low enough to make it competitive with media "broadcast" over satellite or cable systems.

There are several mechanisms and protocols involved in IPv4 multicast:

- An IP multicast group address (one of the IPv4 "class D" addresses described previously)

- A sending node that can convert some kind of data such as audio and/or video into digital form and transmit the resulting UDP packets to that multicast group address

- A multicast distribution tree, where every router crossed supports multicast operation

- A new protocol called IGMP (Internet Group Management Protocol) that allows clients to subscribe to a particular multicast transmission

- Another new protocol called PIM (Protocol Independent Multicast) that sets up multicast distribution trees

- Clients that can "subscribe" to specific multicast addresses (receive the data being transmitted by the sender) and process the received digital data into some kind of service, such as audio or video

Assuming there is a multicast program available on a particular multicast address (e.g., 239.1.2.3), a consumer can use a multicast client application to extend the distribution tree associated with that address to reach their computer. This corresponds to selecting a channel on a television. There may be multiple routers between the sender and this subscriber. All those routers must support multicast and be informed to replicate packets from the sender to that recipient. IGMP[62] is used to subscribe to a specific multicast address, and PIM[63] is used to inform all intervening routers to extend the distribution tree to this client. The multicast server does not need to know anything about the recipients and does not get any response from them. The creation of the distribution tree and subscriptions to particular multicast addresses are handled by the clients and intervening multicast routers, not by the multicast server.

---

[62] https://en.wikipedia.org/wiki/Internet_Group_Management_Protocol
[63] https://en.wikipedia.org/wiki/Protocol_Independent_Multicast

Unlike unicast routers, a multicast router does not need to know how to reach all possible distribution trees, only those for which it is passing traffic from a sender to a recipient. If there is no recipient subscribed to a given channel "downstream" from a router (from the sender to recipient), there is no need for it to replicate packets and forward them downstream. If a recipient downstream from that router subscribes to a particular address, then that router will start replicating incoming upstream packets from the multicast address and relay them downstream toward that recipient (or recipients). This is called adding a "graft" onto the tree. If there are recipients downstream on a particular path from a multicast router and the last one "tunes out," then the last router in the path between the server and that node is informed to stop replicating packets along that path. This is called "pruning" the distribution tree. It is possible that one subscriber "tuning out" could result in an entire chain of multicast routers being pruned if there are no other subscribers down that path.

Multicast is often used for services such as IPTV, including applications such as distance learning. Not all IPv4 routers support multicast and the related protocols, so IPv4 multicast works best in "walled garden" networks, for example, within a single ISP's network (e.g., Comcast subscriber accessing multicast content from Comcast). In such a situation, it is possible to ensure that all intervening routers support the necessary protocols (which are optional in IPv4).

It is possible to build a fully IPv4 multicast-compliant router using open source operating systems and an open source package called XORP[64] (eXtensible Open Router Platform, at *www.xorp.org*). XORP was first developed for FreeBSD, but is available on Linux, OpenBSD, NetBSD, and Mac OS X. The XORP technology and team was transferred to a commercial startup backed by VCs (called XORP Inc.[65]). Many modern enterprise-class routers support Ipv4 multicast, but not all do. Not many small office/home office (SOHO)–class routers do. In IPv6, multicast is an integral part of the standard, and support is mandatory in *all* IPv6-compliant devices. It also works in a very different way and is much more scalable.

Internet Relay Chat[66] (IRC) uses a different approach to multicast (not the standard multicast protocols) and creates a spanning tree across its overlay network to all nodes that subscribe to a given chat channel. Unlike multicast-delivered media content, IRC is a two-way channel.

---

[64] https://en.wikipedia.org/wiki/XORP

[65] www.xorp.org/

[66] https://en.wikipedia.org/wiki/Internet_Relay_Chat

Relevant Standards for IPv4 Multicast

RFC 1112, "Host Extensions for IP multicasting," August 1989 (Standards Track)

RFC 2236, "Internet Group Management Protocol, Version 2," November 1997 (Standards Track)

RFC 2588, "IP Multicast and Firewalls," May 1999 (Informational)

RFC 2908, "The Internet Multicast Address Allocation Architecture," September 2000 (Informational)

RFC 3376, "Internet Group Management Protocol, Version 3," October 2002 (Standards Track)

RFC 3559, "Multicast Address Allocation MIB," June 2003 (Standards Track)

RFC 3973, "Protocol Independent Multicast – Dense Mode (PIM-DM)," January 2005 (Experimental)

RFC 4286, "Multicast Router Discovery," December 2005 (Standards Track)

RFC 4541, "Considerations for Internet Group Management Protocol (IGMP) and Multicast Listener Discovery Protocol (MLD) Snooping Switches," May 2006 (Informational)

RFC 4601, "Protocol Independent Multicast – Sparse Mode (PIM-SM): Protocol Specification (Revised)," August 2006 (Standards Track)

RFC 4604, "Using Internet Group Management Protocol Version 3 (IGMPv3) and Multicast Listener Discovery Protocol Version 2 (MLDv2) for Source-Specific Multicast," August 2006 (Standards Track)

RFC 4605, "Internet Group Management Protocol (IGMP)/ Multicast Listener Discovery (MLD)–Based Multicast Forwarding (IGMP/MLD Proxying)," August 2006 (Standards Track)

RFC 4607, "Source-Specific Multicast for IP," August 2006 (Standards Track)

RFC 4610, "Anycast-RP Using Protocol Independent Multicast (PIM)," August 2006 (Standards Track)

RFC 5015, "Bidirectional Protocol Independent Multicast (BIDIR-PIM)," October 2007 (Standards Track)

RFC 5060, "Protocol Independent Multicast MIB," January 2008 (Standards Track)

RFC 5110, "Overview of the Internet Multicast Routing Architecture," January 2008 (Informational)

RFC 5135, "IP Multicast Requirements for a Network Address Translation (NAT) and a Network Address Port Translator (NAPT)," February 2008 (Best Current Practices)

RFC 5332, "MPLS Multicast Encapsulations," August 2008 (Standards Track)

RFC 5374, "Multicast Extensions to the Security Architecture for the Internet Protocol," November 2008 (Standards Track)

RFC 5384, "The Protocol Independent Multicast (PIM) Join Attribute Format," November 2008 (Standards Track)

RFC 5401, "Multicast Negative-Acknowledgement (NACK) Building Blocks," November 2008 (Standards Track)

RFC 5519, "Multicast Group Membership Discovery MIB," April 2009 (Standards Track)

RFC 5740, "NACK-Oriented Reliable Multicast (NORM) Transport Protocol," November 2009 (Standards Track)

RFC 5771, "IANA Guidelines for IPv4 Multicast Address Assignments," March 2010 (Best Current Practice)

RFC 5790, "Lightweight Internet Group Management Protocol Version 3 (IGMPv3) and Multicast Listener Discovery Version 2 (MLDv2) Protocols," February 2010 (Standards Track)

Internet Group Management Protocol (IGMP)

IGMP[67] is an *Internet Layer* protocol that supports IPv4 multicast. It manages the membership of IPv4 multicast groups and is used by network hosts and adjacent multicast routers to establish multicast group membership. There are three versions of it so far. IGMPv1 is defined in RFC 1112, "Host Extensions for IP Multicasting," August 1989. IGMPv2 is defined in RFC 2236, "Internet Group Management Protocol, Version 2," November 1997. IGMPv3 is defined in RFC 3376,[68] "Internet Group Management Protocol, Version 3," October 2002.

Some "layer 2" switches have a feature called "IGMP snooping," which allows them to look at the "layer 3" packet content, to enable multicast traffic to go only to those ports that have subscribers on them while blocking it (and thereby reducing unnecessary traffic) on ports with no subscribers. A switch without IGMP snooping will flood all connected nodes in the broadcast domain with all multicast traffic. This can be used by hackers to "deny service" to clients who are too busy receiving and ignoring multicast traffic to handle useful traffic. This is called a Denial of Service, or DoS, attack. Active IGMP snooping is described in RFC 4541,[69] "Considerations for Internet Group Management Protocol (IGMP) and Multicast Listener Discovery Protocol (MLD) Snooping Switches," May 2006.

Protocol Independent Multicast (PIM)

PIM[70] supports IPv4 multicast. It is called "protocol independent" because it does not include its own network topology discovery mechanism. PIM does not include routing, but provides multicast forwarding by using static IPv4 routes, or routing tables created by IPv4 routing protocols, such as RIP, RIPv2, OSPF, IS-IS or BGP.

PIM Dense Mode is defined in RFC 3973,[71] "Protocol Independent Multicast – Dense Mode (PIM-DM)," January 2005. This uses dense multicast routing, which builds shortest-path trees by flooding multicast traffic domain-wide and then pruning branches where no receivers are present. It does not scale well.

---

[67] https://en.wikipedia.org/wiki/Internet_Group_Management_Protocol
[68] https://tools.ietf.org/html/rfc3376
[69] https://tools.ietf.org/html/rfc4541
[70] https://en.wikipedia.org/wiki/Protocol_Independent_Multicast
[71] https://tools.ietf.org/html/rfc3973

PIM Sparse Mode is defined in RFC 4601,[72] "Protocol Independent Multicast – Sparse Mode (PIM-SM)," August 2006. PIM-SM builds unidirectional shared trees routed at a rendezvous point per group and can create shortest-path trees per source. It scales fairly well for wide-area use.

Bidirectional PIM is defined in RFC 5015,[73] "Bidirectional Protocol Independent Multicast (BIDIR-PIM)," October 2007. It builds shared bidirectional trees. It never builds a shortest-path tree, so there may be longer end-to-end delays, but it scales very well.

ICMPv4: Internet Control Message Protocol for IPv4

ICMPv4[74] is a key protocol in the *Internet Layer* that complements version 4 of the Internet Protocol (IPv4). It was originally defined in RFC 792,[75] "Internet Control Message Protocol," September 1981. There are several ICMPv4 messages defined. Some of these are generated by the network stack in response to errors in datagram delivery. Some are used for diagnostic purposes (to check for network connectivity). Others are used for flow control (source quench) or routing (redirect).

---

[72] https://tools.ietf.org/html/rfc4601

[73] https://tools.ietf.org/html/rfc5015

[74] https://en.wikipedia.org/wiki/Internet_Control_Message_Protocol

[75] https://tools.ietf.org/html/rfc792

An ICMPv4 message consists of an IPv4 packet header, followed by 8 bytes that specify the details for each ICMPv4 message, followed by 32 or more bytes of data (depending on implementation).

ICMP Message

| Bit | 0 | 1 | 2 | 3 | 4 | 5 | 6 | 7 | 8 | 9 | 10 | 11 | 12 | 13 | 14 | 15 | 16 | 17 | 18 | 19 | 20 | 21 | 22 | 23 | 24 | 25 | 26 | 27 | 28 | 29 | 30 | 31 |

Figure 3-8. *ICMPv4 message syntax*

The IP header *Version* field contains the value 4 (for IPv4).

The IP header *Type of Service* contains the value 0.

The IP header *Length* field contains the sum of 20 (header length) + 8 (ICMPv4 header length) + number of bytes of data to be sent in message.

The IP header *Time To Live* field is set to some reasonable count (or very specific counts if used to implement the *traceroute* function).

The IP header *Protocol* field contains the value 1 (ICMPv4).

The IP header *Source IP Address* field contains the IPv4 address of the sending node.

The IP header *Destination IP Address* field contains the IPv4 address of the intended target node.

The ICMPv4 header *Type of Message* field (8 bits) specifies the ICMPv4 message type, such as 8 for *Echo Request*. See the following for the most possible ICMPv4 message types.

The ICMPv4 header *Code* field (8 bits) specifies options for the specified ICMPv4 message. For example, with Message Type 3, the code defines what failed, for example, 0 means "Destination network unreachable," while 1 means "Destination host unreachable."

The ICMPv4 header *Checksum* (16 bits) field is defined the same way as for an IPv4 header but covers the bytes in the ICMPv4 message (not including the IP header bytes).

The ICMPv4 header *Identifier* field (16 bits) can contain an ID, used only in Echo messages.

The ICMPv4 header *Sequence Number* field (16 bits) contains a sequence number, also used only in Echo messages.

```
ICMPv4 Message Type              Code    Description

0 - Echo Reply                   0       Echo reply (in response to Echo request)

3 - Destination Unreachable      0       Net unreachable
                                 1       Host unreachable
                                 2       Protocol unreachable
                                 3       Port unreachable
                                 4       Fragmentation required, and DF flag set
                                 5       Source route failed

4 - Source Quench                0       Source quench for congestion control

5 - Redirect Message             0       Redirect Datagram for the Network
                                 1       Redirect Datagram for the Host
                                 2       Redirect Datagram for the TOS & network
                                 3       Redirect Datagram for the TOS & host

8 - Echo Request                 0       Echo Request

11 - Time Exceeded               0       TTL expired in transit
                                 1       Fragment reassembly time exceeded
```

Figure 3-9. *ICMPv4 header Sequence Number field options*

For a ping diagnostic, the sending node transmits an ICMPv4 Echo Request message (Type = 8). The ID can be set to any value (0–65,535), and the sequence number is set initially to 0 and then is incremented by one for each ping in a sequence. The Data field (following the ICMPv4 header) can contain any data (typically some ASCII string). When the receiving node gets an ICMPv4 Echo Request, it sends an ICMPv4 Echo Reply (Type = 0). The Identifier, Sequence Number, and Data fields in the reply must contain exactly what were sent in the request.

If the destination of a packet is unreachable, your TCP/IP stack will return a Destination Unreachable ICMPv4 packet, with the code explaining what could not be reached.

If a packet cannot be sent by the preferred path (e.g., due to a link specified in a static route being down), an ICMPv4 Redirect message will be sent to the packet sender (typically the previous router), which should then try other paths.

If the TTL in a packet header is decremented all the way to zero, the packet is discarded, and a Time Exceeded ICMPv4 message will be sent to the packet sender.

If a node is receiving packets faster than it can handle them, it can send an ICMPv4 Source Quench message to the sender, who should slow down.

According to the standards, all nodes should always respond to an Echo Request with an Echo Reply. Due to use of this function by many hackers and worms (for network mapping), many sites now violate the standard and do not reply to Echo Requests. Many ISPs now actually block Echo Requests. Note that in IPv6, you cannot just block all ICMPv6 messages, as it is a far more integral part of the protocol.

IPv4 Routing

TCP/IP was designed from the beginning to be an *internetworking*[76] protocol. This term is where the name "Internet" comes from. TCP/IP supports ways to get packets from one node to another, even across multiple networks, by various routes through a possibly complex series of interconnections. If one or more links go down, the packets may travel by another route. Even within a given group of packets (say, ones that constitute a long email message), some of the packets may go by one route and others by another. The process of determining a viable route (or routes) to get traffic from A to B is called *routing*. This is one of the most complex areas of TCP/IP. There are entire long books on the subject. We will be covering only the simplest details, in order to show how routing differs between IPv4 and IPv6.

Some simpler network protocols (such as Microsoft's NetBIOS or NetBEUI) are *non-routing*. They will work only within a single LAN. TCP/IP and NetWare's IPX/SPX[77] support *routing*. You can connect multiple networks together with them and any node in any network can (in general) exchange data with any other node in any connected network. The Internet is simply the largest set of interconnected networks in the world. TCP/IP's flexible routing capabilities are one of the things that make it possible.

---

[76] https://en.wikipedia.org/wiki/Internetworking

[77] https://en.wikipedia.org/wiki/IPX/SPX

There are many components used to create IP-based networks, including NICs, cables, bridges, switches, and gateways. Of these, only gateways (network devices that can *forward* packets from one network segment to another) do routing. There are several kinds of gateways. The simplest case is a *router*, which uses various protocols, such as RIP, OSPF, and BGP, to determine where to forward packets, depending on their destination address. It is possible to build a router from a generic PC (or another computer) if it has multiple network interfaces (NICs), connected to multiple networks and the ability to forward packets between two or more interfaces. Most operating systems with network support can be configured to do packet forwarding (accepting a packet from one network, via one NIC, and then forwarding it on to another network, via a different NIC). Typically, no changes are made to the IP packet other than decrementing the *hop count* in the IP packet header. If NAT is being performed, numerous changes may be made to the IP packet header. If the packet is layered over Ethernet, there may be a new Ethernet frame[78] wrapped around the IP packet for each stage of its journey.

It is also possible for a gateway node to do other processing as the packets flow through it, such as filtering packets on certain criteria (e.g., allow traffic using port 25 to node 172.20.0.11 to pass, but block port 25 traffic to all other nodes). These are called *packet filtering firewalls*. They are really just routers that allow more control over the flow of traffic and can help protect the network from various attacks. Even in a packet filtering firewall, all processing still takes place at the *Internet Layer*. More sophisticated packet filtering firewalls can "inspect" the contents of the packets and maintain a record ("state") of things that really are associated with higher levels of the network stack (e.g., *Transport* or *Application* Layer). This is called *deep packet inspection*, or *stateful inspection*.

It is also possible to have a bastion host that doesn't just forward traffic; it receives protocol connections on behalf of nodes on the Internet network and relays them onward if they are acceptable. They act as a *proxy* for the internal servers. Processing here takes place at the *Application Layer*. Proxy firewalls are much more secure, but also more complex and slower. Typically, a proxy server must be created for *each protocol* handled by the firewall (e.g., SMTP, HTTP, FTP). There can be both incoming proxies (as described previously) and outgoing proxies (your node makes an outgoing connection to a proxy in your firewall, and it makes a further outgoing connection to the node you

---

[78] https://en.wikipedia.org/wiki/Ethernet_frame

really want to connect to). These allow better control than a simple packet filtering firewall. If a firewall both does packet forwarding with stateful inspection and has proxy servers (incoming and/or outgoing) for at least some protocols, it is called a *hybrid* firewall and can provide the best of both worlds.

Relevant Standard for IPv4 Routing

- **RFC 1058, "Routing Information Protocol," June 1988 (Historic)**

- RFC 1142, "OSI IS-IS Intra-domain Routing Protocol," February 1990 (Informational)

- RFC 1195 "Use of OSI IS-IS for Routing in TCP/IP and Dual Environments," December 1990 (Standards Track)

- **RFC 2328, "OSPF Version 2," April 1998 (Standards Track)**

- **RFC 2453, "RIP Version 2," November 1998 (Standards Track)**

- **RFC 4271, "A Border Gateway Protocol 4 (BGP-4)," January 2006 (Standards Track)**

In Windows, you can view all currently known routes with the "route print" command.

```
C:\>route print
================================================================================
Interface List
 11...00 22 15 24 32 9c ......Realtek PCIe GBE Family Controller
  2...00 22 b0 51 37 7c ......D-Link DGE-530T Gigabit Ethernet Adapter
  1...........................Software Loopback Interface 1
 12...00 00 00 00 00 00 00 e0 Microsoft ISATAP Adapter
 14...00 00 00 00 00 00 00 e0 Microsoft ISATAP Adapter #2
 19...00 00 00 00 00 00 00 e0 Teredo Tunneling Pseudo-Interface
================================================================================

IPv4 Route Table
================================================================================
Active Routes:
Network Destination        Netmask          Gateway       Interface  Metric
          0.0.0.0          0.0.0.0      172.20.0.1     192.168.1.8    276
        127.0.0.0        255.0.0.0         On-link       127.0.0.1    306
        127.0.0.1  255.255.255.255         On-link       127.0.0.1    306
  127.255.255.255  255.255.255.255         On-link       127.0.0.1    306
      192.168.0.0    255.255.255.0         On-link     192.168.1.8    276
      192.168.2.1  255.255.255.255         On-link     192.168.1.8    276
  192.168.255.255  255.255.255.255         On-link     192.168.1.8    276
        224.0.0.0        240.0.0.0         On-link       127.0.0.1    306
        224.0.0.0        240.0.0.0         On-link     192.168.1.8    276
  255.255.255.255  255.255.255.255         On-link       127.0.0.1    306
  255.255.255.255  255.255.255.255         On-link     192.168.1.8    276
================================================================================
Persistent Routes:
  Network Address          Netmask  Gateway Address  Metric
          0.0.0.0          0.0.0.0      192.168.0.1  Default
================================================================================
```

Figure 3-10. *Output of the IPv4 route print command*

There are several routing protocols for IPv4 that are typically handled only in the core or where a customer network meets the core, the *edge router*. These include RIP, RIPv2, EIGRP, IS-IS, OSPF, and BGP.

TCP/IP routing is a very deep, complex subject, and we will be touching only on the most obvious aspects in this book, to give a rough idea of the differences in routing between IPv4 and IPv6.

RIP: Routing Information Protocol,[79] version 1. Defined in RFC 1058,[80] "Routing Information Protocol," June 1988. This protocol is very old and of primarily historic interest, since it does not support address blocks based on CIDR[81] (it is a *classful* routing protocol). It is used to exchange routing information with gateways and other hosts. It is based on the *distance vector* algorithm,[82] which was first used in the ARPANET, circa 1967. RIP is a UDP-based protocol, using port 520.

---

[79] https://en.wikipedia.org/wiki/Routing_Information_Protocol
[80] https://tools.ietf.org/html/rfc1058
[81] https://en.wikipedia.org/wiki/Classless_Inter-Domain_Routing
[82] https://en.wikipedia.org/wiki/Distance-vector_routing_protocol

RIPv2: Routing Information Protocol, version 2.[83] Defined in RFC 2453,[84] "RIP Version 2," November 1998. Although OSPF and IS-IS are superior, there were so many implementations of RIP in use it was decided to try to improve on it. Extensions were made to incorporate the concepts of autonomous systems (ASs), IGP/EGP interactions, subnetting and authentication, as well as address blocks based on CIDR (it is a "classless" routing protocol). The lack of subnet masks in RIPv1 was a particular problem. RIPv2 is limited to networks whose longest routing path is 15 hops. It also uses fixed "metrics" to compare alternative routes, which is an oversimplification. However, RIPv2 becomes unstable if you try to account for different metrics. See RFC for details.

EIGRP: Enhanced Interior Gateway Routing Protocol.[85] This is *not* an IETF protocol, but a Cisco proprietary routing protocol based on their earlier IGRP.[86] EIGRP is able to deal with addresses allocated via CIDR (it is a *classless* routing protocol), including use of variable-length subnet masks. It can run separate routing processes for IPv4, IPv6, IPX, and AppleTalk protocols, but does not support translation between protocols. For details, see Cisco documentation. There is an RFC that covers a subset of the full Cisco EIGRP, RFC 7868,[87] "Cisco's Enhanced Interior Gateway Routing Protocol (EIGRP)," May 2016.

IS-IS: Intermediate System to Intermediate System routing protocol.[88] IS-IS (pronounced "eye-sys") was originally developed by Digital Equipment Corporation (DEC) as part of DECnet Phase V and formally defined as part of ISO/IEC 10589:2002 for the Open System Interconnection reference design. It is *not* an Internet standard, although the details are published as *Informational* RFC 1142,[89] "OSI IS-IS Intra-domain Routing Protocol," February 1990 (since reclassified as *historic* by RFC 7142[90] in 2014). Another RFC specifies how to use IS-IS for routing in TCP/IP and/or OSI environments: RFC 1195,[91] "Use of OSI IS-IS for Routing in TCP/IP and Dual Environments," December 1990. IS-IS is an *Interior Gateway Protocol*, for use within an administrative domain or network. It is not intended for routing between autonomous systems, which is the role

---

[83] https://community.cisco.com/t5/networking-documents/
ripv2-routing-information-protocol/ta-p/3117425
[84] https://tools.ietf.org/html/rfc2453
[85] https://en.wikipedia.org/wiki/Enhanced_Interior_Gateway_Routing_Protocol
[86] https://en.wikipedia.org/wiki/Interior_Gateway_Routing_Protocol
[87] https://tools.ietf.org/html/rfc7868
[88] https://en.wikipedia.org/wiki/IS-IS
[89] https://tools.ietf.org/html/rfc1142
[90] https://tools.ietf.org/html/rfc7142
[91] https://tools.ietf.org/html/rfc1195

of BGP. It is not a distance vector algorithm; it is a *link-state* protocol.[92] It operates by reliably flooding network topology information through a network of routers, allowing each router to build its own picture of the complete network. OSPF (developed by the IETF about the same time) is more widely used, although it appears that IS-IS has certain characteristics that make it superior in large ISPs.

OSPFv2: Open Shortest Path First,[93] version 2. Unlike EIGRP and IS-IS, OSPF is an IETF standard. OSPFv2 is defined in RFC 2328,[94] "OSPF Version 2," April 1998. OSPF is the most widely used Interior Gateway Protocol today (as opposed to BGP, which is an Exterior Gateway Protocol). Like IS-IS, OSPF is a link-state protocol.[95] It gathers link-state information from available routers and builds a topology map of the network. It was designed to support variable-length subnet masking (VLSM) or CIDR addressing models. Changes to the network topology are rapidly detected, and it converges on a new optimal routing structure within seconds. It allows specification of different metrics ("cost of transmission" in some sense) for various links to allow better modeling of the real world (where some links are fast and some slow). OSPF does not layer over UDP or TCP but uses IP datagrams with a protocol number of 89. This is very different from RIP or BGP. OSPF uses multicast, including the special addresses:

```
224.0.0.5      All SPF/link state routers      AllSPFRouters
224.0.0.6      All Designated Routers          ALLDRouters
```

For routing IPv4 multicast traffic, there is MOSPF (Multicast Open Short Path First), defined in RFC 1584,[96] "Multicast Extensions to OSPF," March 1994. However, this is not widely used. Instead, most people use PIM[97] in conjunction with OSPF or other Interior Gateway Protocols.

BGP-4: Border Gateway Protocol 4.[98] Defined in RFC 4271,[99] "A Border Gateway Protocol 4 (BGP-4)," January 2006. This version supports routing only IPv4. There are defined multiprotocol extensions (BGP4+) that support IPv6 and other protocols, which will be described in Chapter 5.

---

[92] https://en.wikipedia.org/wiki/Link-state_routing_protocol
[93] https://en.wikipedia.org/wiki/Open_Shortest_Path_First
[94] https://tools.ietf.org/html/rfc2328
[95] https://en.wikipedia.org/wiki/Link-state_routing_protocol
[96] https://tools.ietf.org/html/rfc1584
[97] https://en.wikipedia.org/wiki/Protocol_Independent_Multicast
[98] https://en.wikipedia.org/wiki/Border_Gateway_Protocol
[99] https://tools.ietf.org/html/rfc4271

BGP is an Exterior Gateway Protocol[100] (compare with IS-IS and OSPFv2, which are Interior Gateway Protocols[101]). It is not used within networks, but only between autonomous systems.[102] Its primary function is to exchange AS network reachability information with other AS networks. This includes information on the list of autonomous systems (ASs) that reachability information traverses. This is sufficient for BGP to construct a graph of AS connectivity from which routing loops can be pruned, and, at the AS level, certain policy decisions may be enforced.

BGP-4 includes mechanisms for supporting CIDR. They can advertise a set of destinations as an IP prefix, eliminating the concept of network "class," which was present in early BGP implementations. BGP-4 also has mechanisms that allow aggregation of routes and AS paths. Most networks that obtain service from ISPs never deploy BGP themselves. It is mostly for exchange of information *between* ISPs, especially if they are multihomed (obtain upstream service from more than one source). This would be referred to as *Exterior Border Gateway Protocol* or EBGP. Enormous networks that are too large for OSPF could deploy BGP themselves as a top level linking multiple OSPF routing domains (this would normally be referred to as *Interior Border Gateway Protocol* or IBGP).

BGP is a *path vector* protocol.[103] It does not use IGP metrics, but makes routing decisions based on path, network policies, and/or rulesets. It replaces the now defunct Exterior Gateway Protocol (EGP), which was formally specified in RFC 904,[104] "Exterior Gateway Protocol Formal Specification," April 1984.

Network Address Translation (NAT)

It is also possible for a gateway to do Network Address Translation[105] (NAT) as packets are forwarded. One form of this ("Full Cone" or "Static" NAT) allows multiple internal nodes (which use private addresses, such as 10.1.2.3) to be translated to *globally routable* addresses (like 123.45.67.89) on the way out. It also can translate the globally routable destination address of packets sent in reply to an outgoing packet back to the private

---

[100] https://en.wikipedia.org/wiki/Exterior_Gateway_Protocol

[101] https://en.wikipedia.org/wiki/Interior_gateway_protocol

[102] https://en.wikipedia.org/wiki/Autonomous_system_(Internet)

[103] https://en.wikipedia.org/wiki/Path_vector_routing_protocol

[104] https://tools.ietf.org/html/rfc904

[105] https://en.wikipedia.org/wiki/Network_address_translation

address of the originating node, so that the internal node can complete a query/response transaction. The port numbers in outgoing packets are shifted by a NAT gateway in such a way that it can figure out which internal node to send reply packets to. This allows many internal nodes to "share" (hide behind) a single globally routable Ipv4 address (necessary now that we are running out of these). NAT will be covered in more detail in the next chapter.

Relevant Standard for IPv4 NAT

RFC 1918, "Address Allocation for Private Internets," February 1996 (Best Current Practices)

RFC 2663, "IP Network Address Translation (NAT) Terminology and Considerations," August 1999 (Informational)

RFC 2694, "DNS Extensions to Network Address Translations (DNS_ALG)," September 1999 (Informational)

RFC 2709, "Security Model with Tunnel-mode IPsec for NAT Domains," October 1999 (Informational)

RFC 2993, "Architectural Implications of NAT," November 2000 (Informational)

RFC 3022, "Traditional IP Network Address Translation (Traditional NAT)," January 2001 (Informational)

RFC 3235, "Network Address Translation (NAT)-Friendly Application Design Guidelines," January 2002 (Informational)

RFC 3519, "Mobile IP Traversal of Network Address Translation (NAT) Devices," April 2003 (Standards Track)

RFC 3715, "IPsec-Network Address Translation (NAT) Compatibility Requirements," March 2004 (Informational)

RFC 3947, "Negotiation of NAT-Traversal in the IKE," January 2005 (Standards Track)

RFC 4008, "Definitions of Managed Objects for Network Address Translations (NAT)," March 2005 (Standards Track)

RFC 4787, "Network Address Translation (NAT) Behavioral Requirements for Unicast UDP," January 2007 (Best Current Practices)

RFC 4966, "Reasons to Move the Network Address Translation – Protocol Translator (NAT-PT) to Historic Status," (Informational)

RFC 5128, "State of Peer-to-Peer (P2P) Communication Across Network Address Translations (NATs)," March 2008 (Informational)

RFC 5207, "NAT and Firewall Traversal Issues of Host Identity Protocol (HIP) Communication," April 2008 (Informational)

RFC 5382, "NAT Behavioral Requirements for TCP," October 2008 (Best Current Practices)

RFC 5389, "Session Traversal Utilities for NAT (STUN)," October 2008 (Standards Track)

RFC 5508, "NAT Behavioral Requirements for ICMP", April 2009 (Best Current Practices)

RFC 5597, "Network Address Translation (NAT) Behavioral Requirements for the Datagram Congestion Control Protocol," September 2009 (Best Current Practices)

RFC 5684, "Unintended Consequences of NAT Deployments with Overlapping Address Space," February 2010 (Informational)

It should be obvious from the number of RFCs that explain how NAT affects other things that NAT has a heavy impact on almost every aspect of networks. There are also a lot of "Informational" RFCs required to explain exactly *how* it impacts these things. Removing NAT has no downside (given sufficient public addresses) and vastly simplifies network architecture and management in addition to lowering costs. It also vastly simplifies application design and implementation. The removal of NAT and restoration of the flat address space is one of the main benefits of moving to IPv6. Unfortunately, we have an entire generation of network engineers who have assumed that NAT is "the way networks are done" and don't realize it was created only as a temporary crutch to extend the life of the IPv4 address space until IPv6 could be completed and deployed. Before NAT, the IPv4 Internet was "flat," and firewalls had very effective security without NAT

(I call this "classic firewall architecture"). In IPv6, we are simply returning to the original concept of "any node to any node connectivity" that characterized the pre-NAT IPv4 Internet. Protocols like SIP, IPsec, IKE, and Mobile IP will work far better without NAT in the way. DNS is also greatly simplified in the absence of NAT (no internal vs. external "views" are required).

Unfortunately, there is *no* possible way to remove NAT from the current Internet. There are far too many users to handle with the possible public addresses, and essentially all the routable addresses are already in use. The only possible way now to remove NAT is by migrating to IPv6.

Most routers and firewalls typically include NAT for IPv4 as part of their functionality, although it would be possible to have a NAT gateway without any filtering or routing capabilities that does *only* NAT.

In general, any gateway that modifies the source and/or destination addresses in a packet (possibly also the source port number) is doing NAT. There are several forms of it, the most popular being *address masquerading* (*hide-mode* NAT) and *one-to-one* (BINAT, or *static* NAT).

Most IPv4 networks today make use of *private addresses* as defined in RFC 1918,[106] "Address Allocation for Private Internets," February 1996. Basically, three blocks of addresses (10.0.0.0/8, 172.16.0.0/12, and 192.168.0.0/16) were permanently removed from the available Internet allocation pool, marked as "unroutable" on the Internet, and reserved for use as something similar to telephone extension numbers in an office (hiding behind a single company phone number, via a Private Branch Exchange). It is possible for any company to use addresses from any or all of these ranges to number the nodes inside their networks. However, these addresses cannot be routed on the Internet from anyone, since they are no longer globally unique. Hence, if the users of nodes with those addresses want to use the Internet, there must be address translation to and from "real" (globally unique) addresses at the gateway that connects them to the Internet, which is what NAT does.

One thing that confuses people is that internal telephone extensions don't look like public telephone numbers (e.g., 100, 101, 1125 vs. 9472-4173). However, private IP addresses look just like public addresses (except for the address ranges) and in fact used to be public addresses that were repurposed.

The RFC 1918 private addresses are in the following ranges:

---

[106] https://tools.ietf.org/html/rfc1918

```
10/8:          10.0.0.0 to 10.255.255.255    (16,777,216 addresses)
172.16/12:     172.16.0.0 to 172.31.255.255  (1,048,576 addresses)
192.168/16:    192.168.0.0 to 192.168.255.255 (65,536 addresses)
```

Figure 3-11. *RFC 1918 address ranges*

More recently another range was reserved for CGN (Carrier-Grade NAT[107]). These addresses cannot be used by end users, only ISPs deploying CGN:

```
100.64/10:      100.64.0.0 to 100.127.255.255   (4,194,304 addresses)
```

Figure 3-12. *New private IP address block for CGN*

Note that the most popular form of NAT is more properly called NAPT ("Network Address Port Translation"), which involves the translation of *both* IP addresses and port numbers.

NAT is defined in RFC 3022,[108] "Traditional IP Network Address Translation (Traditional NAT)," January 2001. Some aspects of NAT are defined in RFC 2663,[109] "IP Network Address Translation (NAT) Terminology and Considerations," August 1999.

One form of NAT traversal (STUN) is defined in RFC 5389,[110] "Session Traversal Utilities for NAT (STUN)," October 2008. STUN is a protocol that serves as a tool for other protocols in dealing with Network Address Translation (NAT) traversal. It can be used by an endpoint to determine the IP address and port allocated to it by a NAT. It can also be used to check connectivity between two endpoints and as a *keepalive* protocol to maintain NAT bindings. STUN works with many existing NATs and does not require any special behavior from them.

Connection Without NAT (Inside the LAN)

Say you have two nodes (Alice and Bob) on your LAN. Alice has the address 10.50.3.12, and Bob has the address 10.50.3.75 (both private addresses). They can make connections within their LAN (to any address in the 10.0.0.0/8 network) with no problem. Say there is a web server (port 80) at 10.1.20.30.

---

[107] https://en.wikipedia.org/wiki/Carrier-grade_NAT

[108] www.ietf.org/rfc/rfc3022.txt

[109] www.ietf.org/rfc/rfc2663.txt

[110] www.ietf.org/rfc/rfc5389.txt

In the following, we will specify the port number appended to the IP address, separated by a colon (e.g., 10.50.3.12:12345). When Alice makes a connection to the web server, the *destination* port is 80, but her *source* port is a randomly chosen value greater than 1024 that is not already in use (e.g., 12345 or 54321). The same source port would be used for the duration of the connection. Replies from the server would be sent using Alice's source address and port as the *destination* address and port in the reply packets. See the following for example.

	Source Addr:Port	Destination Addr:Port
Traffic from Alice to Server:	10.50.3.12:12345	10.1.20.30:80
Replies from Server to Alice:	10.1.20.30:80	10.50.3.12:12345
Traffic from Bob to Server:	10.50.3.75:54321	10.1.20.30:80
Replies from Server to Bob:	10.1.20.30:80	10.50.3.75:54321

Figure 3-13. *Port mapping for IPv4 NAT*

Note: The preceding behavior is somewhat simplified. Such a server could accept only one connection at a time, which would have to complete before anyone else could connect. This is because a given address:port can only handle one connection at any given time. A real-world server would have a parent process listening for connections on a well-known port (e.g., 80). When some client connects to the well-known port, the parent process would create a child process (or thread), which would accept the connection (using yet another unused port number) and process it. Meanwhile, the main process would go back to listening for further connections on the well-known port. If ten users were connected at a time, there would be 11 processes running, one main process and ten child processes (one for each connection). From the viewpoint of the client (e.g., with "netstat –na"), it would appear that the remote port (the one on the server) was the original well-known port (e.g., 80).

Connection Through Hide-Mode NAT

But how do Alice and Bob connect to *www.ipv6.org*? That node happens to have an IPv4 address of 194.63.248.52, and we're still in Chapter 3 (about Ipv4), so they don't have IPv6 yet! Let's say there is a NAT gateway where their LAN (or ISP) connects to the Internet. It has an "outside" address (which must be a valid, routable Ipv4 address) of 12.34.56.137. If *either* Alice or Bob connects to `www.ipv6.org` (over Ipv4), the web page there will indicate to both of them that they are connecting over IPv4, from the address 12.34.56.137, not from their respective private addresses, even if the connections are made at the same instant. The web server log will show that both are connected from that one public address. How can `www.ipv6.org` reply with the correct web page to each of them?

With hide-mode NAT, the gateway is translating the source address in Alice's packets from 10.50.3.12 to 12.34.56.137. It is *also* translating the source address in Bob's packets from 10.50.3.75 to 12.34.56.137. The destination address is 194.63.248.52:80 for both Alice and Bob. Their browsers would each choose a random source port. Let's say Alice's chose 10123 and Bob's chose 20321. The NAT gateway would not only translate the source address from both Alice and Bob; it would also shift the source ports and keep track of that shift in a table, which contains the source address, the original source port, and the shifted source port (for each connection). Let's say Alice's port is shifted to 30567 and Bob's to 40765. The new source address for outgoing connections and the old destination address for incoming connections will always be the same (the outside address of the NAT gateway), so it does not need to keep those in the table. The resulting NAT table would look like the following.

```
                          Source Addr:Port              Shifted Port
                          10.50.3.12:10123              30567
                          10.50.3.75:20321              40765

                                Src Addr:Port            Xlat Src Addr :Port    Dest. Addr :Port
Traffic from Alice to server:   10.50.3.12:10123         12.34.56.137:30567     194.63.248.52:80

                                Src Addr:Port            Dest Addr:Port         Xlat Dest Addr:Port
Replies from server to Alice:   194.63.248.52:80         12.34.56.137:30567     10.50.3.12:10123

                                Src Addr:Port            Xlt Src Addr :Port     Dest. Addr :Port
Traffic from Bob to server:     10.50.3.75:20321         12.34.56.137:40765     194.63.248.52:80

                                Src Addr:Port            Dest Addr:Port         Xlt Dest Addr:Port
Replies from server to Bob:     194.63.248.52:80         12.34.56.137:40765     10.50.3.75:20321
```

Figure 3-14. *Example of outgoing and incoming port mapping for NAT*

Alice's connection to `www.ipv6.org` appears to be coming from 12.34.56.137:30567. When `www.ipv6.org` replies to Alice, it is sent from 194.63.248.52:80 to 12.34.56.137:30567. The NAT gateway looks up that port in its table and sees that it was used for an outgoing connection from 10.50.3.12:10123, so it translates the destination address and port to Alice's private address and port, thereby forwarding the packets correctly to Alice.

Bob's connection to `www.ipv6.org` appears to be coming from 12.34.56.137:40765. When `www.ipv6.org` replies to Bob, it is sent from 194.63.248.52:80 to 12.34.56.137:40765. The NAT gateway looks that port up in its table and sees that it was used from a connection from 10.50.3.75:20321, so it translates the destination address and port to Bob's private address and port, thereby forwarding those packets correctly to Bob.

BINAT (One-to-One NAT)

If you are doing NAT at your gateway, most routers or firewalls support another form of NAT, which is known as BINAT (bidirectional NAT) or one-to-one NAT (sometimes also *static* NAT). This works much the same as regular (hide-mode) NAT, except there is no port shifting involved. This means there can only be *one* internal node associated with each globally routable external address. This is used only for servers that must be accessible from the outside world.

Typically, a server has both an *internal* (private) address (e.g., 10.0.0.13) and an *external* (unique, globally routable) address (e.g., 12.34.56.131). With outgoing connections, the gateway rewrites the source address of each packet to be the external address for that node (but does not shift the port). For incoming connections, the gateway rewrites the destination address to be the internal address for that node. Internally, the node will have only the internal address. However, if you connected to `www.ipv6.org` from such a node, the resulting web page would show a connection not from the *internal* address of the server, but from the unique external address associated with that node. This is similar to hide-mode NAT except that there is exactly one internal node per external address (rather than many), there is no port shifting, and the mapping can be done in both directions (incoming and outgoing connections).

There is a minor problem of the "*missing ARP*" that must be solved in some way for this to work (there is no physical node at the external address, so no node will respond to ARP requests concerning that address). One approach is to configure a static ARP on the gateway that can supply that response. Every operating system or router has some way to do this. Without that, connections from the outside will not work. It is also possible in most cases to assign the external address as an *alias* to the outside interface of the gateway (in addition to its real address). Solving the *missing ARP* problem is one of the most difficult things for firewall administrators to master. This problem only exists in IPv4. As no NAT is needed or done in IPv6, there is no missing ARP (actually in IPv6, it would be a missing ND response).

BINAT at least allows incoming connections but uses up one globally routable IPv4 address for each server node. Most SOHO gateways do not support BINAT. Many do have a simpler mechanism called *port redirection*, which allows incoming connections to the hide-mode external address. At most one internal server can be configured as the target for any given port. So you could configure an internal mail server and redirect ports 25 (SMTP), 110 (POP3), and 143 (IMAP) to it. However, if you have two internal web servers both configured for port 80, you could not redirect port 80 on the gateway to *both* servers.

Ramifications of Using NAT

When Network Address Translation happens, the NAT gateway is actually rewriting new values into the address and port number fields in the IP and TCP (or UDP) packet headers of all packets flowing through the NAT gateway, according to the rule just specified. For outgoing packets, it is rewriting the source address and source port. For incoming packets, it is rewriting the destination address and destination port. Obviously, this would invalidate the IP and TCP header checksums (the IP header contains source and destination addresses; the TCP header contains the source and destination port numbers). Therefore, the NAT gateway *also* has to recalculate both IP and TCP header checksums and rewrite *those* as well.

Packet fragmentation is a real complication for TCP and UDP via NAT. A NAT gateway must reassemble an entire packet, in order to be able to recalculate the TCP checksum (which covers all bytes in the payload, plus the pseudo header, which contains the source and destination addresses). It typically must then re-fragment the packet for further transmission.

What about the IPsec Authentication header (AH)? (Note: IPsec will be discussed in detail in Chapter 6.) The IPsec AH algorithm works like a checksum, but there is a key that only the sender has, required to generate the cryptographic checksum. All this address and port rewriting invalidates the existing AH cryptographic checksum, and the NAT gateway does **not** have the necessary key to regenerate a correct new AH for the modified packet headers. Because of this IPsec does not work through a NAT gateway. Actually, AH is performing its function very effectively; it **is** detecting tampering with the contents of the packet header! It just happens that this tampering is done by a NAT gateway, not a hacker. It's kind of like getting hit by "friendly fire" in a war zone (getting shot by your own side). If any node other than the original sender could generate a new valid AH checksum, then AH would not be very useful! IPsec and NAT are mutually exclusive (although IPsec VPNs can be made to work in conjunction with NAT traversal).

Another ramification involves FTP (File Transfer Protocol). FTP is a *very* old protocol (RFC 765[111] is from 1980, back in the days of the First Internet). In active mode, FTP uses separate connections for control traffic (commands) and for data traffic. The initiating host identifies the corresponding data connection with its Network Layer and Transport Layer addresses. Unfortunately, NAT invalidates this. Fortunately, here, it is possible to create a *reverse FTP proxy* (included on most firewalls) that solves this problem. Without such a proxy though, FTP will not work if NAT is in place, even for outgoing connections. My company early on ported a popular one for IPv4 to IPv6. That allowed FTP connections to dual-stack networks such as freebsd.org to work from our own dual-stack network.

"Peer-to-peer" (like Kazaa, not "real" peer-to-peer) applications have the same kinds of problems with NAT. You must somehow provide a way for your peers to connect to *you* for these applications to work. All participants really need a real, globally routable IP address. This is not easy to arrange on the Second Internet. All such "fake" peer-to-peer applications must use NAT traversal.

SIP (Session Initiation Protocol[112]) is used with many things, including VoIP and video conferencing. It also has major problems with NAT. SIP may use multiple ports to set up a connection and transmit the analog stream over RTP (Real-Time Transport Protocol). IP addresses and port numbers are encoded in the payload and must be known prior to the traversal of NAT gateways (this was bad protocol design, but now we are stuck with it). Again, a SIP proxy on the gateway can help resolve this problem. Another solution is to use NAT traversal, such as STUN. Unfortunately, in these days of widespread NAT, both the caller and the callee are typically behind NAT, so VoIP must overcome problems with NAT both *going out* from the caller and *coming in* to the callee. If this sounds like an ugly mess, *it is*.

Another problem with NAT is the limit of 65,536 ports on the NAT gateway. When NAT was first deployed, most network applications used only one or two ports. Some recent applications (Apple's *iTunes* and Google *Maps*) use 200–300 ports at a time for better performance. If each node is using 300 ports, then there can be *at most* 200 nodes behind a given external IPv4 address. If the NAT gateway runs out of ports, there can be very mysterious failures in network applications. For example, in Google Maps, some areas of the map never get drawn. There is no way for end users (or typically even the network administrator) to determine that this has happened other than by seeing

---

[111] https://tools.ietf.org/html/rfc765
[112] https://en.wikipedia.org/wiki/Session_Initiation_Protocol

mysterious failures in some applications. This means that a larger number of NAT gateways (and valid external IPv4 addresses) are required today than in the past, for a given number of users behind NAT. Just as we are running out of public IPv4 addresses!

Some legacy applications (like web surfing and email) work okay through one layer of NAT. Even with chat, today there must be an intermediary system that two or more chatters connect to via outgoing connections from their nodes (e.g., AOL Instant Messenger). In a flat address space (especially with working multicast), much better connectivity models are possible that may require little or no central facilities.

As the IPv4 addresses run out, it will become more common to have *multiple layers* of NAT (CGN). This can happen today, if you deploy a Wi-Fi access point with NAT behind a DSL modem that *also* has NAT. If you think a single layer of NAT causes problems, you should try dealing with multiple layers of it!

With the wide-scale deployment of NAT, we have lost the original end-to-end model of the early Second Internet, which was a core feature. We've also broken one of the fundamental rules of protocol design: *never tamper with source or destination addresses or ports in an IP packet.*

Today users are either *content producers* who can publish information or videos (e.g., cnn.com, youtube.com) or *content consumers* who can view the content published by the producers. It is much more complicated and expensive to be a producer in the current Second Internet (with NAT in the way) than to be a consumer. There are relatively few producers and millions of consumers. This was not that much of a problem when most people were running mainly web browsers and email clients on their nodes. As newer applications emerge (VoIP, IPTV, multiplayer games, peer-to-peer), this new "digital divide" between producers and consumers is becoming more of a problem. Today, many people would like to be *prosumers* (both producers and consumers of content). With IPv6 that is simple.

Another problem is that since the first implementation of networking on smartphones (WAP), there were not enough public IPv4 addresses for phones, so historically there have never been public IP addresses on phones. Phones could only be used to make outgoing connections – you could not deploy a server on your phone, and Alice's phone could not connect directly to Bob's phone. With IPv6 for the first time, phones have *public addresses* and hence can run servers or do end-to-end connections.

All these problems go away with a flat address space (no NAT). Unfortunately, there is no way to restore the flat address space of the early (pre-NAT) Second Internet. The Second Internet is now permanently broken (there are not enough addresses to allow even the existing users to have access without NAT, even if we use all the remaining unallocated addresses today). The only real solution is to switch to IPv6 (at least for protocols such as VoIP, P2P, multiplayer games, IPTV, and IPsec VPNs).

Basic IPv4 Routing

In the simplest case, where two nodes (A and B) are on the same network segment (not separated by any router), no routing is required. Let's say node A wants to send a packet to node B. Node A determines if node B is in the same network segment by examining B's IP address and the network subnet mask. If node B is in the same subnet as A's IP address, then B is a *local node*. Node A simply uses B's MAC address from its ARP table to send the packet to B. If there is no entry for B's IP address, then node A does *address resolution* (obtains the MAC address for B), as described earlier.

If B's address is *not* in the local subnet, B is *not* a local node, and the packet (with B's correct IP address as the destination) is sent to the node that serves as the *default gateway* for A's subnet (A may first have to do an ARP to obtain the MAC address of the default gateway). The default gateway is a node with multiple network interfaces that knows how to forward the packet on toward the network in which B's IP address is found. Note that by default, *packet forwarding* (relaying packets from one interface to another on a multihomed system) is not enabled. It must be specifically enabled for each protocol (IPv4 and IPv6). The address of a network's default gateway is known to every node in a subnet, either through manual configuration or via DHCPv4. Once the default gateway receives the packet, it may already have the necessary routing information to know where to send that packet (either via *static routes* or via a routing protocol, such as RIP, OSPF, and/or BGP). In the case of a home network, your SOHO router typically just knows how to forward packets for the outside world to yet another gateway at the ISP, where the real routing takes place (via its *own* default gateway, which is a node at the ISP).

Once your traffic gets to your default gateway, that node typically uses an Interior Gateway Routing Protocol (RIP, RIPv2, or OSPF) to route that traffic to the edge of your overall network (e.g., the place your organization's or ISP's network connects to the rest of the Internet). At that point, an Exterior Gateway Routing Protocol (typically BGP-4) is used to determine the best route to the correct edge router for the destination address. Once your traffic arrives there, once again an Interior Gateway Routing Protocol (RIP, RIPv2, or OSPF) takes over and gets the packets to the default gateway of the subnet where the destination node lives. From there, ARP is used to forward the packets to the actual destination node, because the default gateway and the destination node are now on the same subnet. And all this takes place in the blink of an eye, billions of times a day, just like clockwork.

TCP: The Transmission Control Protocol

TCP, the Transmission Control Protocol,[113] is defined in RFC 793,[114] "Transmission Control Protocol," September 1981. This is a *Transport Layer*[115] protocol. TCP implements a *reliable, connection-oriented*[116] model. When we say *reliable*, we aren't talking about a "well-designed" or "robust" protocol. With respect to TCP, "reliable" simply means that the protocol includes error detection and recovery (via retransmission). The term *connection oriented* refers to the fact that TCP is designed to handle potentially large streams of data (typically larger than a single packet). It does this by breaking the large object up into multiple packet-sized chunks and sending those packets out and to the recipient. For example, a large email message or a JPEG photograph might require quite a few packets. Software that uses TCP typically *opens* (initiates) a connection for I/O, reads and/or writes potentially a lot of data from/to it, and then, when done, *closes* (terminates) the connection. This is very similar to the process for reading and writing files, and in fact in UNIX, network streams are just a special kind of file.

Standards Relevant to TCP

> **RFC 793, "Transmission Control Protocol," September 1981 (Standards Track)**
>
> RFC 896, "Congestion Control in IP/TCP Internetworks," January 1984 (Unknown)

---

[113] https://en.wikipedia.org/wiki/Transmission_Control_Protocol
[114] https://tools.ietf.org/html/rfc793
[115] https://en.wikipedia.org/wiki/Transport_layer
[116] https://en.wikipedia.org/wiki/Connection-oriented_communication

RFC 1001, "Protocol Standard for a NetBIOS Service on a TCP/UDP Transport: Concepts and Methods," March 1987 (Standards Track)

RFC 1002, "Protocol Standard for a NetBIOS Service on a TCP/UDP Transport: Detailed Specifications," March 1987 (Standards Track)

RFC 1006, "ISO Transport Service on Top of the TCP Version: 3," May 1987 (Standards Track)

RFC 1085, "ISO Presentation Services on Top of TCP/IP-Based Internets," December 1998

RFC 1086, "ISO-TP0 Bridge Between TCP and X.25," December 1988

RFC 1144, "Compressing TCP/IP Headers for Low-Speed Serial Links," February 1990 (Standards Track)

RFC 1155, "Structure and Identification of Management Information for TCP/IP-Based Internets", May 1990 (Standards Track)

RFC 1180, "TCP/IP Tutorial," January 1991 (Informational)

RFC 1213, "Management Information Base for Network Management of TCP/IP-Based Internets: MIB II," March 1991 (Standards Track)

RFC 1323, "TCP Extensions for High Performance," May 1992 (Standards Track)

RFC 2018, "TCP Selective Acknowledgement Options," October 1996 (Standards Track)

RFC 2126, "ISO Transport Service on Top of TCP (ITOT)," March 1997 (Standards Track)

RFC 2873, "TCP Processing of the IPv4 Precedence Field," June 2000 (Standards Track)

RFC 2883, "An Extension to the Selective Acknowledgement (SACK) Option for TCP," July 2000 (Standards Track)

RFC 2988, "Computing TCP's Retransmission Timer," November 2000 (Standards Track)

RFC 3042, "Enhancing TCP's Loss Recovery Using Limited Transport," January 2001 (Standards Track)

RFC 3293, "General Switch Management Protocol (GSMP) Packet Encapsulation for Asynchronous Transfer Mode (ATM), Ethernet and Transmission Control Protocol (TCP)," June 2002 (Standards Track)

RFC 3390, "Increasing TCP's Initial Window," October 2002 (Standards Track)

RFC 3517, "A Conservative Selective Acknowledgement (SACK)-Based Loss Recovery Algorithm for TCP," April 2003 (Standards Track)

RFC 3782, "The New Reno Modifications to TCP's Fast Recovery Algorithm," April 2004 (Standards Track)

RFC 3821, "Fiber Channel over TCP/IP (FCIP)," July 2004 (Standards Track)

RFC 4015, "The Eifel Response Algorithm for TCP," February 2005 (Standards Track)

RFC 4022, "Management Information Base for the Transmission Control Protocol (TCP)," March 2005 (Standards Track)

RFC 4614, "A Roadmap for Transmission Control Protocol (TCP) Specification Documents," September 2006 (Informational)

RFC 4727, "Experimental Values in IPv4, IPv6, ICMPv4, ICMPv6, UDP and TCP Headers," November 2006 (Standards Track)

RFC 4898, "TCP Extended Statistics MIB," May 2007 (Standards Track)

RFC 4996, "Robust Header Compression (ROHC): A Profile for TCP/IP (ROHC-TCP)," July 2007 (Standards Track)

RFC 5348, "TCP Friendly Rate Control (TFRC): Protocol Specification," September 2008 (Standards Track)

RFC 5482, "TCP User Timeout Option," March 2009 (Standards Track)

RFC 5681, "TCP Congestion Control," September 2009 (Standards Track)

RFC 5682, "Forward RTO-Recovery (F-RTO): An Algorithm for Detecting Spurious Retransmission Timeouts with TCP," September 2009 (Standards Track)

RFC 5734, "Extensible Provisioning Protocol (EPP) Transport over TCP," August 2009 (Standards Track)

TCP Packet Header

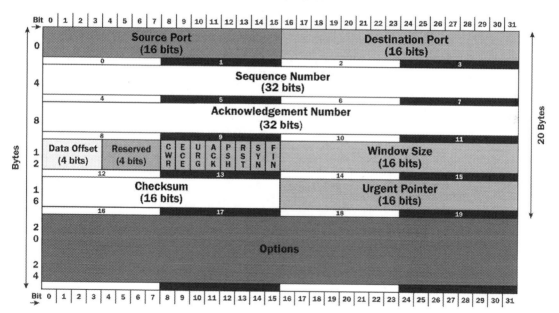

Figure 3-15. *TCP packet header*

Source Port (16 bits): Specifies the port that the data was written to on the sending node.

Destination Port (16 bits): Specifies the port that the data will be read from on the receiving node.

Sequence Number (32 bits): Meaning depends on the value of the SYN flag:

- If the SYN flag is set, this field contains the initial sequence number. The sequence number of the actual first data byte (and the acknowledgment number in the resulting ACK) will then be that value plus 1.

- If the SYN flag is clear, this field contains the accumulated sequence number of the first data byte of this packet for the current session.

Acknowledgement Number (32 bits): Used to acknowledge receipt of data:

- If the ACK flag is set, this field is the next sequence number that the receiver is expecting. This acknowledges receipt of all previous bytes.

- If the ACK flag is clear, this field is not used.

Data Offset (4 bits): Specifies the size of the TCP header in 32-bit words. The minimum value is 5 words (20 bytes), and the maximum value is 15 words (60 bytes), allowing for up to 40 bytes of options.

Reserved (4 bits): Not currently used and must be zeros.

There are eight 1-bit flags (8 bits total) as follows (in order from most significant bit to least significant bit):

- *CWR*: Congestion Window Reduced. If set by the sender, it indicates it has received a TCP segment with the ECE flag set and has responded in a congestion control mechanism.

- *ECE*: ECN Echo. If the SYN flag is set, then ECE set indicates that the TCP peer is ECN capable. If the SYN flag is clear, then the ECE flag set indicates that a Congestion Experienced flag in the IP header set was received during normal transmission.

- *URG*: Indicates whether the Urgent Pointer field is significant.

- *ACK*: If set, indicates that the Acknowledgement Number field is significant. All packets after the initial SYN packet sent by a node should have this flag set.

- *PSH*: Push flag. If set, asks to push any buffered data to the receiving application.

- *RST*: Reset flag. If set, resets the connection.

- *SYN*: Synchronize flag. If set, synchronizes sequence numbers. Only the first packet sent from each end should have this flag set.

- *FIN*: Finished flag – if set, no more data is coming.

Window Size (16 bits): Size of the receive window, which is the number of bytes that the receiver is willing to receive.

Checksum (16 bits): Used for error checking of the TCP header and data.

Urgent Pointer (16 bits): If the URG flag is set, this is the offset from the sequence number indicating the last urgent data byte.

Options (from zero to ten 32-bit words): Optional, not commonly used – see RFC for details.

Protocol Operation

1. Connection is established using a three-way handshake, which creates a virtual circuit.

2. Data is transferred over the virtual circuit until connection is terminated.

3. Connection termination closes the established virtual circuit and releases allocated resources.

TCP operation is controlled by a state machine, with 11 states:

1. *LISTEN*: Wait for a connection request from a remote client.

2. *SYN-SENT*: Wait for the remote peer to send back a segment with SYN and ACK flags set.

3. *SYN-RECEIVED*: Wait for the remote peer to send back acknowledgment after sending back a connection.

4. *ESTABLISHED*: The port is ready to exchange data with the remote peer.

5. FIN-WAIT-1

6. FIN-WAIT-2

7. CLOSE-WAIT

8. CLOSING

9. LAST-ACK

10. *TIME-WAIT*: Ensure the remote peer has received acknowledgment of the termination request (< 4 minutes).

11. CLOSED

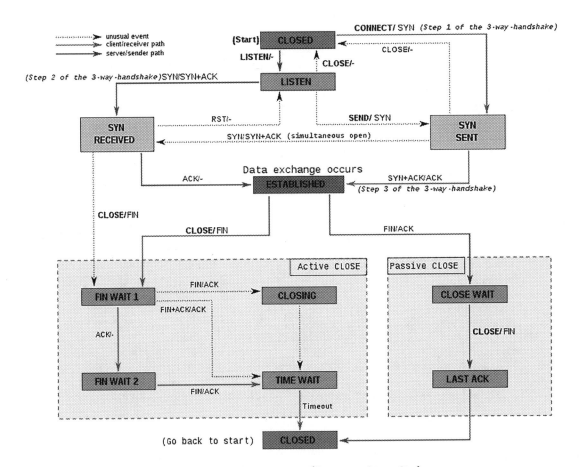

Figure 3-16. *TCP state transition diagram (from Wikipedia)*

Notes

TCP uses sequence numbers to detect lost packets and/or reorder packets that arrive out of order. The cumulative acknowledgment scheme informs the sender that all packets up to the acknowledged sequence number have been received. Selective acknowledgment (RFC 2018[117]) allows for optimization of this feature. Lost data is automatically retransmitted by the sender. End-to-end flow control provides for a mismatch in performance between sender and receiver. A sliding window algorithm allows multiple packets to be in progress, which increases efficiency. Recently, congestion control has been added into TCP to avoid network congestion.

TCP is very complicated. The good news is that when used over IPv6, TCP works essentially the same way. The very minor changes will be covered later.

UDP: The User Datagram Protocol

The User Datagram Protocol is defined in RFC 768,[118] "User Datagram Protocol," August 1980. Like TCP, it is also a *Transport Layer*[119] protocol. Unlike TCP, UDP implements an *unreliable, connectionless* model. When we say *unreliable,* we just mean that error detection and recovery are not built into the protocol, so it is up to the application to do any desired error detection and recovery. By *connectionless*, we mean that each transmission consists of a single (but complete) packet. In IPv4, a packet is typically 1508 bytes, but can be more or less. If you send a big packet, it will likely be fragmented somewhere along the way and reassembled at the other end. Each datagram is an atomic event, not connected to any other datagram. UDP does not handle *streams* of data (as is done with the *connection-oriented* model). Software that uses UDP does not need to *open* or *close* a connection; it can simply read or write datagrams at any time, and each operation sends or receives one packet. This is a much simpler model than TCP, with less overhead. However, when using UDP you are responsible for doing certain things that TCP does for you, such as error detection and retransmission. UDP is often used for things like streaming audio or video. It is also used for DNS queries and responses and for TFTP[120] (Trivial File Transfer Protocol).

---

[117] https://tools.ietf.org/html/rfc2018
[118] https://tools.ietf.org/html/rfc768
[119] https://en.wikipedia.org/wiki/Transport_layer
[120] https://en.wikipedia.org/wiki/Trivial_File_Transfer_Protocol

Standards Relevant to UDP

RFC 768, "User Datagram Protocol," August 1980 (Standards Track)

RFC 2508, "Compressing IP/UDP/RTP Headers for Low-Speed Serial Links," February 1999 (Standards Track)

RFC 3095, "Robust Header Compression (ROHC): Framework and Four Profiles: RTP, UDP, ESP and Uncompressed," July 2001 (Standards Track)

RFC 3828, "The Lightweight User Datagram Protocols (UDP-Lite)," July 2004 (Standards Track)

RFC 4019, "Robust Header Compression (ROHC): Profiles for User Datagram Protocol (UDP) Lite," April 2005 (Standards Track)

RFC 4113, "Management Information Base for User Datagram Protocol (UDP)," June 2005 (Standards Track)

RFC 4362, "RObust Header Compression (ROHC): A Link-Layer Assisted Profile for IP/UDP/RTP," January 2006 (Standards Track)

RFC 4727, "Experimental Values in IPv4, IPv6, ICMPv4, ICMPv6, UDP and TCP Headers," November 2006 (Standards Track)

RFC 4815, "Robust Header Compression (ROHC): Corrections and Clarifications to RFC 3095," February 2007 (Standards Track)

RFC 5097, "MIB for the UDP-Lite Protocol," January 2008 (Standards Track)

RFC 5225, "RObust Header Compression Version 2 (ROHCv2): Profiles for RTP, UDP, IP, ESP and UDP-Lite," April 2008 (Standards Track)

UDP Packet Header

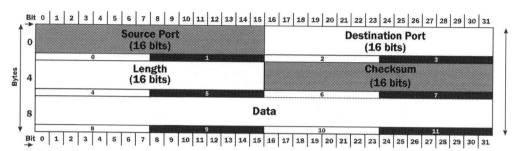

Figure 3-17. UDP packet header

The *Source Port* field (16 bits) specifies which port number the data is being written to on the sending computer. This field is optional (if not used, fill with zeros).

The *Destination Port* field (16 bits) specifies which port number the data is being read from on the receiving computer.

The *Length* field (16 bits) is the number of bytes in the datagram, including the UDP header and the data. Therefore, the minimum value is 8 (the length of the UDP header). The maximum value in theory is 65,536 bytes, but this value is limited by the maximum packet size, typically 1508.

The Checksum field (16 bits) is the 16-bit one's complement sum of the 16-bit words in the following items:

1. A "pseudo header," which contains the source and destination IP addresses, the protocol number, and the UDP length (from the IP header)

2. The UDP header itself

3. The data, padded with a zero byte if required to make an even number of bytes

The Checksum field is optional (if not used, fill with zeros).

The *Data* field begins immediately after the *Checksum* field. It is not really part of the header, but it is factored into the checksum.

DHCPv4: Dynamic Host Configuration Protocol for IPv4

One of the network services that is really useful in network configuration is the Dynamic Host Configuration Protocol (DHCP[121]). The version that works with IPv4 is now called DHCPv4 (to distinguish it from the one for IPv6, which is called DHCPv6). DHCPv4 is specified in RFC 2131,[122] "Dynamic Host Configuration Protocol," March 1997. Without DHCPv4 running on your network, someone must manually configure all IPv4 network settings on every computer. This can be very complicated and error-prone. It also requires at least some expertise, which many users don't possess. It is possible to accidently configure two computers with the same address or mistype a DNS server's address on the 35th computer you install that day. These kinds of errors can cause tricky problems. With a DHCPv4 server, you can configure all the client computers to do "autoconfiguration." When such a computer powers up, it will search for a DHCPv4 server (or a relay agent, connected to a real DHCPv4 server in another network). When it finds one, it will request configuration data (including the default gateway, the IP addresses of the DNS servers, the Internet domain name, and other items, including a lease on an IPv4 address, which should be unique within your network). This makes it easier to change things. If you move a DNS server or make other changes, you need only update your DHCPv4 server configuration and terminate all client leases (all nodes will request new configuration information).

DHCPv4 is widely used by ISPs, especially ones that have lots more customers than valid (globally routable) IPv4 addresses. They can set very short lease times. Then when someone disconnects, the address they had been using can be reused by another customer. Of course, these days, most people want 7×24 Internet connectivity, as opposed to perhaps 1 hour a day or dial-up access. Many ISPs now provide their customers with RFC 1918 private addresses, unless for some reason they specifically require a globally routable address. Some ISPs charge more for a globally routable address and a *lot* more for multiple globally routable addresses. I have one real public IPv4 address for my home network, so I can run email and other services, in addition to using one to tunnel IPv6 into my network over IPv4. DHCPv4 can provide autoconfiguration with private addresses just as easily as with globally routable

---

[121] https://en.wikipedia.org/wiki/Dynamic_Host_Configuration_Protocol
[122] https://tools.ietf.org/html/rfc2131

addresses, so they still use DHCPv4 to assign those. Basically, all their users are now "hiding" behind a single public address, via NAT. More recently, users are behind *two* layers of NAT – they don't even have *one* public IPv4 address anymore. This is done with CGNAT (Carrier-Grade NAT[123]), with one mapping from a public address at the ISP to one private address from 100.64/10 at the ISP and another from that private address to multiple RFC 1918 private addresses in their network.

DHCPv4 uses broadcast (which doesn't exist in IPv6) and can only deliver 32-bit addresses (for the assigned IP address or things like DNS IP addresses), so it had to be completely rewritten for IPv6. The differences will be covered in Chapter 6.

Most client operating systems in use today (especially on personal computers) include a DHCPv4 client, including all versions of Windows, FreeBSD, Linux, Solaris, Mac OSX, etc. Even smartphones with Wi-Fi include a DHCPv4 client. In practice, DHCPv6 may not be widely used, as IPv6 addresses and even discovery of IPv6 addresses for DNS are more likely to be done via Stateless Address Autoconfiguration (SLAAC) than via DHCPv6 (see RFC 6106,[124] "IPv6 Router Advertisement Options for DNS Configuration", November 2010).

Most server operating systems (such as Windows Server, FreeBSD, Linux, etc.) include a DHCPv4 server. The most common one for UNIX and UNIX-like servers is *dhcpd* from the Internet Systems Corporation (ISC). It is configured by editing some complex ASCII text configuration files (with a text editor). This type of configuration has not changed appreciably in 50 years (and you thought *IPv4* was old). The DHCPv4 server included with Windows Server at least has a GUI configuration tool, which is much easier to use. Most appliances that provide DHCPv4 service include a GUI web-based configuration tool (as a "front end" to dhcpd, in most cases).

When you configure a DHCPv4 server, you typically configure one or more pools of addresses to be managed by that server. You can have more than one DHCPv4 server in a given network subnet, but the managed address ranges must not overlap. DHCPv4 clients cannot contact DHCPv4 servers on another subnet (on the other side of a router) directly (since DHCPv4 servers are found via broadcast). So you either need to have a DHCPv4 server (or at least a *DHCPv4 relay agent*) in every subnet ("broadcast domain"). You can create a "scope" on the server and configure the "stateless" items that it will use to autoconfigure clients, including the domain name, the subnet mask, the address of the default gateway, the IP addresses of two DNS servers, etc. There are dozens of

---

[123] https://en.wikipedia.org/wiki/Carrier-grade_NAT
[124] https://tools.ietf.org/html/rfc6106

things you can autoconfigure with DHCPv4. You also specify a range of addresses (e.g., 192.168.5.100 to 192.168.5.199) as a *pool* from which to lease addresses. You should not manually assign any of these addresses to other nodes. If you do for some reason, you can exclude that address from the available pool.

Once such a server is installed and configured, just set up your client computers to "Obtain an IP address automatically" and to "Obtain DNS server address automatically." As soon as you specify that or anytime the computer powers up, it will obtain all necessary information (including a unique IPv4 node address) from the DHCPv4 server. In Windows, you can use the "ipconfig /all" command (in a DOS prompt window) to view the obtained settings (look for the interface named Local Area Connection).

By default, addresses are assigned on a "first come, first served" basis. If you want a given node to be assigned a specific address each time, you can make an *address reservation* by associating one of the pool addresses with that node's MAC address. Any time that node requests configuration data from the DHCPv4 server, it will be assigned the reserved address for that MAC address, rather than a random one from the pool.

The DHCPv4

The DHCPv4 lives in the *Application* Layer. It uses port 67 for data from client to server and port 68 for data from the server to the client (both over UDP). There are four phases in a DHCPv4 network configuration:

1. IP Discovery

2. IP Lease Offer

3. IP Request

4. IP Lease Acknowledgment

Let's say our network uses 192.168.0.0/16. That means the subnet mask is 255.255.0.0. Our DNS servers are at 192.168.0.11 and 192.168.0.12. The DHCPv4 server is also running on 192.168.0.11. The default gateway is 192.168.0.1. We have created a pool of addresses from 192.168.1.0 to 192.168.1.255.

In the *Discover IP* phase, the client sends a DHCPDISCOVER request, as follows:

- Source address = 0.0.0.0, source port = 68

- Destination address = 255.255.255.255, destination port = 67

- *DHCP option 50*: IP address 192.168.1.100 is requested.

- *DHCP option 53*: Message is DHCPDISCOVER.

- Request subnet mask, default gateway, domain name, and domain name server(s).

In this case, the node is requesting its last known IP address. Assuming it is still connected to the same network and the address is not already leased to someone else, the server may grant the request. Otherwise, the client will have to negotiate for a new address.

In the *DHCP Lease Offer* phase, the server will reserve an IP address for the client (in this case it is accepting the request for the last known address) and send a DHCPOFFER message to the client, as follows:

- Source address = 192.168.0.11, source port = 67

- Destination address = 255.255.255.255, destination port = 68

- *DHCP option 01*: Subnet mask is 255.255.0.0.

- *DHCP option 03*: Default gateway is 192.168.0.1.

- *DHCP option 06*: IP addresses of DNS servers are 192.168.0.11 and 192.168.0.12.

- *DHCP option 51*: Lease duration is 86400 seconds (1 day).

- *DHCP option 53*: Message is DHCPOFFER.

- *DHCP option 54*: IP address of the DHCP server is 192.168.0.11.

In the *IP Request* phase, the client accepts the offer and sends a DHCPREQUEST message as follows:

- Source address = 0.0.0.0, source port = 68

- Destination address = 255.255.255.255, destination port = 67

- *DHCP option 50*: IP address 192.168.1.100 is requested.

- *DHCP option 53*: Message is DHCPREQUEST.

- *DHCP option 54*: IP address of the DHCP server is 192.168.0.11.

In the *IP Acknowledgement* phase, the server officially registers the assignment and notifies the client of the configuration values:

- Source address = 192.168.0.11, source port = 67

- Destination address = 255.255.255.255, destination port = 68

- *DHCP option 01*: Subnet mask is 255.255.0.0.

- *DHCP option 03*: Default gateway is 192.168.0.1.

- *DHCP option 06*: IP addresses of DNS servers are 192.168.0.11 and 192.168.0.12.

- *DHCP option 51*: Lease duration is 86400 seconds (1 day).

- *DHCP option 53*: Message is DHCPACK.

- *DHCP option 54*: IP address of the DHCP server is 192.168.0.11.

At this point, the client actually configures those values for its network interface and can begin using the network.

Useful Commands Related to DHCPv4

In Windows, there are some commands available in a DOS prompt box related to DHCPv4:

> ipconfig /release: Release the assigned IPv4 address and de-configure network.
>
> ipconfig /renew: Do a new configuration request for IPv4.
>
> ipconfig /all: View all network configuration settings.

This is an example of the output from "ipconfig /all".

```
C:> ipconfig /all
...
Ethernet adapter Local Area Connection:

        Connection-specific DNS Suffix   . : redwar.org
        Description . . . . . . . . . . . : Realtek PCIe GBE Family Controller
        Physical Address. . . . . . . . . : 00-22-15-24-32-9C
        DHCP Enabled. . . . . . . . . . . : Yes
        IPv4 Address. . . . . . . . . . . : 192.168.1.8(Preferred)
        Subnet Mask . . . . . . . . . . . : 255.255.0.0
        Default Gateway . . . . . . . . . : 192.168.0.1
        DNS Servers . . . . . . . . . . . : 192.168.0.11
                                            192.168.0.12
        NetBIOS over Tcpip. . . . . . . . : Enabled
```

Figure 3-18. *Output of the ipconfig /all command*

IPv4 Network Configuration

Let's assume our LAN has the following configuration:

```
Network Address:    192.168.0.0/16 (mask = 255.255.0.0)
Default Gateway:    192.168.0.1
DHCPv4 Address:     192.168.0.11
DNS Server Address: 192.168.0.11, 192.168.0.12
Domain Name:        redwar.org
```

Furthermore, assume the DHCPv4 server is correctly configured with this information and is managing the address range 192.168.1.0–192.168.1.255 (and that some leases have already been granted).

Any node connected to a network with IPv4 must have certain items configured, including

- IPv4 node address

- Subnet mask (or, equivalently, CIDR subnet mask length)

- IPv4 address of the default gateway

- IPv4 addresses of DNS servers

- Nodename

- DNS domain name

Manual Network Configuration

It is possible to perform IPv4 network configuration on a node manually, either by editing ASCII configuration files, as in FreeBSD or Linux, or via GUI configuration tools, as in Windows. If you have understood the material in this chapter, it should be fairly easy for you to configure your node(s). In most cases, if you have ISP service, the ISP will give you all the information necessary to configure your node(s).

Let's configure a FreeBSD 7.2 node manually. Assign it the nodename "us1.redwar.org" and the IP address 192.168.0.13. The interface we are configuring has the FreeBSD name "vr0".

You need to edit the following files (you will need root privilege to do this):

/etc/rc.conf

```
...
hostname="us1.redwar.org"
ifconfig_vr0="inet 192.168.0.13 netmask 255.255.0.0"
defaultrouter="192.168.0.1"
...
```

/etc/resolv.conf

```
domain      redwar.org
nameserver  192.168.0.11
nameserver  192.168.0.12
```

If you make these changes and reboot, you can check the configuration as shown:

```
$ ifconfig vr0
vr0: flags=8843<UP,BROADCAST,RUNNING,SIMPLEX,MULTICAST> metric 0 mtu 1500
        options=2808<VLAN_MTU,WOL_UCAST,WOL_MAGIC>
        ether 00:15:f2:2e:b4:1c
        inet 192.168.0.13 netmask 0xffff0000 broadcast 192.168.255.255
        media: Ethernet autoselect (100baseTX <full-duplex>)
        status: active
$ uname -n
us1.redwar.org
$ nslookup
> server
```

```
Default server: 192.168.0.11
Address: 192.168.0.11#53
Default server: 192.168.0.12
Address: 192.168.0.12#53
> exit

$ netstat -rn
Routing tables

Internet:
Destination          Gateway              Flags    Refs      Use  Netif Expire
default              192.168.0.1          UGS         0        5   vr0
...
```

Auto Network Configuration Using DHCPv4

It is also possible for a node to be automatically configured if a DHCPv4 server (or relay agent) is available somewhere on the LAN (or possibly from the ISP). If you are deploying several nodes on a home network, it is likely that there is a DHCPv4 server in your home gateway/DSL modem.

Let's configure a FreeBSD 7.2 node automatically using DHCPv4. Assign it the nodename "us1.redwar.org" and any IP address from DHCPv4. The interface we are configuring has the FreeBSD name "vr0".

You need to edit the following file (you will need root privilege to do this): */etc/rc.conf*

```
...
hostname="us1.redwar.org"
ifconfig_vr0="DHCP"
...
```

If you make these changes and reboot, you can check the configuration as shown:

```
$ ifconfig vr0
vr0: flags=8843<UP,BROADCAST,RUNNING,SIMPLEX,MULTICAST> metric 0 mtu 1500
        options=2808<VLAN_MTU,WOL_UCAST,WOL_MAGIC>
        ether 00:15:f2:2e:b4:1c
```

```
        inet 192.168.1.9 netmask 0xffff0000 broadcast 192.168.255.255
        media: Ethernet autoselect (100baseTX <full-duplex>)
        status: active
```

$ **uname -n**

```
us1.redwar.org
```

$ **nslookup**

> **server**

```
Default server: 192.168.0.11
Address: 192.168.0.11#53
Default server: 192.168.0.12
Address: 192.168.0.12#53
```

> **exit**

$ **netstat -rn**

```
Routing tables
```

```
Internet:
Destination         Gateway          Flags    Refs      Use  Netif Expire
default             192.168.0.1       UGS        0         5   vr0
```

Step 1 – IPv4 Network Configuration

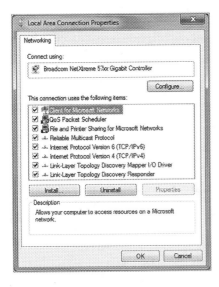

- To bring this up this dialog:
    - Click on *Windows Start Icon* (lower left)
    - Click on *Control Panel*
    - Click on *Network and Sharing Center*
    - Click on *Change Adapter Settings*
    - Right click on *Local Area Connection*
    - Select *Properties* from pull down menu
- If IPv4 is not currently enabled (a *check* in the square box at start of line), click in that box to enable it.
- Double click on the *Internet Protocol Version 4 (TCP/IPv4)* item.

Figure 3-19. *TCP/IP network configuration – main tab*

Step 2 – IPv4 (TCP/IPv4) Properties

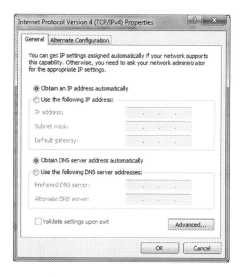

- To obtain an IPv4 node address, subnet mask and default gateway automatically via DHCPv4, select *Obtain an IP address automatically*

- To manually specify them, select *Use the following IP address*

- To obtain the IPv4 addresses of available DNS servers via DHCPv4, select *Obtain DNS server addresses automatically*

- To manually specify them, select *Use the following DNS server addresses*

Figure 3-20. *IPv4 network configuration – TCP/IPv4 Properties dialog*

Step 3 – Manual IPv4 Network Configuration

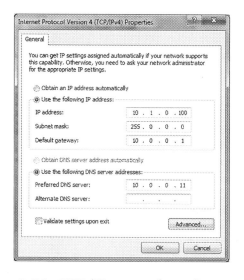

- Here, an IPv4 node address, a subnet mask, a default gateway and the IPv4 address of a DNS server have been manually specified.

- Note that when you choose manual configuration of the IPv4 address, that also selects manual configuration of the DNS server addresses. With automatic address configuration you can still select manual configuration of DNS server addresses.

- When done, click on *OK*.

Figure 3-21. *TCP/IP network configuration – IPv4 manual configuration dialog*

Step 4 – IPv6 Network Configuration

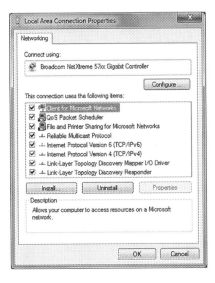

* To bring this up this dialog:
    * Click on *Windows Start Icon* (lower left)
    * Click on *Control Panel*
    * Click on *Network and Sharing Center*
    * Click on *Change Adapter Settings*
    * Right click on *Local Area Connection*
    * Select *Properties* from pull down menu

* If IPv6 is not currently enabled (a *check* in the square box at start of line), click in that box to enable it.

* Double click on the *Internet Protocol Version 6 (TCP/IPv6)* item.

Figure 3-22. *TCP/IP network configuration main tab – select IPv6*

Step 5 – IPv6 (TCP/IPv6) Properties

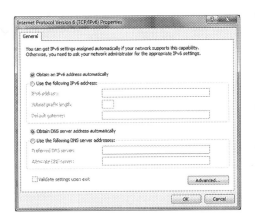

* To obtain an IPv6 node address, subnet prefix length and default gateway automatically via SLAAC and/or DHCPv6, select *Obtain an IPv6 address automatically*.

* To manually specify them, select *Use the following IPv6 address*

* To obtain the IPv6 addresses of available DNS servers via DHCPv6, select *Obtain DNS server address automatically*.

* To manually specify them, select *Use the following DNS server addresses*

Figure 3-23. *TCP/IP network configuration – TCP/IPv6 automatic configuration*

Step 6 – Manual IPv6 Network Configuration

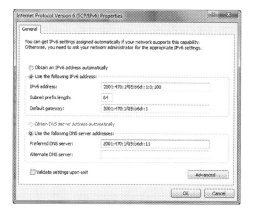

- Here, an IPv6 node address, a subnet prefix length, a default gateway and the IPv6 address of a DNS server have been manually specified.

- Note that when you choose manual configuration of the IPv6 address, that also selects manual configuration of the DNS server addresses. With automatic address configuration you can still select manual configuration of DNS server addresses.

- When done, click on *OK*.

Figure 3-24. *TCP/IP network configuration – TCP/IPv6 manual configuration*

Step 7 – Verify Network Configuration

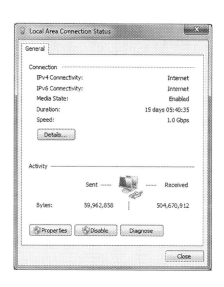

- To get this dialog, right click on the *Local Area Connection* item in the *Network Connections* window, and select *Status* from the menu.

- This shows the status of connectivity to the IPv4 Internet, connectivity to the IPv6 Internet, whether the interface has a carrier (from being connected to a working switch), how long the connection has been working, and the network speed of the connection (10 Mbps, 100 Mbps, or 1.0 Gpbs).

- To see details click on *Details...*

Figure 3-25. *TCP/IP network configuration – verify configuration*

Step 8 – Check Network Settings

- This Details dialog shows:
  - the node's MAC (physical) address
  - whether DHCP was enabled
  - IPv4 node address
  - IPv4 subnet mask
  - IPv4 default gateway address
  - IPv4 address of DNS server(s)
  - One or more IPv6 global unicast node address(es)
  - Link-local IPv6 node address
  - Link-local IPv6 default gateway address
  - Global IPv6 default gateway address
  - IPv6 address of DNS server(s)

Figure 3-26. *TCP/IP network configuration – check network configuration details*

Summary

In this chapter, we reviewed the technical aspects of IPv4. First off, some of you might not be familiar with the details of this protocol or how it has evolved over the many years since 1981, when it was specified.

In particular, we covered how we have "broken the Internet" by introducing NAT and private addresses and why it was done. This allowed us to keep using IPv4 well past its real shelf life, but at a very high cost (more complex network software design and limits on what most uses can do in terms of connections to and from other nodes). We have splintered the original IPv4 "monolithic address space" into millions of tiny "private Internets" loosely coupled together through NAT gateways.

There are many RFCs that specify how IPv4 works, going back to 1981 (RFCs 791 and 792).

Since IPv6 is heavily based on IPv4, you need to understand IPv4 in order to understand what is new in IPv6.

Likely, IPv4 will be phased out (at least at the international backbone level) in the next few years. Until then, we will have a global Internet that is partly IPv4 and partly IPv6. The two can exist in parallel, but it is not really practical to translate between them in either direction (NAT64 is very limited to allow translation from IPv6 addresses to external IPv4 servers).

If you are like most network engineers and developers today, you only know IPv4. If so, you are rapidly becoming obsolete. The future is IPv6. This book will help you make the leap from IPv4 to IPv6, so you will still have a job in a few years. Today, when I run into a corporate network that is IPv4-only, I feel like I am being asked to ride on a horse instead of my much more powerful and fast car (IPv6). The four-octet IPv4 addresses now look quaint and primitive to me, like the one-octet NDP addresses from ARPANET look to you now. Get used to 128-bit addresses in hexadecimal. The future is here.

A good analogy is when Novell NetWare was being replaced by TCP/IP some years ago. Many people were very tied to NetWare, with multiple certifications and extensive expertise, but soon there were no jobs for them. All networks were being converted over to TCP/IP because that was the native protocol of the Internet. They had to learn TCP/IP to be useful. The next generation has arrived.

If you want to have a good dose of reality, check out "Sunset IPv4" – the working group of the IETF whose charter was to figure out how to finally put IPv4 to sleep for good, like NetWare or OSI. So let us say a fond farewell to IPv4. The king is dead. Long live the king.

CHAPTER 4

The Depletion of the IPv4 Address Space

Some people today are aware that the folks in charge of the Internet are running out (or have already run out) of public IPv4 addresses. Most of them are *not* aware that this is not the first time we've faced this or just how low that pool of addresses is today. The majority of Internet users are either completely oblivious to what is going on and think that the Internet will go on like it has, forever. If they have heard any rumors about an address shortage, they have a blind faith that the people in charge can simply work some magic and the problem will go away. Well, they did once, in the mid-1990s (with NAT and private addresses), and they have found another trick with Carrier-Grade NAT to extend the lifetime of IPv4 even longer. However, each of these stopgap measures has caused major new problems. IPv4 is simply at its end of life, and it is time to start using its successor, IPv6.

L. E. Hughes, *Third Generation Internet Revealed*, https://doi.org/10.1007/978-1-4842-8603-6_4

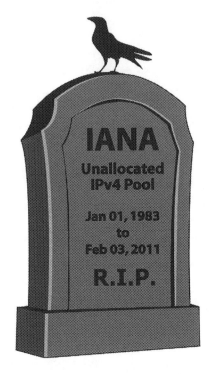

Figure 4-1. *RIP, IPv4 public address allocation pool*

OECD IPv6 Report, March 2008

The best study on this done to date (in my opinion) is in the OECD report presented at the OECD Ministerial Meeting on the Future of the Internet Economy, in Seoul, Korea, June 17–18, 2008. I was a speaker at the concurrent Korean IPv6 Summit. The full name of the OECD is *Organisation for Economic Co-operation and Development*. It was established in 1961 and currently has 30 member nations, including most members of the EU, plus Australia, Canada, Japan, Korea, Mexico, New Zealand, Turkey, the United Kingdom, and the United States. It had a 2009 budget of EUR 320 million. Their goals are to

- Support sustainable economic growth.

- Boost employment.

- Raise living standards.

- Maintain financial stability.

- Assist other countries' economic development.

- Contribute to growth in world trade.

Unlike the IETF or ISO, the OECD is not specifically concerned with technology. They are primarily concerned with the economies of their member countries. However, they have determined that the imminent exhaustion of the IPv4 address space will have a major impact on most of their goal areas. Because of this, they did a major study, the results of which are presented in Ministerial Background Report DSTI/ICCP(2007)20/FINAL, "Internet Address Space: Economic Considerations in the Management of IPv4 and in the Deployment of IPv6." The report[1] is available free to download over the Internet. You should actually read the entire report, but I will summarize the most important aspects of it in this chapter.

Let me quote one paragraph from the "Main Points" section:

> There is now an expectation among some experts that the currently used version of the Internet Protocol, IPv4, will run out of previously unallocated address space in 2010 or 2011, as only 16% of the total IPv4 address space remains unallocated in early 2008. The situation is critical for the future of the Internet economy because all new users connecting to the Internet, and all businesses that require IP addresses for their growth, will be affected by the change from the current status of ready availability of unallocated IPv4 addresses.

As of early 2010, only 8% of the addresses remained unallocated. The IANA pool was officially exhausted in February 2011.[2] All five RIRs have all reached "end of normal IPv4 allocation" since then.

Another key passage from this section follows:

> As the pool of unallocated IPv4 addresses dwindles and transition to IPv6 gathers momentum, all stakeholders should anticipate the impacts of the transition period and plan accordingly. With regard to the depletion of the unallocated IPv4 address space, the most important message may be that there is no complete

---

[1] www.oecd.org/internet/ieconomy/40605942.pdf

[2] www.computerworld.com/article/2512924/update--icann-assigns-its-last-ipv4-addresses.html

solution and that no option will meet all expectations. While the Internet technical community discusses optional mechanisms to manage IPv4 address space exhaustion and IPv6 deployment and to manage routing table growth pre- and post-exhaustion, governments should encourage all stakeholders to support a smooth transition to IPv6.

IPv6 adoption is a multi-year, complex integration process that impacts all sectors of the economy. In addition, a long period of co-existence between IPv4 and IPv6 is projected during which maintaining operations and interoperability at the application level will be critical. The fact that each player is capable of addressing only part of the issue associated with the Internet-wide transition to IPv6 underscores the need for awareness raising and co-operation.

Basically, there is no good or lasting solution for those wanting to remain with IPv4. It is going to take *multiple years* to make the transition. We are now in 2022, and there are still a lot of people and organizations who have not completed their transition to IPv6 (and some who haven't even *begun* it). Such transitions are usually not done well when rushed. And once the addresses are gone, that's it. The IETF assumed that the transition would be *done* by 2010 *before* IPv4 public addresses ran out. One problem is that the best transition mechanism (6in4 tunneling) requires one public IPv4 address at the customer site, and today those are *very* hard to come by. For example, in the Philippines, *no* personal ISP accounts include a public IPv4 address, and even business accounts have a very small number (five or maybe just one). The ISPs just don't have any more to allocate. The tunneling schemes that work through NAT are much more problematic and unstable.

The OECD report acknowledges that in the early phases of a major technology transition such as this, there may be little or no incentive to shift to the new technology. However, once a critical mass of users adopt the new technology, there is often a *tipping point* after which adoption grows rapidly until it is widespread. In theory this tipping point is reached when the marginal cost, for an ISP or an organization, of implementing the next device with IPv4 becomes higher than the cost of deploying the next device with IPv6. For an ISP, there are costs associated with deploying IPv4 nodes such as the cost of obtaining the addresses themselves and the costs of designing and deploying network infrastructure that uses fewer and fewer public (globally routable) addresses (by using NAT). When these become higher than the cost of deploying IPv6, they will begin migration in earnest. Reaching this tipping point depends on a number of factors, including customer demand, opportunity costs, emerging markets, the introduction of new services, government incentives, and regulation.

For mobile telco service providers (especially in the United States), they have already passed this tipping point. It was far cheaper for them to migrate to IPv6 than to keep IPv4 alive for one more year. Even with private addresses, the largest block (10/8) only has 16.7M addresses in it, and many mobile telcos have far more than 16.7M customers. So multiple /8 blocks must be deployed and somehow "stitched together" into a single network. This is very difficult. Many mobile operators are ditching IPv4 altogether and providing *only* IPv6 service. This is viable due to something called "464XLAT," (RFC 6877[3] – 464XLAT: Combination of Stateful and Stateless Transition," April 2013). This allows legacy smartphone apps that only support IPv4 to still work. 464XLAT has been in Android since 4.3 in October 2013. On iPhones, since iOS 10, all apps in the App Store must work in an IPv6-only environment. Today many mobile phone service providers have migrated to IPv6 – typical companies are at 90+% migrated.

As 5G is rolled out, this will likely be mostly IPv6 based. Apart from the higher speed, 5G is supposed to support end-to-end direct connection, which is only possible with IPv6.

One of the key requirements for migrating to IPv6 is technical expertise in the subject. This is necessary to provide countries and companies with competitive advantage in the area of technology products and services and the benefit from ICT-enabled innovation. Countries who are early adopters and provide training and incentives for their companies to embrace it or even help fund the necessary infrastructure (as in China) will have significant competitive advantages in years to come over countries that are laggards in this transition. India has taken the lead in this by requiring all ISPs to deploy IPv6. They are now at 60% migration nationwide (the highest in the world).

Increasing scarcity of IPv4 addresses can raise competitive concerns in terms of barriers to new entry and strengthening incumbent positions. There has been much discussion over how to manage previously allocated IPv4 addresses once the free pool

---

[3] `https://tools.ietf.org/html/rfc6877`

has been exhausted. A global market for IPv4 addresses has emerged. Desperate ISPs and cloud service providers can still buy previously allocated IPv4 addresses for about $16 per address (early 2019). Today, you only *borrow* (lease?) addresses from an ISP for so long as you have service with that ISP. If you terminate that service, the addresses are reclaimed by the ISP for allocation to other customers. You don't really *own* those addresses, so you can't *sell* them. Even the ISP doesn't own them. If an ISP goes out of business, their address pool probably returns to the RIR they got them from. Some of these situations are not currently well defined, but they will be as the IPv4 address space nears exhaustion. Notably, the situation on the early class A block allocations is not quite so well defined. Those blocks *are* owned by those early adopter companies.

One of the companies that got a class A block (Nortel), when it closed down, sold off 666,000 IPv4 public addresses to Microsoft[4] for USD 7.5 million (primarily for use in Azure Cloud).

There is also discussion of how existing and increasing use of NAT requires developers of network-aware products and applications to build increasingly complex central gateways or NAT traversal mechanisms to allow clients (who are in most cases *both* behind NAT gateways) to communicate. This is creating barriers to innovation and to the development of new services. It is also causing problems with the overall performance and stability of the Internet.

There is a risk of some parts of the world deploying IPv6, while others continue running IPv4 with multiple layers of NAT. Such decisions would impact the economic opportunities offered by the Internet with severe repercussions in terms of stifled creativity and deployment of generally accessible new services. Also, there could be serious issues of interoperation between people in the IPv6 world and those left behind in the IPv4 world. This could lead to a fragmentation of the Internet.

The five sections of the report cover the following topics:

- Overview of the major initiatives that have taken place in Internet addressing to date and the parallel development of institutions that manage Internet addressing.

- Summary of proposals under consideration for management of the remaining IPv4 addresses.

---

[4]www.networkworld.com/article/2228854/microsoft-pays-nortel--7-5-million-for-ipv4-addresses.html

- Overview of the drivers and challenges for transitioning to IPv6 through a dual-stack (IPv4 + IPv6) environment. It reviews factors that influence IPv6 adoption, drawing on available information.

- Economic and public policy considerations and recommendations to governments.

- Lessons learned from several IPv6 deployments.

OECD Follow-Up Report on IPv6, April 2010

In April 2010, the OECD released a follow-up report to the IPv6 report mentioned previously. It is called "Internet Addressing: Measuring Deployment of IPv6."[5] They still expected IPv4 addresses to run out in 2012. As of March 2010, only 8% of the full IPv4 address space was available for allocation at the IANA level. At that time, IPv6 use was growing faster than IPv4 use, albeit from a still small base. Several large-scale deployments were taking place or were in planning. Some of the key findings, all as of March 2010, were as follows:

- 5.5% of the networks on the Internet (1,800 networks) could handle IPv6 traffic.

- IPv6 networks have grown faster than IPv4-only networks since mid-2007.

- Demand for IPv6 address blocks has grown faster than demand for IPv4 address blocks.

- One out of five transit networks (i.e., networks that provide connections through themselves to other networks) handled IPv6. This means that Internet infrastructure players were actively readying for IPv6.

---

[5] www.oecd.org/internet/ieconomy/44953210.pdf

- As of January 2010, over 90% of installed operating systems were IPv6 capable, and 25% of end users ran an operating system that enabled IPv6 by default (e.g., Windows Vista or Mac OS X). This percentage has probably increased since the release of Windows 7, but no measurement is available.

- As of January 2010, over 1.45% of the top 1000 websites were available over IPv6, but as of March 2010 (when Google IPv6 enabled their websites), this jumped to 8%.

- Over 4,000 IPv6 prefixes (address blocks) had been allocated. Of these 2,500 (60%) showed up as routed on the Internet backbone (were actually in use).

- At least 23% of Internet Exchange Points explicitly supported IPv6.

- Seven out of 13 DNS root servers were accessible over IPv6.

- 65% of top-level domains (TLDs) had IPv6 records in the root zone file.

- 80% of TLDs had name servers with an IPv6 address.

- 1.5 million domain names (about 1% of the total) had IPv6 DNS records.

Operators in the RIPE NCC and APNIC service areas were given a survey in 2009. The results showed the following:

- 7% of APNIC respondents claimed to have equal or more IPv6 traffic than IPv4 traffic.

- 2% of RIPE respondents claimed to have equal or more IPv6 traffic than IPv4 traffic.

- Of those respondents not deploying IPv6, 60% saw cost as a major barrier.

- Of those respondents deploying IPv6, 40% considered lack of vendor support the main obstacle.

OECD Second Follow-Up Report on IPv6, November 2014

Since the 2010 book, the OECD has released another report on IPv6: The Economics of Transition to Internet Protocol version 6 (IPv6).[6]

Citation: OECD (2014), "The Economics of Transition to Internet Protocol version 6 (IPv6)," *OECD Digital Economy Papers*, No. 244, OECD Publishing. DOI: `https://doi.org/10.1787/5jxt46d07bhc-en`[7]

- As of April 2014, worldwide traffic over IPv6 was roughly 3.5%.

- The adoption of IPv6 has differed from that of other technologies, for the following reasons:

    - The primary benefit to adopters is access to the larger IP address space.

    - Since most people deploying IPv6 are implementing dual stack, they still must cope with the lack of public addresses and NAT on the IPv4 side – many of the benefits will not come until users can turn off IPv4.

    - Implementation may involve solving new and unexpected technical challenges, and there has been a lack of skills for implementing IPv6.

    - ISPs and vendors have invested heavily in alternative solutions such as Carrier-Grade NAT, despite many negative aspects.

- Deployment by mobile service providers has been much stronger than by wired service providers, due to certain technical factors, which make it less expensive to deploy IPv6 than to keep IPv4 alive in that space.

- Not transitioning to IPv6 has a range of economic implications:

---

[6] `www.oecd.org/officialdocuments/publicdisplaydocumentpdf/?cote=DSTI/ICCP/CISP%282014%293/FINAL&docLanguage=En`

[7] `www.oecd-ilibrary.org/science-and-technology/the-economics-of-transition-to-internet-protocol-version-6-ipv6_5jxt46d07bhc-en`

- Alternative solutions break some applications and disrupt the modularity of the Internet.

- Economic costs of IPv4 depletion are asymmetric, affecting some products, services, and providers more than others.

- A market has emerged to trade in unused IPv4 addresses.

- The World IPv6 Launch (June 6, 2006) was effective in promoting adoption.

- IPv6 is infrastructure, not a product in its own right.

- The net benefits from adoption are not distributed equally across the stakeholders.

- As of April 2014, IANA, APNIC, RIPE, and LACNIC had already reached end of normal IPv4 allocation. ARIN was expected to end allocation in February 2015 and AfriNIC in June 2020. [ARIN actually ended normal allocation in September 2015 but left no buffer in stock – other RIRs ended allocation when they reached one "/8" of addresses (16.7M).]

- IPv6 connections to Google search website had reached 3.5% [it is now at 26.7%, just 5 years later].

- IPv6 adoption is based on the "Probit model," where (1) the benefits of adoption exceed the costs and (2) it is better to adopt now compared with any other time.

- To many users, the benefits are uncertain and will occur primarily in the future.

- As with many technologies, there is a "network effect," where as more people adopt it, the incentives to adopt increase.

- Adoption of early users can influence adoption by later users. Early ISP adopters include Comcast, Verizon Wireless, AT&T, Free, Deutsche Telekom, and KDDI. Large content providers include Google, Facebook, and Yahoo.

- Another OECD report in 2013 estimated that a family with two teenagers could have as many as 50 devices connected in their home by 2022. They further estimated 50 billion devices by 2020–2030. These volumes cannot be handled by IPv4.

- Support for IPv6 by network equipment vendors was "excellent," while CPE vendors were not as good.

- Costs of deploying CGN were estimated at US$ 90,000 per 10,000 users, plus US$ 10,000 per year ongoing.

- Costs of a provider transitioning to IPv6 are lower if stretched over several years than if done all at once (replace IPv4-only gear with IPv6-compliant gear during normal replacement cycles).

- Mobile providers do not have to consider CPE costs and complexity, and 40% of LTE handsets in 2014 supported IPv6 [virtually all do now]. The presence of 464XLAT on almost all Android handsets makes the migration to IPv6 especially easy for mobile providers.

- In enterprise, current use of NAT and private addresses reduces pressure to have more IP addresses, but complicates the transition to IPv6.

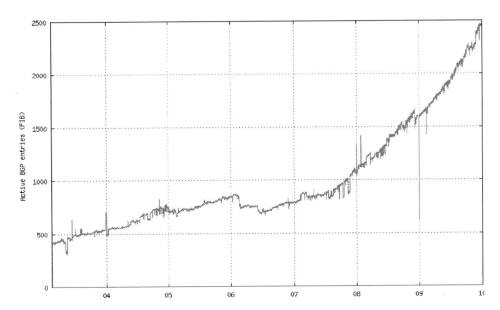

Figure 4-2. *Routed IPv6 refixes, 2003 to end of 2009*

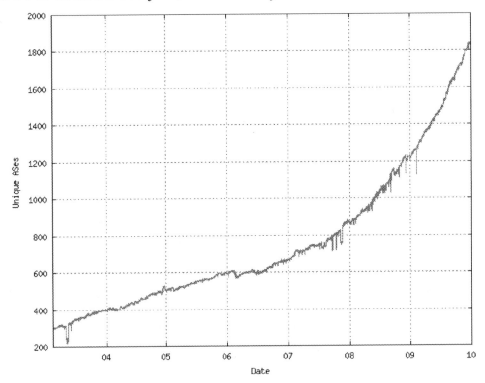

Figure 4-3. *IPv6 unique autonomous systems, 2003 to end of 2009*
Source: ITAC/NRO Contribution to the OECD, Geoff Huston and George Michaelson, data from end of year 2009

Since 2008, the ratio of routed IPv6 prefixes to IPv4 prefixes had climbed from 0.45% to 0.8%, which indicates that the number of routed IPv6 prefixes was increasing more rapidly than that of routed IPv4 prefixes. The ratio of IPv6 to IPv4 AS entities actively routing went from about 3.2% in 2008 to 5.5% in 2010.

The compound annual growth rate from February 24, 2009, to November 5, 2009, for dual-stack AS networks was 52%, for IPv6-only AS networks was 13%, and for IPv4-only AS networks was 8%. At year-end 2009, there were 31,582 AS networks using IPv4-only, there were 1806 AS networks using dual stack, and there were 59 AS networks using IPv6-only.

One trend is that service providers, corporations, public agencies, and end users were using IPv6 for advanced and innovative activities on private networks. IPv6 was also being used in 6LoWPANs (IPv6 over Low-Power Wireless Personal Area Networks), as specified in RFC 4944, "Transmission of IPv6 Packets over IEEE 802.15.4 Networks," September 2007.

How IPv4 Addresses Were Allocated in the Early Days

In the early days, before IANA and the RIRs were created, IPv4 addresses were actually allocated manually by a single individual, Jon Postel.[8] He never dreamed how large the Internet would grow or that it would be a worldwide phenomenon that had a major impact on most world economies. He is the one responsible for allocating large chunks ("class A" blocks) to a few early adopters (e.g., HP, Apple, and MIT). Unfortunately, those allocations are very difficult to undo today, so about one-third of all the addresses allocated in the United States belong to fewer than 50 organizations. The IANA now just considers those *legacy allocations* and tried to do the best they could with the address space remaining at the time they took over allocation.

Original "Classful" Allocation Blocks

The first 50% of the full IPv4 address space (0.0.0.0–127.255.255.255) was divided up into 128 "class A" blocks (now known as "/8" or "slash-8" blocks). Each of these contained 2^{24}-2, or some 16.8 million usable addresses. Here is a list of some of the lucky

---

[8] https://en.wikipedia.org/wiki/Jon_Postel

organizations that own these blocks today, either from the original allocation or by buying other companies that owned them:

General Electric	3.x.x.x
Level 3 Communications	4.x.x.x
U.S. Army Info Systems Center	6.x.x.x
(formerly DoD, now ARIN)	7.x.x.x
Level 3 Communications	8.x.x.x
IBM	9.x.x.x
U.S. DoD Intel Info Systems	11.x.x.x
AT&T Worldnet	12.x.x.x
Xerox Corp.	13.x.x.x
HP	15.x.x.x
DEC (now HP)	16.x.x.x
Apple Computer	17.x.x.x
Mass. Inst. Of Technology	18.x.x.x
Ford Motor Company	19.x.x.x
Computer Sciences Corp	20.x.x.x
DoD DISA DDN-RVN	21.x.x.x
U.S. DoD DISA	22.x.x.x
U.K. Ministry of Defense	25.x.x.x
U.S. DoD DISA	26.x.x.x
U.S. DoD DSI-North	28.x.x.x
U.S. DoD DISA	29.x.x.x
U.S. DoD DISA	30.x.x.x
AT&T Global Network Svcs	32.x.x.x
U.S. DoD DLA Sys Auto Ctr	33.x.x.x
Halliburton	34.x.x.x
InterOp Show	45.x.x.x
Bell-Northern (now Nortel)	47.x.x.x
Prudential Insurance	48.x.x.x
E.I. DuPont de Nemours	52.x.x.x
Daimler AG	53.x.x.x
U.S. DoD Network Info Ctr	55.x.x.x
U.S. Postal Service	56.x.x.x

Another 25% of the full address space (128.0.0.0–191.255.255.255) was divided up into 16,384 "class B" blocks (now known as "/16" blocks). Each of these contained 2^{16}-2, or 65,534 usable addresses.

Another 12.5% of the full address space (192.0.0.0–224.255.255.255) was divided up into about 2.1 million "class C" blocks (now known as "/24" blocks). Each of these contained 2^8-2, or 254 usable addresses.

Another 6.25% of the full address space (224.0.0.0–239.255.255.255) was reserved for multicast (these are known as class D addresses). There is no way to "recover" any of this address space.

The final 6.25% of the full address space (240.0.0.0–255.255.255.255) was reserved for future use, experimentation, and limited broadcast. These are known as class E addresses. These addresses cannot be "recovered" without modifications to essentially every router in the world (most routers block them by default).

The sub-block of class E from 255.0.0.0 to 255.255.255.255 is actually used for "limited broadcast" (limited because it will not cross routers). A packet sent to any of these addresses will be received by all nodes on your LAN. Of these, normally only the address 255.255.255.255 is used. There is no broadcast in IPv6 (although there is a multicast address that has much the same effect).

The US Department of Defense has ten "/8" blocks, for about 168 million addresses. This is almost 4% of the total IPv4 address space. One entire "/8" block (127.x.x.x) has only one address used, which is 127.0.0.1 (the IPv4 "loopback" address, used to address your own node). A small block at 169.254.0.0/16 is reserved for IPv4 link-local usage (similar to IPv6 link-local addresses). For details, see RFC 5735, "Special Use IPv4 Addresses," January 2010.

One "/8" block (10.0.0.0/8), one "/12" block (172.16.0.0/12), and one "/16" block (192.168.0.0/16) were reserved for use as "private" addresses by RFC 1918, "Address Allocation for Private Internets," February 1996. These addresses can be used by any organization for any internal network but should never be routed onto the Internet (although in practice you can sometimes find these addresses on the backbone due to misconfigured routers). These would correspond to internal phone "extensions" such as 101, 102, etc. Every company with a PBX might use that same set of extensions.

As of June 4, 2010, only 16 of the possible 256 "/8" blocks (about 6.25% of the full address space) were still unallocated. Here is a map of the status of all 256 "/8" blocks. By February 2011, there weren't any dots left. All the blocks with dots (unallocated "/8"s) in the chart today were allocated to one of the RIRs (ARIN, RIPE, APNIC, LACNIC, or AfriNIC).

	+0	+1	+2	+3	+4	+5	+6	+7	+8	+9
000	R	AP	RN	L	L	.	L	L-AR	L	L
010	R	L	L	L	AP	L	L	L	L	L
020	L	L	L	.	AR	L	L	AP	L	L
030	L	RN	L	L	L	L	.	.	L	.
040	L	AF	.	L-AP	L	L	RN	L	L	.
050	AR	L	L	L	L	L	L	L	AP	AP
060	AP	AP	RN	AR	AR	AR	AR	AR	AR	AR
070	AR	AR	AR	AR	AR	AR	AR	RN	RN	RN
080	RN	RN	RN	RN	RN	RN	RN	RN	RN	RN
090	RN	RN	RN	RN	RN	RN	AR	AR	AR	AR
100	.	.	.	.	.	.	.	AR	AR	RN
110	AP	AP	AP	AP	AP	AP	AP	AP	AP	AP
120	AP	AP	AP	AP	AP	AP	AP	R	L-AR	L-AR
130	L-AR	L-AR	L-AR	L-AP	L-AR	L-AR	L-AR	L-AR	L-AR	L-AR
140	L-AR	L-RN	L-AR	L-AR	L-AR	L-RN	L-AR	L-AR	L-AR	L-AR
150	L-AP	L-RN	L-AR	L-AP	L-AF	L-AR	L-AR	L-AR	L-AR	L-AR
160	L-AR	L-AR	L-AR	L-AP	L-AR	L-AR	L-AR	L-AR	L-AR	L-AR
170	L-AR	L-AP	L-AR	AR	AR	AP	RN	LA	RN	.
180	AP	LA	AP	AP	AR	.	LA	LA	L-RN	LA
190	LA	L-LA	L-AR	RN	RN	RN	L-AF	AF	L-AR	AR
200	LA	LA	AP	AP	AR	AR	AR	AR	AR	AR
210	AP	AP	RN	RN	L	L	AR	RN	AP	AP
220	AP	AP	AP	AP	R-MC	R-MC	R-MC	R-MC	R-MC	R-MC
230	R-MC	R-MC	R-MC	R-MC	R-MC	R-MC	R-MC	R-MC	R-MC	R-MC
240	R-FU	R-FU	R-FU	R-FU	R-FU	R-FU	R-FU	R-FU	R-FU	R-FU
250	R-FU	R-FU	R-FU	R-FU	R-FU	R-FU				

Key and Analysis

AR	ARIN allocated	33	72	28.13%	(ARIN total)
L-AR	Legacy, admin by ARIN	39			
AP	APNIC allocated	38	44	17.19%	(APNIC total)
L-AP	Legacy, admin by APNIC	6			
RN	RIPE NCC allocated	33	37	14.45%	(RIPE total)
L-RN	Legacy, admin by RIPE NCC	4			
LA	LACNIC allocated	8	9	3.52%	(LACNIC total)
L-LA	Legacy, admin by LACNIC	1			
AF	AfriNIC allocated	2	4	1.56%	(AfriNIC total)
L-AF	Legacy, admin by AfriNIC	2			
L	Legacy, early allocation	39	39	15.23%	(Legacy total)
R	Reserved	3	3	1.17%	
R-MC	Reserved, Multicast	16	16	6.25%	
R-FU	Reserved, Future Use	16	16	6.25%	
.	Unallocated	16	16	6.25%	(Unallocated)
		---	---	-------	
		256	256	100.00%	

Almost all the "Legacy, early allocation" blocks are in the United States, so ARIN's real share of the total IPv4 address space is over 40% (for less than 5% of the world's population).

Of course, today, all remaining /8 blocks at IANA have been allocated, but the percentages are not that different from those shown.

Classless Inter-Domain Routing (CIDR)

The original allocation block sizes (classes A, B, and C) did not fit all organizations. For many organizations, even the smallest block (class C) was too big. If we had stuck with the original allocation block sizes, we would have run out of addresses around 1997. When this was realized, the IETF introduced Classless Inter-Domain Routing as defined in RFC 1518, "An Architecture for IP Address Allocation with CIDR," September 1993, and RFC 1519, "Classless Inter-Domain Routing (CIDR): An Address Assignment and Aggregation Strategy," September 1993. CIDR allowed the two parts of an address to be split along any of the 30 possible places to divide them, not just at multiples of 8 bits. Some useful CIDR allocation block sizes are

Size	Subnet Mask	Number of usable addresses in block
/30	255.255.255.252	2
/29	255.255.255.248	6
/28	255.255.255.240	14
/27	255.255.255.224	30
/26	255.255.255.192	62
/25	255.255.255.128	126
/24	255.255.255.0	254 (old Class C)
/23	255.254.0.0	510
/22	255.252.0.0	1,022
/21	255.248.0.0	2,046
/20	255.240.0.0	4,094
/19	255.224.0.0	8,190
/18	255.192.0.0	16,382
/17	255.128.0.0	32,766
/16	255.255.0.0	65,534 (old Class B)
/8	255.0.0.0	16,777,214 (old Class A)

CIDR allows a closer fit to actual organization size than the old classful "three-sizes-fit-all" scheme. However, each allocated block requires an entry in the core routing tables. As we allocate smaller and smaller blocks, the number of entries in the core routing tables is growing *very* rapidly. Many things are beginning to go wrong as we get closer and closer to an empty barrel.

In the mid-1990s, there were steps taken (NAT and private addresses) to further limit the number of public IPv4 addresses being allocated to each organization. NAT was only ever envisioned by its creators as a "quick fix" that would buy us a few years to really solve the problem. They understood all the problems NAT would cause and were willing to live with them for a short time, when the alternative was to run out of IPv4 addresses somewhere around 1997. For the real long-term fix, the IETF also began working on the next-generation Internet Protocol with a much larger address space. That next-generation Internet Protocol is complete, mature, and being deployed globally today. It is called IPv6.

Problems Introduced by Customer Premises Equipment NAT (CPE NAT)

Since the mid-1990s, we have been living with problems created by the introduction of Network Address Translation doing conventional "hide-mode" (cone) NAT at the customer premise (CPE NAT). These include the following:

- Difficulty for internal nodes to accept incoming connections, for VoIP (SIP), peer-to-peer (P2P), running your own mail (SMTP), web (HTTP/HTTPS), file transfer (FTP/SSH), or other servers.

- Problems with protocols that embed IPv4 addresses in packet transmissions (SIP, many games).

- Problems with protocols that detect tampering to IP and/or TCP/UDP header fields (e.g., IP addresses, port numbers), such as the IPsec Authentication header (AH).

- Problems due to advances in web technology (primarily Web 2.0/AJAX) that use large numbers of connections, each over a different port, such as iTunes and Google Maps. This can be as high as 200–300 ports per application. Since NAT systems share the 65,536 possible ports associated with a single "real" IPv4 address among the nodes hidden behind each address, each internal user on average can use at most 65,536 divided by the number of users behind that address. In enterprise networks, this might (until recently) have been thousands or tens of thousands of nodes behind one real address. For 1,000 nodes, on average each user could use no more than 65 ports. For 10,000 nodes, on average each user could use no more than 6 ports. To allow each user up to 200 ports, no more than 300 users should be hidden behind each IPv4 address. Currently, the *average* number of ports used per user is actually quite low (less than 10), but this is expected to grow rapidly as more users begin using Web 2.0-/AJAX-type applications. If possible, NAT schemes should use ports on a *first come, first served* basis, rather than allocating 1/n of the possible ports to each node.

- Difficulty tracking abuse to specific users behind a NAT. This requires keeping large amounts of information including source IP address, destination IP address, port number(s), and accurate timestamps for *every* connection. This may have to be kept for up to 1 year. A year's worth of such data for a single user can be tens of gigabytes to terabytes in size. Multiplied by the number of users, this is a staggering amount of storage that ISPs are required to keep. Hackers love to "hide behind" NAT gateways.

Essentially, private IP addresses behind "hide-mode" NAT are good only for outgoing connections using the simplest connectivity paradigms (e.g., client to server, using a small total number of ports per user). Note that since the start of accessing the Internet from phones, we have had *only* private addresses. With IPv6 on phones, for the first time, we have *public* addresses on phones. That means you can run a server on your phone that will be accessible from anywhere, or your phone can connect directly to any other phone in the world (so long as both of you have IPv6 addresses and nothing is blocking the ports involved). That is very exciting.

It is possible to allow *at most* one internal node to accept incoming connections on a given port (e.g., port 80 for HTTP) for the gateway external IPv4 address, using *port forwarding*. For example, your NAT gateway can be configured to forward any incoming connection to its external IPv4 address on port 25 to the private address of a single internal node where an email server is running. The gateway could *also* forward incoming connections to its external IPv4 address on port 80 to the same (or a different) internal node's private address where a web server is running. This limits the entire LAN (or that part of it behind a given real IPv4 address) to a single server for any given port number (when using port redirection). This still translates the destination IPv4 address on the way in and the source IPv4 address on the way out (but port numbers are left unchanged). This still causes many of the problems listed previously. One-to-one NAT (BINAT) does not have this limitation, but one valid external IPv4 address (in addition to the valid external IPv4 address used for hide-mode NAT and port redirection) is required for each internal server.

If you tried to map incoming port 80 traffic to *two* different internal addresses with port redirection, your browser would be very confused by receiving responses from two different web servers simultaneously. A good firewall or router should flag the attempt to do this as an error.

Some firewalls (or other NAT gateways) in addition to "hide-mode" (cone) NAT for outgoing connections, and port forwarding, also support *bidirectional NAT* (called BINAT, symmetric NAT, and "1-to-1" NAT, among other names). This type of NAT makes a two-way address translation between a single external IP address and a single private internal address (hence "1-to-1"). The full 65,536 possible ports may be used on the internal node, but a distinct real IPv4 address is required for *each* such BINAT mapping. This would allow deployment of multiple web servers within a LAN or an easy way to provide access to many services on a single node (e.g., a Windows Server–based computer). This still translates the destination IP address of packets on the way in and the source IP address of packets on the way out (again, port numbers are not affected), still causing many of the problems listed previously. In addition, it uses up one real address per internal server and requires addressing the "missing ARP" problem (caused by the fact that there is no physical node at the external address to respond to ARP queries). This can be solved by configuring a static ARP for the external IPv4 address on the NAT gateway or various other solutions. Solving the "missing ARP" problem is one of the most difficult and least widely understood aspects of managing a NAT gateway (or firewall).

There is in fact an external interface on the NAT gateway, with a valid external address (say 123.45.67.81), which will reply as usual to an ARP request to its *primary* address (e.g., 123.45.67.81) with the MAC address of the external interface. With BINAT, however, you also assign an additional *alias* address for each BINAT mapping (e.g., 123.45.67.82, 123.45.67.83, etc.) to the *gateway's* external interface. If an ARP is done for two of these *alias addresses*, by default the external interface will not respond to them, hence the "missing ARP" (actually, "missing ARP *response*"). The ARP request is not translated to the internal node, and even if it was, the node doing the ARP doesn't want the MAC address of the internal node – it wants the MAC address of the external interface of the gateway. To get the external interface to respond with its MAC address to ARP requests for an *alias* address, you must configure a *proxy ARP* on the external interface *for that alias address*. The commands for configuring alias addresses on the external interface, and proxy ARPs for them, vary widely from one OS to another. See the labs in Chapter 10 for an example of this with m0n0wall (based on FreeBSD). In some cases, another mechanisms may be used to solve the "missing ARP" problem, such as configuring a static route for each alias address. This eliminates the need for other nodes to do an ARP request.

There are several *NAT traversal* protocols (STUN, TURN, SOCKS, NAT-T, etc.) that allow incoming connections to internal nodes that have only private addresses (without any port forwarding or BINAT support in the NAT gateway). These typically require an outside server to assist (this alone should raise security and reliability concerns). STUN uses an outside server only to *establish* the connection, while TURN also routes all traversing traffic through an outside gateway. All NAT traversal schemes involve encapsulating traffic over UDP, which complicates error detection and recovery and intrusion detection, as well as supporting the "connection-oriented" nature of TCP traffic. All require extensive modifications to the source code of clients, which is quite complex and very specific to the NAT traversal algorithm used. Usually, the external servers used are not under control of the network, leading to security issues. One of the most popular network applications (Skype) uses standard UDP-encapsulated "hole punching" traversal, which causes *many* security issues. Anyone with access to the external server can easily track who you are calling (and who is calling you) and even *listen in* or redirect the call. With IPv6, there is no NAT, hence no need for NAT traversal.

Note that there are many variants of NAT, and a given implementation of NAT traversal may work with only one or two of them. Also, many schemes fail if there are two or more NAT mappings in series (say your ISP doing a NAT44 mapping to one private address and then your CPE router/modem doing a *second* NAT44 mapping of that private address to yet another private address – this is sometimes called NAT444).

Here are RFCs related to NAT traversal:

RFC 1928, "SOCKS Protocol Version 5," March 1996 (Standards Track)

RFC 3947, "Negotiation of NAT-Traversal in the IKE," January 2005 (Standards Track)

RFC 3948, "UDP Encapsulation of IPsec ESP Packets," January 2005 (Standards Track)

RFC 5389, "Session Traversal Utilities for NAT (STUN)," October 2008 (Standards Track)

RFC 5766, "Traversal Using Relays Around NAT (TURN): Relay Extensions to Session Traversal Utilities for NAT (STUN)," March 2010 (Standards Track, awaiting final approval)

Implementing NAT at the Carrier: Carrier-Grade NAT (CGN)

As we have progressed from the "end times" for IPv4 to "life *after* IPv4" (beyond the depletion date for IPv4), those who have not already migrated to IPv6 will face even greater problems, as ISPs deploy *Carrier-Grade NAT*[9] solutions in their networks, as opposed to doing NAT only in the Customer Premises Equipment (CPE NAT). The reason for this is to try to make optimal use of an even smaller number of globally routable IPv4 addresses than is possible with CPE NAT. Essentially the ISP will have a very small pool of real IPv4 addresses (far less than the number of customers). They will share single real IPv4 addresses *across customers*. This will make the problems associated with CPE NAT dramatically worse. There is excellent coverage of the issues associated with deploying NAT in the carrier in RFC 6269,[10] "Issues with IP Address Sharing," June 2011 (Informational):

- Dual-Stack Lite, RFC 6333[11]

- Carrier-Grade NAT (CGN), RFCs 6888[12] and 6598[13]

- NAT64, RFC 6146[14]

- IVI, RFC 6219[15]

- Address+Port (A+P), RFC 6346

Of these, only Dual-Stack Lite makes dual-stack service available to users. It provides direct IPv6 service (no NAT, no tunneling). It provides IPv4 service tunneled over IPv6 (called 4in6 tunneling) with only one level of NAT44 (which takes place at the carrier). Customers will get *only* private IPv4 addresses. It is possible that some ISPs may provide a few precious "real" (globally routable) IPv4 addresses to business customers at a

---

[9] https://en.wikipedia.org/wiki/Carrier-grade_NAT
[10] https://tools.ietf.org/html/rfc6269
[11] https://tools.ietf.org/html/rfc6333
[12] https://tools.ietf.org/html/rfc6888
[13] https://tools.ietf.org/html/rfc6598
[14] https://tools.ietf.org/html/rfc6146
[15] https://tools.ietf.org/html/rfc6219

significant price premium (*all the market will bear*, which could easily reach thousands of dollars per address per year). All the NAT schemes extend the address space by adding port information. They differ in the way they manage the port value.

With CPE NAT, a given public IPv4 address covers only one legal entity (a home, a company, etc.). With Carrier-Grade NAT, multiple legal entities will be behind most real IPv4 addresses, which will vastly complicate the legal issues (such as tracking down a source of network abuse or being able to prove who really did something).

You will see the terms *NAT444* and *NAT464* in discussions of carrier-based NAT. The existing NAT that is widely deployed now is called *NAT44* (NAT from IPv4 to IPv4). There is also *NAT46* (NAT from IPv4 to IPv6) and *NAT64* (NAT from IPv6 to IPv4).

NAT444[16] essentially leaves the CPE NAT44 (the existing one-layer NAT that is widely deployed today) intact at the customer premise, while the carrier deploys a *second* layer of NAT44 before it ever reaches the customer (using the new reserved block 100.64/10). It is really just two NAT44 mechanisms in series. The CPE NAT44 will map the private addresses supplied from the carrier NAT44 onto yet another set of internal private addresses. The transport from carrier to customer is also over IPv4. The difference from existing systems is that today the CPE NAT usually has one real IPv4 address, which it shares among multiple internal nodes. In NAT444 systems, there won't be *even one* real IPv4 address at the customer premise. It will be quite difficult (and probably very expensive) to host servers with public IPv4 addresses (e.g., web, mail, VoIP) at customer sites – most will have to be hosted at a colocation facility.

For an analogy, imagine deploying *nested* telephone PBXes. There would be an outer PBX, with a real telephone number, and behind that other PBXes with internal extensions from the outer PBX. Behind each internal PBX, you would have sets of internal phones. To call an internal phone, you would dial the real phone number of the outer PBX and have to do something to select an internal PBX (dial the internal PBX's extension number?). Then once connected to the internal PBX, you would need to interact with *it* to select an internal phone (e.g., dial the first three characters of the phone owner's name). This is the kind of complexity that IPv4 applications will now have to cope with. It will be much simpler to just convert them directly to IPv6.

---

[16] https://chrisgrundemann.com/index.php/2011/nat444-cgn-lsn-breaks/

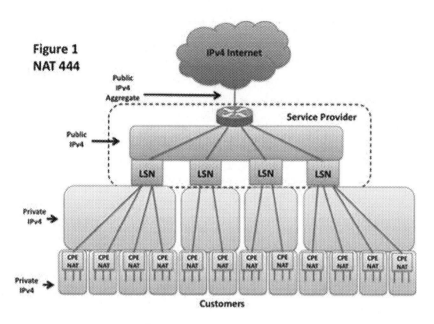

Figure 4-4. *How NAT444 works*

NAT464 is similar but involves doing one layer of NAT46 (from IPv4 to IPv6) at the carrier, followed by a second layer of NAT64 (from IPv6 to IPv4) at the customer premise. This allows the transport from carrier to customer to be over IPv6, which is a good thing, but involves upgrading or replacing all Customer Premises Equipment to ones that are NAT64 compliant (few are today). Also, address translation between IP families (IPv4 to IPv6 and IPv6 to IPv4) has even more problems than address translation within a single IP family (only IPv4 to IPv4 – *there is no IPv6-to-IPv6 NAT!*).

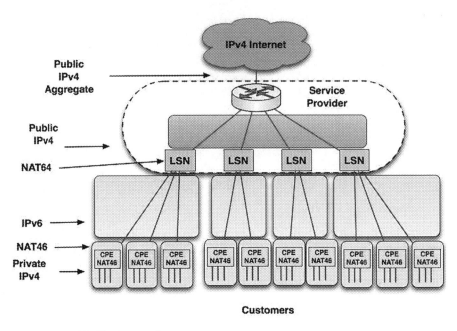

Figure 4-5. *How NAT464 works*

In either case (NAT444 or NAT464), there are some protocols that will work across one layer of NAT, but fail when there is more than one layer of NAT. Both NAT444 and NAT464 will introduce these kinds of issues, since both involve *at least* two layers of NAT. Some home or small business users may unintentionally introduce *even more* layers of NAT due to lack of understanding, for example, by deploying a firewall/NAT box behind a modem/NAT gateway.

The following problems are made worse by Carrier-Grade NAT compared even with CPE NAT. Some affect only the end user, some affect third parties (e.g., law enforcement), and many affect both:

- The number of ports available per node will be even less, so Web 2.0/AJAX applications such as iTunes and Google Maps will fail in unpredictable ways, especially with schemes that divide the available ports into equally sized port ranges per customer.

- Incoming port negotiations may fail – for example, Universal Plug and Play (UPnP).

- Incoming connections to well-known ports will not work (e.g., SMTP, HTTP, SIP, etc.).

- Reverse DNS pretty much breaks down completely.

- Inbound ICMP will fail in most cases.

- Security issues are even worse than with CPE NAT.

- Packet fragmentation requires special handling.

- There are more single points of failure and decreased network stability.

- Port randomization is affected (especially in schemes that restrict ports to ranges).

- Penalty boxes no longer work.

- Spam blacklisting will affect many other nodes that use the same address.

- Geolocation services may not be reliable or particularly specific.

- Load balancing algorithms are impacted.

- Authentication mechanisms are impacted.

- IPv6 transition mechanisms will be affected (Dual-Stack Lite is the exception here).

- Frequent keep-alives will reduce battery life in mobile nodes.

Applications that had to be modified to support NAT traversal to work through NAT44 will have to be modified *once again*, with even more complicated schemes, to traverse multiple layers of NAT. Application Layer gateway (ALG) workarounds now have to be implemented at the carrier, not just at the customer premise. ALGs that have to deal with port-range restrictions will have an even harder job.

Blocking incoming access to services based on IPv4 addresses will likely affect many "innocent bystanders" that happen to share the same real IPv4 address. One obvious example is spam blacklists. A less obvious example is that some secure devices restrict access by source IP address (only *this* node can connect to my firewall). Now, many other nodes, even in different organizations, will be sharing that same IP address legitimately, so may be able to access such nodes.

With reverse DNS, you publish the nodename associated with a given IP address. With CPE NAT this affects many nodes, but this will be completely meaningless for nodes behind carrier-based NAT. There is no way to publish thousands of nodenames for a single IP address, nor is there any way for someone asking for the reverse lookup to interpret the response correctly.

IPv6 transition mechanisms such as 6to4 will not work at all behind carrier-based NAT, but Teredo might. Likewise, IPv4 multicast and Mobile IPv4 will have to be modified extensively for carrier-based NAT.

Summary

In this chapter we covered the inevitable depletion of public IPv4 addresses. First, IANA ran out of addresses to allocate to RIRs in 2011. Over the next few years, all five RIRs have reached end of normal allocation. Even companies, telcos, and ISPs are now pretty much out of public IPv4 addresses.

They have tried to continue operation of IPv4 via various schemes, but those are causing even more problems now.

IPv4 is at end of life. It's time for everyone to migrate to IPv6.

CHAPTER 5

IPv6 Deployment Progress

This chapter presents the progress to date in the deployment of IPv6. There are many sources of information on this. We are now in the rapid adoption phase (finally).

Cisco's 6lab site[1]

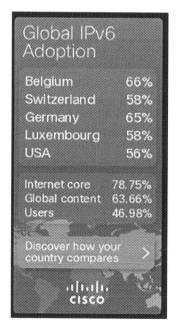

Figure 5-1. *Global IPv6 adoption*

This widget shows a summary of the deployment percentages in the top five countries (all already above 50%). The "Internet core" shows readiness of the Internet backbones for IPv6. The "Global content" shows how much of the popular content is available over IPv6 (almost always over both IPv4 and IPv6 at this time – there is very little content available over just IPv6). The "Users" indicates what percentage of users can access IPv6 content.

---

[1] https://6lab.cisco.com/stats/

L. E. Hughes, *Third Generation Internet Revealed*, https://doi.org/10.1007/978-1-4842-8603-6_5

There are many world maps on the 6lab site with frequently updated information about each country.

Figure 5-2. *Deployment by country, Cisco*

If you hover over a country, it will show a number of items about IPv6 deployment in that country. For example, on April 1, 2019, the United States shows

> *IPv6 deployment*: 50.2%
>
> *Prefixes*: 33.69%
>
> *Transit AS*: 67.47%
>
> *Content*: 56.66%
>
> *Users*: 34.9%

The "Prefixes" value is from measurements done by APNIC and Eric Vyncke. It shows the percentage of "allocated" prefixes that are "live" (actually have traffic going to them or at least have entries in the BGP routing tables). Typically, a lot of organizations have made the first step of obtaining a block of addresses from their RIR (Regional Internet Registry), but only some of those have gone live at this time.

The "Transit AS" value has to do with autonomous systems used only for "transit" – information being source and destination ASs. This indicates the readiness of the Internet core.

The "Content" value has to do with how many of the popular ("Alexa 500") websites are available over IPv6.

The "Users" value is the percentage of users that connect to Google's search page over IPv6. Note that language and national blocking of google.com (e.g., in China) affect this number.

The overall deployment value is a weighted combination of the above values.

Here are some per-country stats ("IPv6 deployment") from the Cisco site as of March 17, 2019. Cisco is measuring very different things from Google, including infrastructure, and is not as influenced by language:

Country	IPv6 %
Belgium	63.7
Germany	59.07
Uruguay	55.26
Greece	53.93
Malaysia	52.81
Finland	51.13
Vietnam	50.85
United States	50.36
Japan	49.13
Thailand	49.02
India	49.01
Brazil	48.66
United Kingdom	48.25
France	47.95
Estonia	47.42
Portugal	46.89
Canada	46.40

Country	IPv6 %
Mexico	45.85
Hungary	44.14
Norway	42.80
Ireland	40.42
Peru	39.52
Australia	38.96
Singapore	38.68
Saudi Arabia	38.31
Czech Republic	37.30
Sweden	36.81
Romania	33.78

So How Did IPv4 Depletion Go at RIRs?

The following widget on the IPv6 Forum website tells the story.

Figure 5-3. *IPv4 exhaustion counter, toward the end of IPv4 available in RIRs*

- IANA ran out of IPv4 allocable (public) addresses on February 3, 2011. After that, they could not provide any more IPv4 blocks to the RIRs. It didn't take long for the RIRs to go through their remaining inventory.

- APNIC (AsiaPac + Japan) was the first RIR to reach the "end of normal allocation," on April 15, 2011.

- RIPE NCC (EU and Middle East) was next on September 14, 2012.

- LACNIC (Latin America) stopped normal allocation on June 10, 2014.

- ARIN (North America except for Mexico) ran completely out on September 24, 2015.

- AfriNIC (African continent) stopped normal allocation on January 13, 2020.

Apart from ARIN, the RIRs decided to stop allocating IPv4 addresses when they reached their final /8 (16.7 million) addresses. After that they could only be given out in small blocks (e.g., 1024 addresses), for special purposes (such as migration to IPv6). ARIN chose to keep doing allocation until the barrel was completely empty.

Some ISPs, telcos, and cloud providers tried to get as many IPv4 addresses as they could, but many of those have now run out. They can buy a few on the IP address market, but even those will run out at some point, and even now the price of public IPv4 addresses continues to rise.

Google Statistics

IPv6 Adoption

We are continuously measuring the availability of IPv6 connectivity among Google users. The graph shows the percentage of users that access Google over IPv6.

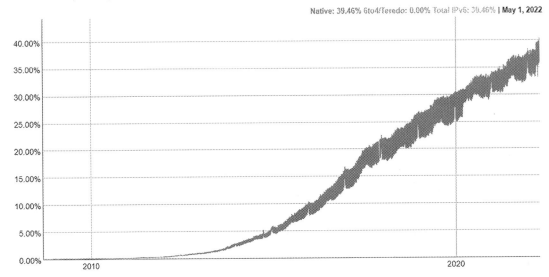

Figure 5-4. *Percentage of connections to google.com globally over IPv6*

Google tracks the number of connections to its search engine over both IPv4 and IPv6. The preceding chart shows the growth of IPv6 as a percentage of total usage. As of May 1, 2022, connections over IPv6 represent about 40% of all connections, globally. Note that the actual percentage is likely higher, due to distortion caused by Happy Eyeballs.[2] This is a modification to all recent browsers that will attempt to connect IPv4 or IPv6, depending on which one connected in the shortest time. This can result in a situation where both the user and the website both support IPv6, but they also both support IPv4, and for whatever reason, connection over IPv4 was faster. You can see the current version of this chart at `www.google.com/intl/en/ipv6/statistics.html#tab=ipv6-adoption`.[3]

For full details on Happy Eyeballs, see RFC 8305,[4] "Happy Eyeballs Version 2: Better Connectivity Using Concurrency," December 2017.

---

[2] `https://en.wikipedia.org/wiki/Happy_Eyeballs`

[3] `www.google.com/intl/en/ipv6/statistics.html#tab=ipv6-adoption`

[4] `https://tools.ietf.org/html/rfc8305`

You can see the incredible increase in deployment globally from when *The Second Internet* book was written. At that time, the Google stats showed .25% adoption. In the intervening years, adoption has increased by over 100 times, even with Happy Eyeballs.

The reason the line is wide vertically is actually quite interesting. It seems that usage is highest on weekends and drops on weekdays. This is because more people have IPv6 at home than they do at work.

Google also breaks down their statistics per country. You can see the current version of this information at `www.google.com/intl/en/ipv6/statistics.html#tab=per-country-ipv6-adoption`.[5]

Here are a few selected countries from the Google per-country stats as of March 17, 2019:

Country	IPv6 %
Belgium	53.17
Germany	41.72
Greece	35.99
United States	35.24
Malaysia	34.13
Uruguay	34.07
India	32.59
Vietnam	32.19
Japan	29.66
Switzerland	28.94
Brazil	27.68
Taiwan	25.97
France	24.62
Estonia	24.35
Finland	23.80
Mexico	22.92

---

[5] `www.google.com/intl/en/ipv6/statistics.html#tab=per-country-ipv6-adoption`

Country	IPv6 %
Canada	22.77
United Kingdom	22.71
Hungary	20.44
Portugal	20.06
Ecuador	19.79
Thailand	18.82
Ireland	18.70
Trinidad and Tobago	17.65
New Zealand	17.11
Netherlands	16.75
Peru	16.62
Australia	15.37
Romania	12.95
Norway	12.27
Puerto Rico	12.26
Bolivia	11.90
Czechia	10.87
Slovenia	10.46
Guatemala	9.51
Saudi Arabia	8.97

Note that these values are based on actual measurements but are heavily influenced by English-speaking ability (the Google site is in English) and per-country restrictions (China blocks Google). Most of the other countries are below 1%.

Another good source of statistics on IPv6 adoption from actual measurements can be found at www.vyncke.org/ipv6status/.[6] This site breaks it down by websites, email, and DNS (from Alexa).

Predictions for Future Years

Cisco had a chart to predict future IPv6 deployment based on past data from Google. It predicts that we will reach 100% by 2028. That is just 6 years from now. Perhaps I will do another update to this book in 2028 and see what really happened.

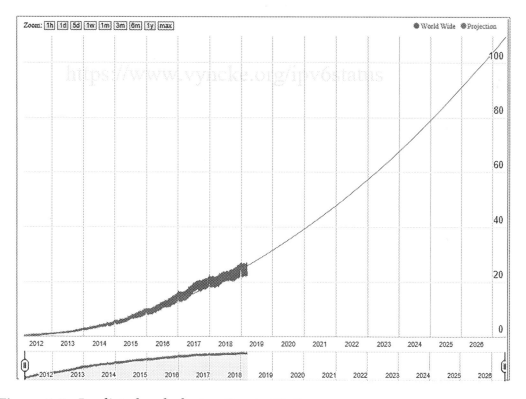

Figure 5-5. *Predicted end of migration to IPv6*

---

[6]www.vyncke.org/ipv6status/

Summary

IPv6 deployment is proceeding well globally. Most developed countries are already over 50% of all traffic going over IPv6. The adoption curve for IPv6 from Google (for percentage of connections to google.com over IPv6) is a strong indication of how this is progressing. That is a classic adoption curve.

CHAPTER 6

IPv6 Core Protocols

This chapter introduces the new concepts and technical specifics of IPv6, the foundation of the Third Internet. Since IPv6 is based heavily on IPv4, the approach will be to describe the differences between the two. This will help those who already are familiar with IPv4 to make the leap to IPv6. The subchapter headings are intentionally similar to those in Chapter 3, to allow you to compare the old and the new, topic by topic. Again, there is no intent to be comprehensive. There is a lot of content available on all aspects of IPv6 listed in the bibliography and/or available online. The ultimate references are the RFCs, so this chapter includes hyperlinks to the relevant ones, for those who want to drill deeper on specific topics.

In other chapters we will discuss topics such as advanced aspects of IPv6 (IPsec, IKEv2), the new things that IPv6 makes possible, who is involved in making it happen, and how we get from the Second Internet to the Third Internet (migration). This chapter covers the core protocols of IPv6.

Network Hardware

Essentially the same network hardware that was used to deploy IPv4 networks is being used to deploy the IPv6 networks, with some notable exceptions, primarily hardware that implements things at the Internet Layer or above, such as smart ("layer 3") switches, routers, and firewalls. Also, DNS and DHCP servers must be updated or replaced with ones that support IPv6 (more typically both IPv4 and IPv6, or "dual stack"). As IPv6 is deployed, Virtual Private Networks (VPNs) will likely move away from "SSL/VPN"[1] to IPsec-based VPNs, which are the only IETF-approved technology for VPNs. There are no RFCs for SSL/VPN because it is not considered to be a viable approach. Unfortunately, IPsec is incompatible with NAT, which is now endemic in the Second Internet. VoIP and

---

[1] https://openvpn.net/faq/why-ssl-vpn/

L. E. Hughes, *Third Generation Internet Revealed*, https://doi.org/10.1007/978-1-4842-8603-6_6

IPTV appliances will probably be upgraded to (or replaced with) IPv6-based systems. Any device with TCP/IP hardware acceleration (such as high-end routers) will probably need to be redesigned or replaced. Simply upgrading the firmware will not be sufficient on such products (there are hardware dependencies on IPv4). There are some routers that only have hardware acceleration for the IPv4 stack (IPv6 is done entirely with software), which has led some people to think there are performance issues with IPv6. Already there are hardware acceleration chips that support both IPv4 and IPv6 and are available and being used in new product designs.

The hardware of most computing *nodes* does not need to change, especially client and server computers. Replacement or upgrade of the operating system and applications is all that is needed. The good news is that almost all operating systems and many network applications that run on client computers are *already* fully compliant with IPv6, and those are widely deployed. Those that aren't yet compliant can be upgraded or configured to support it with very reasonable effort and cost. Many server applications (especially open source ones) are already compliant as well. Virtually everything Microsoft makes fully (except Azure Cloud VMs, Skype, and Teams) supports IPv6 today. For client computers, Windows Vista and Windows 7 had fairly complete support. Later versions of Windows (8.1, 10, and 11) have very complete support. Windows XP had some support but was missing some key features (like GUI configuration of IPv6 addresses and DNS queries over IPv6). For server computers, Windows Server 2008 and Exchange Server 2007 (and most other server software since 2007) have full support for IPv6. Most open source operating systems (Linux, FreeBSD, OpenBSD, and NetBSD) have had full support for IPv6 for many years. Most open source network applications (Apache, Nagios, Postfix, Dovecot, etc.) also have full support (although in some cases, documentation may be hard to find).

NICs (Network Interface Connectors) do not need to change unless they have IPv4-specific hardware acceleration, and even those will typically run IPv6 with no problem, but the IPv6 part won't be accelerated (it will run at "software" performance levels, in terms of packets or bytes processed per second). There are already many chips available to build hardware-accelerated NICs that fully support both IPv4 *and* IPv6, so soon, even NICs with hardware acceleration will be no problem. They will accelerate IPv4 and/or IPv6 traffic. For the most part, NICs work at the *Link Layer* and hence are IP version agnostic (except for hardware acceleration).

Existing *Wi-Fi NICs* are also IP version agnostic (they work at the *Link Layer*), and every Wi-Fi NIC that I've tried has worked with IPv6 with no upgrades or workarounds required. **Wi-Fi routers** are another matter, because they include higher layer functionality such as IPv4 routing, often including IPv4 NAT and a DHCPv4 server. Even here, there is a simple workaround. Most Wi-Fi access points have a "WAN" connector, which is the input to the NAT gateway, and one or more "LAN" connectors that are on the *client* side of the NAT gateway. The LAN connectors are intended to plug in wired client nodes, which are peers to the wireless client nodes (both wired and wireless client nodes obtain configuration information and translated IP addresses from the DHCPv4 server and NAT gateway built into the Wi-Fi access point). Of course, the existing IPv4 routing, IPv4 NAT, and DHCPv4 in such devices are not compatible with IPv6. There are dual-stack Wi-Fi access points available now from companies like D-Link, but some of the products available today do not have routing, firewall, or DHCP support for IPv6. Any device listed on the IPv6-ready list of certified products[2] fully supports IPv6.

However, if you plug the cable from your ISP DSL modem (or from an existing Ethernet network) into *one of the LAN connectors* on your Wi-Fi access point, instead of into the WAN connector as you are supposed to, you can simply ignore the IPv4-specific parts of the Wi-Fi access point. You are now using the router in "bridge mode." The actual Wi-Fi transmitter part is IP agnostic, and if there are both IPv4 and IPv6 on the feed you connect, they will both be broadcast on wireless, and all existing nodes with Wi-Fi NICs will receive it (assuming each OS supports IPv6 and you have configured it). Of course, if you want your Wi-Fi nodes to obtain IPv4 addresses automatically, you must have a DHCPv4 server somewhere in your network (properly configured). Your Wi-Fi access point is no longer performing this function. Likewise, if you want Wi-Fi clients to obtain IPv6 addresses through stateless autoconfiguration, there must be a Router Advertisement Daemon in your network (just as for wired IPv6). If your wireless node has a DHCPv6 client and you have a DHCPv6 server in your network, stateful autoconfiguration will work over Wi-Fi as well. Of course, you can manually configure IPv6 addresses for Wi-Fi nodes just as you can with wired nodes. No NAT is required (or needed) for IPv6. For IPv4, no NAT will be performed in the Wi-Fi access point, so if you need it, it must be performed at the outside gateway (e.g., a wired DSL modem from your ISP). Your wireless nodes will be *peers* to your wired nodes. All of them (wired and wireless) will get IPv4 addresses from the same DHCPv4 pool (if you use DHCP), and all will be in the same subnet. Normally if you connected a Wi-Fi gateway with NAT inside an existing NATted network, your wireless nodes would be behind *two* levels of NAT, which can cause some problems. One level of NAT is bad enough – two levels are even worse.

---

[2] `www.ipv6ready.org/db/index.php/public`

You will also find that some consumer devices that support Wi-Fi already have support for IPv6, such as Android and iOS. It's kinda cool to deploy dual-stack Wi-Fi and show people the dancing turtle at www.kame.net[3] on your phone.

The KAME project

1998.4 - 2006.3

Dancing kame by atelier momonga

The KAME project was a joint effort of six companies in Japan to provide a free stack of IPv6, IPsec, and Mobile IPv6 for BSD variants.

Our products are available in:

- FreeBSD 4.0 and beyond
- OpenBSD 2.7 and beyond
- NetBSD 1.5 and beyond
- BSD/OS 4.2 and beyond

The project officially concluded in March 2006 (see press release from the WIDE project). Almost all of our implemented code has been merged to FreeBSD and NetBSD. The historical archive of the KAME repository is available at github.

| | Google |

[Top] [Old info]

Figure 6-1. *The dancing kame*

With some of today's phones, however, the only thing that works over IPv6 today (if anything) is Wi-Fi Internet access, not the voice traffic or "dataplan" service. Some mobile phone service providers are including IPv6 today (most in the United States are). In theory you could add a dual-stack softphone (VoIP client) and do voice communications over IPv6, but only via the Wi-Fi connection through a Wi-Fi access point connected to the main Internet, not over your wireless telephone carrier's Internet service via WAP, GPRS, EDGE, HSDPA, or whatever else they provide. Someday even these services will be dual stack (probably primarily LTE).

---

[3] www.kame.net/

There are now dual-stack Wi-Fi access points that fully support routing for IPv4 and IPv6, NAT for IPv4, and a Router Advertisement Daemon to enable IPv6 stateless auto-configuration. D-Link in Taiwan has several that fully support IPv6, as do other vendors.

Network cables are totally IP version agnostic. You will not need to rewire your network just for IPv6.

All conventional ("layer 2") *hubs and switches* are IP version agnostic, although "layer 3" features of some switches (such as web management, SNMP, and VLANs) must be upgraded to support IPv6. In most cases, this will be possible simply with new downloaded firmware. No hardware changes are needed (assuming there is sufficient RAM and ROM to handle the more complex firmware). Contact your switch vendor and demand that they add support for IPv6. There are already a few layer 3 switches on the market that support IPv6. I have an SMC 8848M 48-port gigabit managed switch in my home network that has quite a bit of IPv6 support, including web management over IPv6, IPv6-based VLANs, SNMP over IPv6, etc. Unfortunately, traffic statistics do not break out IPv4 and IPv6 traffic; just the total is reported. D-Link also has a dual-stack smart switch series. They are already *IPv6 ready* certified. One example is their DGS-3627 XSTACK managed 24-port gigabit stackable L3 switch.

Many *enterprise-grade routers* and *firewalls* already support IPv6, although in some cases you must pay extra for the IPv6 functionality. Cisco routers used to require "advanced IP services" for IOS (at additional cost), before IPv6 worked. For example, the Cisco 2851 router ($6495) included only the base IOS (no IPv6 support). The Advanced IP Services Feature Pack for it was an additional $1700 (all prices list). When buying or considering using Cisco routers for use in IPv6 networks, make sure they already include advanced IP services or include the additional cost of the feature pack. More recent Cisco routers include IPv6 support for free in the base IOS.

Home network gateways that support IPv6 are further behind, but coming soon, especially from Asian vendors, such as D-Link. A typical one will have all the features of existing IPv4-based gateways, plus 6in4 tunneling (to tunnel in IPv6 from a virtual ISP), a Router Advertisement Daemon (to enable stateless auto-configuration), and firewall rules for IPv6 traffic. They should also be able to accept direct (in addition to tunneled) IPv6 service, for when dual-stack ISP service becomes more widely available. Their DNS relay should support DNS over both IPv4 and IPv6. More advanced gateways might include a DHCPv6 server.

Note that some DSL or cable modems also include IPv4 firewall functionality. Of course, this will not allow you to control IPv6 traffic. Therefore, if you are connecting your LAN to the IPv6 Internet, there must be IPv6 firewalling somewhere, possibly in a 6in4 tunnel endpoint that is routing IPv6 traffic into your LAN. A dual-stack gateway firewall may include routing to accept incoming "direct" IPv6 service and/or a 6in4 endpoint to accept incoming "tunneled" IPv6 service, together with both IPv4 and IPv6 filtering rules, and a Router Advertisement Daemon to support stateless auto-configuration for the internal nodes that support IPv6.

Some **IP phones** in use today support IPv6, such as those from Snom in Germany and Moimstone in Korea. Cisco supports IPv6 on a number of their recent phones, including the 7906G, 7911G, 7931G, 7941G/GE, 7942G, 7945G, 7961G/GE, 7962G, 7965G, 79770G, 7971G/GE, and 7975G. Most of the older Cisco IP phones currently in use do not support IPv6, and their firmware cannot be upgraded for various reasons (e.g., insufficient RAM or ROM).

When looking for hardware products that already support IPv6, an excellent source of information is the *IPv6-ready approved products list*. If possible, choose products that have passed the phase 2 (gold-level) testing. This ensures full compliance with all relevant RFCs and interoperability with many other products. Phase 2 testing *also* ensures compliance with all items denoted SHOULD in the relevant RFCs (a much more comprehensive set of functionalities). These lists are updated and maintained by the IPv6 Ready Logo Committee of the IPv6 Forum. They can be found at the IPv6-ready list of certified products.[4]

RFCs: A Whole Raft of New Standards for IPv6

There are many new RFCs that define the protocols, addressing and routing schemes, as well as migration issues for IPv6. I will cover the most important of those in this chapter.

You can trace the beginnings and evolution of IPv6 in some early RFCs. In 1990, when the IETF first realized that a successor to IPv4 was going to be needed (and soon), the fun began. One key RFC related to this is RFC 1752,[5] "The Recommendation for the IP Next Generation Protocol," January 1995. Prior to this, people referred to the successor protocol as IPng (IP next generation), but in this RFC the term IPv6 was used. RFC 1752 says that the IETF started its effort to select a successor in late 1990 and that several parallel efforts were started. Among these proposals were "CNAT," "IP Encaps," "Nimrod," "Simple CLNP," the "P Internet Protocol," the "Simple Internet Protocol," and "TP/IX." None of these ever made it past the Internet Draft stage.

---

[4] www.ipv6ready.org/db/index.php/public
[5] https://tools.ietf.org/html/rfc1752

By late 1993, an IPng Working Group was formed, and the various proposals still around were reviewed. These included CATNIP, TUBA, and SIPP. Relevant RFCs (now of only historical interest) are

RFC 1347, "TCP and UDP with Bigger Addresses (TUBA)," June 1992 (Informational)

RFC 1526, "Assignment of System Identifiers for TUBA/CLNP Hosts," September 1993 (Informational)

RFC 1561, "Use of ISO CLNP in TUBA Environments," December 1993 (Experimental)

RFC 1707, "CATNIP: Common Architecture for the Internet," October 1994 (Informational)

RFC 1710, "Simple Internet Protocol Plus White Paper," October 1994 (Informational)

The CLNP referred to in several of these was the "Connectionless-mode Network Layer Protocol," defined in ISO/IEC 8473, which did not make it into the final IPv6 specification. By 1995 a consensus had emerged, with the best features of all the contenders. The consensus was summarized in RFC 1752. Before the end of the year (barely), the first real IPv6 specifications were published:

RFC 1883, "Internet Protocol, Version 6 (IPv6) Specification," December 1995, obsoleted by RFC 2460 and then by RFC 8200

RFC 1884, "IP Version 6 Addressing Architecture," December 1995, obsoleted by RFC 2373, then by RFC 3513, and then by RFC 4291

RFC 1885, "Internet Control Message Protocol (ICMPv6) for the Internet Protocol Version 6 (IPv6) Specification," December 1995, obsoleted by RFC 2463 and then by RFC 4443

RFC 1886, "DNS Extensions to Support IP Version 6," December 1995, obsoleted by RFC 3596

RFC 1887, "An Architecture for IPv6 Unicast Address Allocation," December 1995

Four of these have been replaced (multiple times in some cases) since then, and there are quite a few new ones since 1995, but this is where it really started. Yes, IPv6 is 27 years old in 2022 and has finally grown up.

IPv6

The software that is making the Third Internet (and virtually all Local Area Networks) possible will be around for quite some time. Like its predecessor, IPv4, it is a suite (family) of protocols. Once again, the core protocols are TCPv6 (Transmission Control Protocol version 6) and IPv6 (Internet Protocol version 6). TCPv6 has changes from TCPv4, but only a few, due to the larger addresses that require more storage and the odd method of calculating the checksum defined in TCPv4 (this involves a "pseudo header" that includes the source and destination addresses from the IP header, which of course are different in IPv4 and IPv6).

There is no new RFC specifically about TCPv6, but there are several RFCs that include details about the new features.

UDP has only very minor changes to work over IPv6, primarily to provide more storage for IPv6 addresses. The UDP packet header checksum also includes the IP addresses, once again using the new pseudo header.

The following standards are relevant to IPv6 in general.

RFCs specific to IPv4-IPv6 transition can be found here.

RFCs specific to IPv6 and DNS can be found here.

RFC 1809, "Using the Flow Label Field in IPv6," June 1995 (Informational)

RFC 1881, "IPv6 Address Allocation Management," December 1995 (Informational)

RFC 1887, "An Architecture for IPv6 Unicast Address Allocation," December 1995 (Informational)

RFC 2428, "FTP Extensions for IPv6 and NATs," September 1998 (Standards Track)

RFC 2474, "Definition of the Differentiated Service Field (DS Field) in the IPv4 and IPv6 Headers," December 1998 (Standards Track)

RFC 2526, "Reserved IPv6 Subnet Anycast Addresses," March 1999 (Standards Track)

RFC 2675, "IPv6 Jumbograms," August 1999 (Standards Track)

RFC 2711, "IPv6 Router Alert Option," October 1999 (Standards Track)

RFC 2894, "Router Renumbering for IPv6," August 2000 (Standards Track)

RFC 3111, "Service Location Protocol Modifications for IPv6," May 2001 (Standards Track)

RFC 3122, "Extensions to IPv6 Neighbor Discovery for Inverse Discovery Specification," June 2001 (Standards Track)

RFC 3175, "Aggregation of RSVP for IPv4 and IPv6 Reservations," September 2001 (Standards Track)

RFC 3178, "IPv6 Multihoming Support at Site Exit Routers," October 2001 (Informational)

RFC 3306, "Unicast-Prefix-based IPv6 Multicast Addresses," August 2002 (Standards Track)

RFC 3314, "Recommendations for IPv6 in Third Generation Partnership Project (3GPP) Standards," September 2002 (Informational)

RFC 3363, "Representing Internet Protocol version 6 (IPv6) Addresses in the Domain Name System," August 2002 (Informational)

RFC 3364, "Tradeoffs in Domain Name System (DNS) Support for Internet Protocol version 6 (IPv6)," August 2002 (Informational)

RFC 3531, "A Flexible Method for Managing the Assignment of Bits of an IPv6 Address Block," April 2003 (Informational)

RFC 3574, "Transition Scenarios for 3GPP Networks," August 2003 (Informational)

RFC 3582, "Goals for IPv6 Site-Multihoming Architectures," August 2003 (Informational)

RFC 3587, "IPv6 Global Unicast Address Format," August 2003 (Informational)

RFC 3595, "Textual Conventions for the IPv6 Flow Label," September 2003 (Standards Track)

RFC 3701, "6bone (IPv6 Testing Address Allocation) Phaseout," March 2004 (Standards Track)

RFC 3750, "Unmanaged Networks IPv6 Transition Scenarios," April 2004 (Informational)

RFC 3756, "IPv6 Neighbor Discovery (ND) Trust Models and Threats," May 2004 (Informational)

RFC 3769, "Requirements for IPv6 Prefix Delegation," June 2004 (Informational)

RFC 3849, "IPv6 Address Prefix Reserved for Documentation," July 2004 (Informational)

RFC 3879, "Deprecating Site Local Addresses," September 2004 (Standards Track)

RFC 3974, "SMTP Operational Experience in Mixed IPv4/v6 Environments," January 2005 (Informational)

RFC 4007, "IPv6 Scoped Address Architecture," March 2005 (Informational)

RFC 4029, "Scenarios and Analysis for Introducing IPv6 into ISP Networks," March 2005 (Informational)

RFC 4057, "IPv6 Enterprise Network Scenarios," June 2005 (Informational)

RFC 4074, "Common Misbehavior Against DNS Queries for IPv6 Addresses," May 2005 (Informational)

RFC 4135, "Goals of Detecting Network Attachment in IPv6," May 2005 (Informational)

RFC 4147, "Proposed Changes to the Format of the IANA IPv6 Registry," August 2005 (Informational)

RFC 4159, "Depreciation of ip6.in," August 2005 (Best Current Practice)

RFC 4177, "Architectural Approaches to Multihoming for IPv6," September 2005 (Informational)

RFC 4192, "Procedures for Renumbering an IPv6 Network Without a Flag Day," September 2005 (Informational)

RFC 4193, "Unique Local IPv6 Unicast Addresses," October 2005 (Standards Track)

RFC 4215, "Analysis of IPv6 Transition in Third Generation Partnership Project (3GPP) Networks," October 2005 (Informational)

RFC 4218, "Threats Relating to IPv6 Multihoming Solutions," October 2005 (Informational)

RFC 4291, "IP Version 6 Addressing Architecture," February 2006

RFC 4294, "IPv6 Node Requirements," April 2006 (Informational)

RFC 4311, "IPv6 Host-to-Router Load Sharing," November 2005 (Standards Track)

RFC 4339, "IPv6 Host Configuration of DNS Server Information Approaches," February 2006 (Informational)

RFC 4380, "Teredo: Tunneling IPv6 over UDP through Network Address Translations (NATs)," February 2006 (Standards Track)

RFC 4429, "Optimistic Duplicate Address Detection (DAD) for IPv6," April 2006 (Standards Track)

RFC 4443, "Internet Control Message Protocol (ICMPv6) for the Internet Protocol Version 6 (IPv6) Specification," April 2006 (Standards Track)

RFC 4472, "Operational Considerations and Issues with IPv6 DNS," April 2006 (Informational)

RFC 4554, "Use of VLANs for IPv4-IPv6 Coexistence in Enterprise Networks," June 2006 (Informational)

RFC 4659, "BGP-MPLS IP Virtual Private Network (VPN) Extensions for IPv6 VPN," September 2006 (Standards Track)

RFC 4692, "Considerations on the IPv6 Host Density Metric," October 2006 (Informational)

RFC 4727, "Experimental Values in IPv4, IPv6, ICMPv4, ICMPv6, UDP and TCP Headers," November 2006 (Standards Track)

RFC 4779, "ISP IPv6 Deployment Scenarios in Broadband Access Networks," January 2007 (Informational)

RFC 4818, "RADIUS Delegated-IPv6-Prefix Attribute," April 2007 (Standards Track)

RFC 4852, "IPv6 Enterprise Network Analysis – IP Layer 3 Focus," April 2007 (Informational)

RFC 4861, "Neighbor Discovery for IP version 6 (IPv6)," September 2007 (Standards Track)

RFC 4862, "IPv6 Stateless Address Autoconfiguration," September 2007 (Standards Track)

RFC 4864, "Local Network Protection for IPv6," May 2007 (Informational)

RFC 4890, "Recommendations for Filtering ICMPv6 Messages in Firewalls," May 2007 (Informational)

RFC 4919, "IPv6 over Low-Power Wireless Personal Area Networks (6LoWPANs): Overview, Assumptions, Problem Statement and Goals," August 2007 (Informational)

RFC 4941, "Privacy Extensions for Stateless Address Autoconfiguration in IPv6," September 2007 (Standards Track)

RFC 4943, "IPv6 Neighbor Discovery On-Link Assumption Considered Harmful," September 2007 (Informational)

RFC 4968, "Analysis of IPv6 Link Models for 802.16 Based Networks," August 2007 (Informational)

RFC 5095, "Deprecation of Type 0 Routing Headers in IPv6," December 2007 (Standards Track)

RFC 5172, "Negotiation for IPv6 Datagram Compression Using IPv6 Control Protocol," March 2008 (Standards Track)

RFC 5175, "IPv6 Router Advertisement Flags Option," March 2008 (Standards Track)

RFC 5181, "IPv6 Deployment Scenarios in 802.16 Networks," May 2008 (Informational)

RFC 5350, "IANA Considerations for the IPv4 and IPv6 Router Alert Options," September 2008 (Standards Track)

RFC 5375, "IPv6 Unicast Address Assignment Considerations," December 2008 (Informational)

RFC 5453, "Reserved IPv6 Interface Identifiers," February 2009 (Standards Track)

RFC 5533, "Shim6: Level 3 Multihoming Shim Protocol for IPv6," June 2009 (Standards Track)

RFC 5534, "Failure Detection and Locator Pair Exploration Protocol for IPv6 Multihoming," June 2009 (Standards Track)

RFC 5549, "Advertising IPv4 Network Layer Reachability Information with an IPv6 Next Hop," May 2009 (Standards Track)

RFC 5570, "Common Architecture Label IPv6 Security Option (CALIPSO)," July 2009 (Informational)

RFC 5619, "Softwire Security Analysis and Requirements," August 2009 (Standards Track)

RFC 5701, "IP Address Specific BGP Extended Community Attribute," November 2009 (Standards Track)

RFC 5722, "Handling of Overlapping IPv6 Fragments," December 2009 (Standards Track)

RFC 5739, "IPv6 Configuration in Internet Key Exchange Protocol version 2 (IKEv2)," February 2010 (Experimental)

RFC 5798, "Virtual Router Redundancy Protocol (VRRP) Version 3 for IPv4 and IPv6," March 2010 (Standards Track)

RFC 5855, "Nameservers for IPv4 and IPv6 Reverse Zones," May 2010 (Best Current Practices)

RFC 5871, "IANA Allocation Guidelines for the IPv6 Routing Header," May 2010 (Proposed Standard)

RFC 5881, "Bidirectional Forwarding Detection (BFD) for IPv4 and IPv6 (Single Hop)," June 2010 (Proposed Standard)

RFC 5905, "Network Time Protocol Version 4: Protocol and Algorithms Specification," June 2010 (Standards Track)

RFC 5908, "Network Time Protocol (NTP) Server Option for DHCPv6," June 2010 (Proposed Standard)

RFC 5942, "IPv6 Subnet Model: The Relationship Between Links and Subnet Prefixes," July 2010 (Proposed Standard)

RFC 5952, "A Recommendation for IPv6 Address Text Representation," August 2010 (Proposed Standard)

RFC 5963, "IPv6 Deployment in Internet Exchange Points (IXPs)," August 2010 (Informational)

RFC 5970, "DHCPv6 Options for Network Boot," September 2010 (Proposed Standard)

RFC 6036, "Emerging Service Provider Scenarios for IPv6 Deployment," October 2010 (Informational)

RFC 6059, "Simple Procedures for Detecting Network Attachment in IPv6," November 2010 (Proposed Standard)

RFC 6085, "Address Mapping of IPv6 Multicast Packets on Ethernet," January 2011 (Standards Track)

RFC 6088, "Traffic Selectors for Flow Exchange Bindings," January 2011 (Proposed Standard)

RFC 6092, "Recommended Simple Security Capabilities in Customer Premises Equipment (CPE) for Providing Residential Internet Service," January 2011 (Informational)

RFC 6119, "IPv6 Traffic Engineering in IS-IS," February 2011 (Proposed Standard)

RFC 6144, "Framework for IPv4/IPv6 Translation," April 2011 (Informational)

RFC 6156, "Traversal Using Relays Around NAT (TURN) Extensions for IPv6," April 2011 (Proposed Standard)

RFC 6157, "IPv6 Transition in the Session Initiation Protocol (SIP)," April 2011 (Proposed Standard)

RFC 6164, "Using 127-Bit IPv6 Prefixes on Inter-Router Links," April 2011 (Standards Track)

RFC 6177, "IPv6 Address Assignments to End Sites," March 2011

RFC 6204, "Basic Requirements for IPv6 Customer Edge Routers," April 2011 (Informational)

RFC 6214, "Adaptation of RFC 1149 for IPv6," April 1, 2011 (Informational)

RFC 6221, "Lightweight DHCPv6 Relay Agent," May 2011 (Proposed Standard)

RFC 6250, "Evolution of the IP Model", May 2011 (Informational)

RFC 6264, "An Incremental Carrier-Grade NAT (CGN) for IPv6 Transition," June 2011 (Informational)

RFC 6294, "Survey of Proposed Use Cases for the IPv6 Flow Label," June 2011 (Informational)

RFC 6343, "Advisory Guidelines for 6to4 Deployment," August 2011 (Informational)

RFC 6384, "An FTP Application Layer Gateway (ALG) for IPv6-to-IPv4 Translation," October 2011 (Proposed Standard)

RFC 6437, "IPv6 Flow Label Specification," November 2011 (Standards Track)

RFC 6438, "Using the IPv6 Flow Label for Equal Cost Multipath Routing and Link Aggregation in Tunnels," November 2011

RFC 6459, "IPv6 in 3rd Generation Partnership Project (3GPP) Evolved Packet System (EPS)," January 2012 (Informational)

RFC 6540, "IPv6 Support Required for All IP-Capable Nodes," April 2012 (Best Current Practice)

RFC 6556, "Testing Eyeball Happiness," April 2012 (Informational)

RFC 6564, "A Uniform Format for IPv6 Extension Headers," April 2012 (Proposed Standard)

RFC 6568, "Design and Application Spaces for IPv6 over Low-Power Wireless Personal Area Networks (6LoWPANs)," April 2012 (Informational)

RFC 6606, "Problem Statement and Requirements for IPv6 over Low-Power Wireless Personal Area Network (6LoWPAN) Routing," May 2012 (Informational)

RFC 6619, "Scalable Operation of Address Translators with Per-Interface Bindings," June 2012 (Proposed Standard)

RFC 6724, "Default Address Selection for Internet Protocol Version 6 (IPv6)," September 2012 (Standards Track)

RFC 6890, "Special-Purpose IP Address Registries," April 2013 (Best Current Practice)

RFC 7066, "IPv6 for Third Generation Partnership Project (3GPP) Cellular Hosts," November 2013 (Informational)

RFC 7098, "Using the IPv6 Flow Label for Load Balancing in Server Farms," January 2014

RFC 7707, "Network Reconnaissance in IPv6 Networks," March 2016 (Informational)

RFC 8106, "IPv6 Router Advertisement Options for DNS Configuration," March 2017 (Standards Track)

RFC 8200, "Internet Protocol, Version 6 (IPv6) Specification," July 2017 (Standards Track)

RFC 8201, "Path MTU Discovery for IP version 6," July 2017 (Standards Track)

RFC 8504, "IPv6 Node Requirements," January 2019 (Best Current Practices)

Four-Layer IPv6 Architectural Model

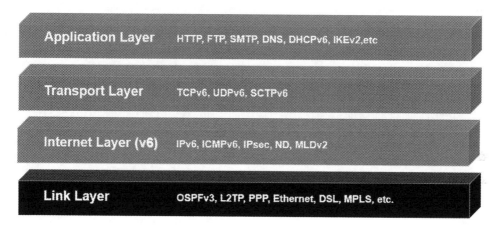

Figure 6-2. *Four-layer IPv6 model*

The major changes from the IPv4 model are as follows:

- *Application Layer*: DHCPv4 replaced with DHCPv6

- *Transport Layer*: TCPv4 replaced with TCPv6, UDPv4 replaced with UDPv6

- *Internet Layer*: IPv4 replaced with IPv6, ICMPv4 replaced with ICMPv6 (which includes ND)

- *Link Layer*: Removed ARP, OSPFv2 replaced with OSPFv3

In the following discussion, traffic really flows both *down* from the Application Layer to the Link Layer (then out the wire) *and* from the wire up through the Link Layer to the Application Layer. For clarity, only the downward path is described in the following. When traffic goes up through the layers, each layer strips off one header and hands off the remaining bytes to the layer above.

The ***Application Layer*** implements the protocols most people are familiar with (e.g., HTTP). The software routines for these are typically contained in application programs such as browsers or web servers that make system calls to subroutines (or "functions" in C terminology) in the "socket API" (an API is an Application Program Interface, or a collection of related subroutines, typically supplied with the operating system or programming language). The application code creates outgoing data streams and then calls routines in the API to actually send the data via TCP (Transmission Control Protocol) or UDP (User Datagram Protocol). Output to the *Transport Layer* is *[DATA]* using IP addresses.

The ***Transport Layer*** implements TCP (the Transmission Control Protocol) and UDP (the User Datagram Protocol). These routines are internal to the socket API. They add a TCP or UDP packet header to the data passed down from the *Application Layer* and then pass the data down to the *Internet Layer* for further processing. Output to the *Internet Layer* is *[TCP HDR [DATA]]*, using IP addresses.

The ***Internet Layer*** implements IPv6 (the Internet Protocol) and various other related protocols such as ICMPv6 (which includes the "ping" function among other things). The IP routine takes the data passed down from the *Transport Layer* routines, adds an IPv6 packet header onto it, and then passes the now complete IPv6 packet down to routines in the *Link Layer*. Output to the Link layer is *[IPv6 HDR [TCP HDR [DATA]]]* using IP addresses. ND (Neighbor Discovery) is actually a part of ICMPv6. It helps locate the Link Layer address of other nodes on the link in addition to other functionality.

The ***Link Layer*** contains routines that actually read and write packets (as fed down to it by routines in the *Internet Layer*) onto the network wire, in compliance with Ethernet or other standards. Output to wire is Ethernet frame using MAC addresses (or the equivalent if other network hardware is used, such as Wi-Fi), which includes the entire IPv6 packet.

Link Layer Issues with IPv6

The following standards are relevant to the Link Layer in IPv6 (primarily the binding mechanisms from IPv6 to the Link Layer):

> **RFC 2464**, "Transmission of IPv6 Packets over Ethernet Networks," December 1998 (Standards Track)
>
> RFC 2467, "Transmission of IPv6 Packets over FDDI Networks," December 1998 (Standards Track)
>
> RFC 2470, "Transmission of IPv6 Packets over Token Ring Networks," December 1998 (Standards Track)
>
> RFC 2491, "IPv6 over Non-Broadcast Multiple Access (NBMA) Networks," January 1999 (Standards Track)
>
> RFC 2492, "IPv6 over ATM Networks," January 1999 (Standards Track)
>
> RFC 2497, "Transmission of IPv6 Packets over ARCnet Networks," January 1999 (Standards Track)
>
> RFC 2590, "Transmission of IPv6 Packets over Frame Relay Networks Specification," May 1999 (Standards Track)
>
> RFC 3146, "Transmission of IPv6 Packets over IEEE 1394 Networks," October 2001 (Standards Track)
>
> RFC 4338, "Transmission of IPv6, IPv4 and Address Resolution Protocol (ARP) Packets over Fibre Channel," January 2006 (Standards Track)
>
> RFC 4392, "IP over InfiniBand (IPoIB) Architecture," April 2006 (Informational)
>
> RFC 4944, "Transmission of IPv6 Packets over IEEE 802.15.4 Networks," September 2007 (Standards Track)
>
> **RFC 5072**, "IP Version 6 over PPP," September 2007 (Standards Track)
>
> RFC 5121, "Transmission of IPv6 via the IPv6 Convergence Sublayer over IEEE 802.16 Networks," February 2008 (Standards Track)

IPv6: The Internet Protocol, Version 6

IPv6 is the foundation of the Third Internet and accounts for many of its distinguishing characteristics, such as its 128-bit address size, its addressing model, and its packet header structure and routing. IPv6 is currently defined in RFC 8200,[6] "Internet Protocol, Version 6 (IPv6) Specification," July 2017, but there are several RFCs that extend the definition.

IPv6 Packet Header Structure

So what are these packet headers mentioned previously? In IPv6 packets, there is an IPv6 packet header, then zero or more packet header extensions, then a TCP or UDP header, and finally the packet data. Each header and header extension is a structured collection of data, including things such as the IPv6 address of the sending node and the IPv6 address of the destination node. Why are we getting down to this level of detail? Because some of the big changes from IPv4 to IPv6 have to do with the new and improved IP packet header architecture in IPv6. In this chapter, we'll cover the IPv6 packet header. Here it is.

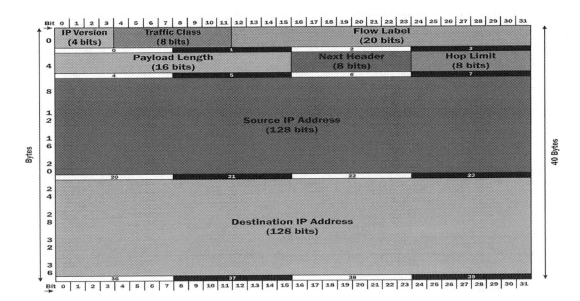

Figure 6-3. *IPv6 packet header*

---

[6]www.rfc-editor.org/rfc/rfc8200.txt

The *IP Version* field (4 bits) contains the value 6 (imagine that!), which in binary is "0110." This field allows IPv4 and IPv6 traffic to be mixed in a single network.

The *Traffic Class* field (8 bits) is available for use by originating nodes and/or forwarding routers to identify and distinguish between different classes or priorities of IPv6 packets, in a manner virtually identical to that of IPv4 "Type of Service."

The *Flow Label* field (20 bits) is something new in IPv6. It can be used to tag up to 2^{20} (1,048,576) distinct traffic flows, for purposes such as fine-grained bandwidth management (QoS). Its use is still experimental. Hosts or routers that do not support this function should set it to zero when originating a packet or ignore it when receiving a packet. The semantics and usage of this field are covered in RFC 8200. Further information is found in RFCs 3595, 6294, 6437, 6438, and 7098. Even today, very few routers actually act on the contents of this field. Until they do, it will be of limited value.

The *Payload Length* field (16 bits) is the length of the IPv6 packet payload in bytes, not counting the standard packet header (as it is in IPv4 *Total Length*), but *counting* the size of any extension headers, which don't exist in IPv4. You can think of packet extension headers as being the first part of the Data field (payload) of the IPv6 packet.

The *Next Header* field (8 bits) indicates the type of header immediately following the standard IPv6 packet header. It uses the same values as the IPv4 *Protocol* field, as defined in RFC 1700, "Assigned Numbers," October 1994. If this value contains the code for *TCP*, then the TCP header and packet payload (data) begins immediately after the IPv6 packet header. Otherwise, one or more IPv6 extension headers will be found before the TCP header and data begins. Since each extension header has another *Next Header* field (and a *Header Length* field), this constitutes a linked list of headers before the final extension header, which is followed by the data. UDP packets can also have extension headers.

The *Hop Limit* field (8 bits) is to prevent packets from being shuttled around indefinitely on a network. Every time a packet crosses a switch or router, the hop count is decremented by one. If it reaches zero, the packet is dropped. Typically, if this happens, an ICMPv6 message ("Time Exceeded") is returned to the packet sender. This mechanism is how the *traceroute* command works.

The *Source IP Address* field (128 bits) contains the IPv6 address of the packet sender.

The *Destination IP Address* field (128 bits) contains the IPv6 address of the packet recipient.

Note   The following fields from the IPv4 packet header have been eliminated in the IPv6 packet header: *Header Length*, *Identification (Fragment ID)*, *Fragmentation Flags*, *Fragment Offset*, *Header Checksum*, and *Options*. The value in the *Payload Length* field no longer includes the length of the standard packet header. The *Flow Label* field has no corresponding field in the IPv4 packet header. Some of the missing fields (e.g., fragmentation information) have been pushed into extension packet headers. For example, in IPv6 only fragmented packets have the fragmentation header extension. Unfragmented packets do not have to carry the unnecessary overhead. In IPv4, all packets have the fragmentation fields in their header, whether they are fragmented or not.

IPv6 Packet Fragmentation and Path MTU Discovery

The fields related to fragmentation are now found in the Fragment extension header, which exists only in fragmented packets (no need to clutter up unfragmented packets, as in IPv4). In IPv6, only the originating node can fragment packets (no intervening node is supposed to do this). The originating node uses *MTU Path Discovery* to determine the "width" of the proposed path (the maximum packet size that it can handle). *MTU* stands for Maximum Transmitted Unit (maximum packet length). Any packets larger than that size must be fragmented before transmission by the originating node and reassembled upon receipt by the destination node. There is a default packet size that any IPv6 node must be able to handle (1280 bytes). MTU Path Discovery allows the sender to determine if larger (more efficient) packets can be used. The originating node assumes the path MTU is the MTU of the first hop in the path. A trial packet of this size is sent out. If any link is unable to handle it, an ICMPv6 *Packet Too Big* message is returned. The originating node iteratively tries smaller packet sizes until it gets no complaints from any node and then uses the largest MTU that was acceptable along the entire path. This process takes place automatically in the *Internet Layer*. There is no corresponding mechanism in IPv4.

Extension Headers (New in IPv6)

After the main header, there can be zero or more *extension headers*, before the payload (actual packet data). This approach makes IPv6 highly extensible, for new functionality in years to come. Several extension headers are already defined, and doubtless more will be defined over time.

The first byte of each extension header contains a *Next Header* field, identical to the same named field in the main IPv6 packet header (using codes from RFC 1700). The second byte of each extension header contains a *Header Extension Length* field, which specifies the length of this header, in 8-byte units, not including the first 8 bytes. Thus, every extension header is at least 8 bytes long and is a multiple of 8 bytes in length. The following header (or data, if no more extension headers) will begin immediately after the end of this extension header. This effectively defines a *linked list* (a data structure familiar to all programmers). Here are some typical packet header sequences to illustrate how each chains to the next.

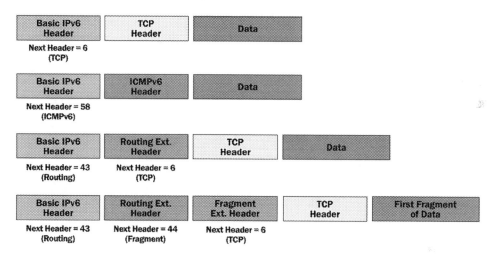

Figure 6-4. *Typical IPv6 packet headers with extensions*

The basic extension headers are defined in RFC 8200,[7] "Internet Protocol, Version 6 (IPv6) Specification," January 2017. These include the following:

- Options extension header

- Hop-by-Hop Options extension header

- Routing extension header

- Fragment extension header

- Destination extension header

---

[7]www.rfc-editor.org/rfc/rfc8200.txt

Two extension headers are used for IPsec (IP Layer security). The IPsec Authentication extension header (IPsec AH) is defined in RFC 2402, "IP Authentication Header," November 1998. The Encapsulating Security Payload header (IPsec ESP) is defined in RFC 2406, "IP Encapsulating Security Payload (ESP)," November 1998.

When multiple extension headers are used in a single packet, the following order should be followed:

- IPv6 basic header

- Hop-by-Hop Options header

- Destination Options header (for options to be processed by more than just the final recipient)

- Routing header

- Fragment header

- Authentication header

- Encapsulating Security Payload header

- Destination Options header (for options to be processed only by the final recipient)

- Upper Layer header (TCP, UDP, or SCTP)

Hop-by-Hop Options header: Used to carry optional information that must be examined by every node along a packet's delivery path. This option is indicated by a *Next Header* value of 0.

Routing header: Used by an IPv6 source node to list one or more intermediate nodes to be "visited on the way" to a packet's destination. This is similar to IPv4's Loose Source and Record Route option. The *Routing header* is identified by a *Next Header* value of 43.

Fragment header: Used by an IPv6 source to send a packet larger than would fit in the path MTU to its destination. In IPv6, packet fragmentation is performed only by the source node, which must use MTU discovery to determine the maximum packet size along the proposed path. The *Fragment header* is identified by a *Next Header* value of 44.

Destination Options header: Used to carry optional information that needs to be examined only by a packet's destination node(s). The *Destination Options header* is identified by a *Next Header* value of 60.

For the specific details on each of the above header extension packets, see RFC 8200. The Authentication header and ESP packet headers will be described later, under IPsec.

IPv6 Addressing Model

In IPv6, addresses are 128 bits in length. They are simply numbers from 0 to about 340 undecillion (340 trillion, trillion, trillion). In exponential notation, that would be 3.40 e+38 (think of it as a 38-digit phone number, where an IPv4 address is a 9-digit phone number). Regardless of how you write it, that's a *really* big number. For the convenience of humans, these numbers are typically represented in what I call *coloned hex* notation (as opposed to the *dotted decimal* notation used with IPv4). This splits the 128-bit addresses into eight 16-bit fields, and each of which is represented with a hexadecimal (base 16) number from 0 to ffff (you can use upper- or lowercase for the hexadecimal digits A–F, but it is common practice in IPv6 to use lowercase). These hexadecimal numbers cover all possible 16-bit binary patterns from 0000 0000 0000 0000 to 1111 1111 1111 1111. The hexadecimal numbers are separated by colons (":"). Leading zeros can be eliminated in each field. At most one run of zeros can be replaced by the double colon, "::". The following are all valid IPv6 addresses written in coloned hex notation:

```
2001:df8:5403:3000:b5ea:976d:679f:30f5    An EUI-64 unicast address
2001:df8:5403:3000::1e                     Manually assigned unicast
fe80::b5ea:976d:679f:30f5                  Link-local EUI-64 address
ff02::1                                    Multicast address
::1                                        Loopback address for IPv6
::                                         The unspecified address
```

Some people are aware that you can use IPv4 addresses instead of nodenames in web URIs, for example: http://123.45.67.89/main.html. You can also use IPv6 addresses, but because colons demark other things in URIs (such as nonstandard port number), you cannot use IPv6 addresses "as is"; enclose them in square brackets ([]). For example, http://[2001:df8:5403:3000::d]/nagios is a valid URI that includes an IPv6 numeric address.

In certain cases, the size of the subnet is specified after the address, similar to CIDR. This is especially common when representing prefixes, for example:

```
2001:df8:5403::/48         An organization's 48 bit network prefix
2001:df8:5403:3000::/52    A /52 block routed into a branch office
2001:df8:5403:3000::/64    The 64 bit prefix for one branch subnet
```

When an RIR (e.g., APNIC) allocates a "/32" block of addresses to an ISP, they assign the first 32 bits of those addresses, based on the next available "/32" block from the unallocated pool at that time. A "/32" block contains 65,536 "/48" blocks to allocate to customers. If the ISP allocates all those, then the RIR will give them a new "/32" block, each address of which will have a completely different first 32 bits from the addresses in the previous "/32" block given to the ISP. The leftmost, or *most significant*, 32 bits of every address in a given "/32" block will all be the same. All addresses from smaller blocks (like a "/48" block or "/64" block) carved out of that "/32" block by the ISP (for allocation to customers) will have the same first 32 bits. For example, many of NTT America's IPv6 allocations include addresses that start with "*2001:418::/32*". No other ISP in the world will *ever* be allocated a "/32" block with those particular first 32 bits. Up to 65,536 of NTT America's customers might get "/48" blocks whose addresses start with those 32 bits.

When an ISP allocates a "/48" block for a customer from their "/32" block, the *next* 16 bits (following the first 32 bits) are chosen by that ISP, so that the first 48 bits will be unique to that customer in the entire world. The first 48 bits of every address in a "/48" block given to an end-user organization will all be the same but will be different from the first 48 bits of the addresses in any other "/48" block in the world. You can think of this 48-bit sequence as the *organization prefix*. All addresses in our "/48" block from HE happens to start with "*2001:470:ed3a::/48*". No other customer of HE has ed3a in the third 16-bit field of their addresses. When a customer deploys subnets, they choose a *further* 16-bit value (unique within their organization) for each subnet, which, together with the organization's 48-bit prefix, creates a globally unique 64-bit prefix for a working subnet. This can be used to manually configure 128-bit addresses for nodes on that subnet, or they can be configured on the Router Advertisement Daemon that supplies prefixes to nodes in that subnet for Stateless Address Autoconfiguration. If using stateful DHCPv6, the administrator can also create pools of addresses for assignment, where each 128-bit address in a pool has that same 64-bit subnet prefix.

IPv6 Packet Transmission Types

In IPv4, there were several packet transmission types (unicast, anycast, and multicast). IPv4 multicast uses class D addresses, while all other addresses are unicast (or reserved). There is no real concept of *scope* in IPv4 (the part of the network in which a given address is valid and unique). IPv4 "private addresses" are a step in this direction, but

IPv6 defines real scope rules for certain kinds of addresses. These concepts are defined in RFC 4291, "IP Version 6 Addressing Architecture," February 2006. Note: In Windows, "ping" is used for both IPv4 and IPv6. In Linux and BSD, the "ping" command is used just for IPv4 – in IPv6, the command is "ping6." In the following, I use just the generic "ping," but be aware that for IPv6 on some platforms, "ping6" would actually be used.

IPv6 Address Scopes

The *scope* of an address specifies in what part of the network it is valid and unique. The defined scopes in IPv6 are as follows:

> *Node Local*: Valid only within the local node (e.g., loopback address).

> *Link Local*: Valid only within a single network link (subnet). All such addresses start with the 10 bits "1111 1110 10" followed by 54 bits of 0 (fe80::/64). When specified in commands, you usually must follow a link-local address with "%" and the interface ID of the link it is connected to. In FreeBSD, this might be something like "fxp0", so to ping a link-local address, you might use the command

```
ping fe80::3c79:b2ca:90ce:5d59%fxp0
```

> In Windows, interface IDs are numbers, so a ping command there might look like

```
ping fe80::3c79:b2ca:90ce:5d59%11
```

> *Site Local*: Valid only within a "site." They start with the 10 bits "1111 1110 11" (fec0::/10). These were intended to be like IPv4 RFC 1918 "private addresses," but are no longer used as of RFC 3878, "Deprecating Site Local Addresses," September 2004.

> *Global*: Valid anywhere on the IPv6 Internet. Global unicast addresses are in the 2000::/3 block. When you specify global addresses, there is no need to append the interface ID, so a ping command for such an address might look like

```
ping 2001:df8:5403:3000::c
```

IPv6 Address Types

A ***unicast address*** specifies a single network interface (destination address). Currently, all *global* unicast addresses are in the 2000::/3 block. There are also *link-local* unicast addresses, in the fe80::/10 block. The global unicast address type is defined in RFC 3587, "IPv6 Global Unicast Address Format," August 2003. This RFC deprecates (makes historic) the "Top-Level Aggregator" and "Next-Level Aggregator" (TLA/NLA) scheme previously defined for global unicast addresses and formalizes the 48-bit organization prefix, 16-bit subnet number, and 64-bit interface identifier concept used today:

```
| 3 |     45 bits        |  16 bits  |            64 bits             |
+---+-------------------+----------+----------------------------+
|001|global routing prefix| subnet ID |         interface ID           |
+---+-------------------+----------+----------------------------+
```

There are two *special* unicast addresses:

> :: *(all bits zero)*: The unspecified address must never be assigned to any node.

> ::1 *(127 zeros followed by a 1)*: The loopback address for IPv6 (corresponds to 127.0.0.1 in IPv4).

When the site-local scope was deprecated, a new address type called ***unique local unicast*** was defined in RFC 4193, "Unique Local IPv6 Unicast Addresses," October 2005. These addresses are in the *fc00::/7* block. The first 7 bits are "1111 110". The eighth bit is called "L". If L = 1, the address is locally assigned (L = 0 is reserved for future use). The next 40 bits are a global ID that ensures the global uniqueness of the overall address. It is generated pseudo-randomly and must not be sequential. The next 16 bits are a subnet ID, and the final 64 bits are an interface ID (just like in global unicast addresses). Perhaps someday there will be a way to reserve specific global IDs from a central authority (to prevent anyone else from using one you have chosen), but no such mechanism exists today. These addresses have much the same semantics as the IPv4 private addresses:

```
| 7 bits |1|  40 bits  |  16 bits  |            64 bits             |
+--------+-+-----------+----------+----------------------------+
| Prefix |L| Global ID | Subnet ID |          Interface ID          |
+--------+-+-----------+----------+----------------------------+
```

An ***anycast address*** can specify any of a group of addresses (usually on different nodes). A packet sent to an anycast address will be delivered to exactly one of those interfaces, typically the "nearest" one (in the network sense, not geographic sense). Anycast addresses look just like unicast addresses and differ only in being injected into the routing protocol at multiple locations in the network.

A ***multicast address*** specifies multiple network destinations (multiple nodes can be configured with the same multicast address). A packet sent to a multicast address will be delivered to *all* nodes that have been assigned that address. Multicast addresses all have the special prefix *ff00::/8* (the first 8 bits of multicast addresses are all ones). After the first 8 bits, there are 4 bits of flags (0,0,0,T). If T=0, the address is a "well-known" address assigned by IANA. If T=1, then the address is a non-permanently assigned ("transient") address. The scope is specified in the next 4 bits, followed by 112 bits of group ID:

```
|   8    | 4 | 4 |                 112 bits                      |
+------- -+----+----+-------------------------------------------+
|11111111|flgs|scop|               group ID                     |
+--------+----+----+-------------------------------------------+
```

There are several multicast scopes defined by the four scope bits. All other combinations are unassigned:

0	reserved
1	interface-local scope
2	link-local scope
3	reserved
4	admin-local scope
5	site-local scope
8	organization-local scope
E	global scope
F	reserved

The following multicast groups are "well known" (T=0):

1	node
2	router
5	OSPF IGP router
6	OSPF IGP Designated router
9	RIP router

```
a    EIGRP router
b    mobile agent
d    PIM router
16   MLDv2 capable router
fb   DNS server
101  NTP server
108  NIS+ server
1:2  DHCPv6 relay agent or server
1:3  DHCPv6 server (but not relay agent)
```

As there are 112 bits for *group* ID, there are 2^{112} (about 5.19 e+33) possible multicast groups. That is enough for the entire world, for quite some time to come. You can think of a multicast group as similar to a TV channel number. As examples, the following multicast addresses are all valid (and are all "well known"):

```
ff02::1      All nodes on the local link
ff05::1      All nodes in the organization

ff02::2      All routers on the local link
ff05::2      All routers in the site

ff02::fb     All DNS servers on the local link
ff08::fb     All DNS servers in the organization

ff02::1:2    All DHCPv6 relay agents or servers on local link
             (note, DHCPv6 relay agents can only be reached
             via link local addresses, so wider scope
              addresses for relay agents don't make sense)

ff02::1:3    All DHCPv6 servers on the local link
ff05::1:3    All DHCPv6 servers in the site
```

With the scopes larger than the organization, multicast addresses must be specifically configured on nodes (you have to "subscribe to that channel"). If you ping the multicast address *ff0e::1*, you are not going to get a response from every node on earth, unless you can first talk everyone into adding that address to their nodes. Even then, various routers along the way would probably block *that* packet. An organization's routers enforce the scope rules so that link-local multicast addresses will not cross any routers, organization-local multicast addresses will not cross the organization's border router, but global multicast addresses will cross *any* router (in the real world, this is actually managed by the MLD, the Multicast Listener Discovery protocol, and PIM, the Protocol Independent Multicast protocol).

A ***solicited node multicast address*** is a special multicast address (addressed to all nodes on the local link) created from a global unicast address by appending the least significant (rightmost) 24 bits of the unicast address to the special prefix *ff02:0:0:0:1:ff::/104*. For the global unicast address

`2001:df8:5403:3000:3c79:b2ca:90`**ce:5d59**

the solicited node multicast address is:

`ff02::1:ffce:5d59`

These addresses are used by ND (the Neighbor Discovery protocol) in the process of mapping IPv6 addresses to Link Layer (MAC) addresses.

There is no ***broadcast address*** in IPv6, but a multicast to all *nodes on the local link* multicast group *ff02::1* will have pretty much the same result.

Perhaps someday there will be a central authority to coordinate use (and allow reservation) of multicast group IDs. No such authority currently exists. Once IPv6 multicast broadcasters start making their programming available over large regions (or even worldwide), such coordination will be necessary and corresponds to the FCC's management of broadcast frequencies that prevent stations from interfering with each other. Because the number of potential group IDs is so large (2^{112} or about 5.19 e+33), for now, choosing them randomly is sufficient. The probability of any two randomly generated group IDs being the same is quite low, even with millions of people using this scheme. You might think of these group IDs as being in some sense *channel numbers* as found today on TVs. I can envision a search engine that would allow you to find multicast channels associated with programming that caters to specific tastes, such as Bollywood music videos over IPTV.

Special Case: IPv4-Compatible IPv6 Addresses (Now Deprecated)

The entire 4.3 billion addresses of IPv4 were mapped into the IPv6 address space, not just once, but twice – once as *IPv4-compatible IPv6 addresses* (::w.x.y.z) and a second time as *IPv4-mapped IPv6 addresses* (::ffff:w.x.y.z).

The addresses in the first special block all start with 96 bits of 0, followed by a 32-bit IPv4 address (which can be specified in dotted decimal). When you send traffic to an IPv4-compatible IPv6 address, it is sent as an IPv6 packet, but encapsulated with an IPv4 header, with the Protocol field of the IPv4 packet header set to 41 to indicate that the payload is an IPv6 packet. The IPv4 header allows the traffic to travel across an IPv4-only infrastructure. Upon receipt, the packet payload (the IPv6 packet) is passed to IPv6. This is called automatic IPv6 tunneling over IPv4 networks (defined in RFC 2893, "Transition Mechanisms for IPv6 Hosts and Routers," August 2000).

IPv4-compatible IPv6 addresses were deprecated in RFC 4291, "IP Version 6 Addressing Architecture," February 2006. No current transition mechanism uses them. New implementations are not required to support these addresses. Note however that two special addresses that are widely used actually fall into this range, the "unspecified" address (all zeros, or "::") and the loopback address ("::1").

Special Case: IPv4-Mapped IPv6 Addresses (Still Valid)

The addresses in the second special block of addresses all start with 80 bits of 0 (*0:0:0:0:0*), followed by 16 bits of 1 (*ffff*) and then a 32-bit IPv4 address (which can be, but does not have to be, specified in dotted decimal). When such an address is used on a dual-stack node that supports IPv4-mapped IPv6 addresses, it causes an IPv4 packet to be sent using the last 32 bits of the IPv4-mapped IPv6 address, as the IPv4 address. As an example, on a Windows 7 node configured with dual stack, you can ping an IPv4 node as usual with the command

```
C:\Users\lhughes>ping 10.1.0.14

Pinging 10.1.0.14 with 32 bytes of data:
Reply from 10.1.0.14: bytes=32 time<1ms TTL=64
Reply from 10.1.0.14: bytes=32 time<1ms TTL=64
Reply from 10.1.0.14: bytes=32 time<1ms TTL=64
Reply from 10.1.0.14: bytes=32 time<1ms TTL=64
```

You could ping the same IPv4, by using an IPv4-mapped IPv6 address, as follows. The ping command would first view the address as a valid IPv6 address and create an IPv6 socket as usual. The IPv6 socket would look at the IPv6 address, realize it is an IPv4-mapped IPv6 address, and then hand the operation over to the IPv4 stack to handle, using the low 32 bits of the IPv4-mapped IPv6 address. Normal IPv4 packets would be sent from the IPv6 socket, indistinguishable from the IPv4 packets sent in the preceding example:

```
C:\Users\lhughes>ping ::ffff:10.1.0.14

Pinging 10.1.0.14 with 32 bytes of data:
Reply from 10.1.0.14: bytes=32 time<1ms TTL=64
Reply from 10.1.0.14: bytes=32 time<1ms TTL=64
Reply from 10.1.0.14: bytes=32 time<1ms TTL=64
Reply from 10.1.0.14: bytes=32 time<1ms TTL=64
```

In general, you can do any I/O operation to an IPv4 node using IPv4 packets, *from an IPv6 socket*, by using these IPv4-mapped addresses (on nodes where this is supported). Some operating systems (e.g., OpenBSD) don't support this kind of "cross-stack" operation at all. On some operating systems (Linux, NetBSD, FreeBSD), this mode is disabled by default, but in FreeBSD can be enabled by including the following line in /etc/rc.conf:

```
...
IPv6_IPv4mapping="YES"
...
```

In general, it is best to avoid use of these addresses since support varies from operating system to operating system, behavior is implementation dependent, and there are potential vulnerabilities if it is enabled. It was originally intended as a transition mechanism, but it caused more problems than it solved, so it is better left unused and, ideally, disabled.

Simple IPv6 Address Assignment Scheme (for Manually Assigned Addresses)

The following is not part of any standard, IETF or otherwise. It is a best-practices recommendation, which may help you in migration to IPv6.

Many administrators have adopted a simple scheme for assigning IPv6 addresses manually to nodes, based on existing IPv4 address conventions or actual addresses. It could be argued that it can lead to confusion (by humans) between decimal and hexadecimal. It uses the same numeric digits that are currently used in your IPv4 scheme, to create what are really hexadecimal fields. It is possible to use the numeric digits (0–9) to create up to three hex digits in each of the four 16-bit groups in the IPv6 interface identifier. The resulting address may look strange in binary, but this scheme will make it easier for you to keep track of your IPv6 nodes and is especially useful in dual-stack networks, where you can use what appears to be the "same" address (not counting the prefix) on a given node, in both IPv4 and IPv6.

As an example, say our 48-bit organization prefix is 2001:df8:5403::/48. Let's also say we have four subnets (independent links) for IPv4, so we would also have four subnets for IPv6. Let's arbitrarily assign the IPv6 subnet numbers as 3000, 3100, 3200, and 3300 (all hex) for these subnets. Choose any values you want for subnet numbers (when setting up your network architecture) – you have 65,536 (from 0000 to ffff) to play with. The following IPv4 addresses from these subnets could be assigned the corresponding IPv6 addresses:

```
IPv4 Subnet           IPv4 Address     Corresponding IPv6 Address
123.45.67.00/24       123.45.67.1      2001:df8:5403:3000:123:45:67:1
192.168.0.0/16        192.168.5.13     2001:df8:5403:3100:192:168:5:13
172.16.0.0/12         172.31.25.32     2001:df8:5403:3200:172:31:25:32
10.0.0.0/8            10.30.1.43       2001:df8:5403:3300:10:30:1:43
```

Alternatively, It is also possible to use just the *interface identifier* part of the IPv4 address ("node number within subnet") as the IPv6 interface identifier, in which case, the preceding addresses would be

```
Subnet                IPv4 Address     Corresponding IPv6 Address
123.45.67.00/24       123.45.67.1      2001:df8:5403:3000::1
192.168.0.0/16        192.168.5.13     2001:df8:5403:3100::5:13
172.16.0.0/12         172.31.25.32     2001:df8:5403:3200::15:25:32
10.0.0.0/8            10.30.1.43       2001:df8:5403:3300::30:1:43
```

The mapping for the 172.31.25.32 address may confuse you – this is because a /12 subnet mask length divides the second 8-bit field right in the middle (4 bits of it are the network address, and 4 bits are the interface identifier). This is why using dotted decimal for IPv4 was a bad idea and hexadecimal is used in IPv6. This can get even more confusing with very odd subnet lengths, like /19. The following should clear things up:

```
172       31        25        32        Full address, dotted decimal

A    C    1   F    1    9    2    0      Full address, hex

1010 1100 0001 1111 0001 1001 0010 0000  Full address, binary

          F    1    9    2    0          Interface identifier, hex

          15        25        32         Interface identifier,
                                         dotted decimal
```

Using only the IPv4 interface identifier is less likely to produce addresses that collide with automatically generated addresses but requires a good understanding of IPv4 subnetting (see the preceding discussion). Use whichever scheme makes the most sense to you but try to be consistent.

The Simple IPv6 Address Assignment Scheme can also be used to manually assign link-local addresses. In this case, there is no IPv6 subnet number, because each address is valid only within a subnet. The following link-local addresses could be assigned to the preceding nodes:

Subnet	IPv4 Address	Corresponding IPv6 Address
123.45.67.00/24	123.45.67.1	fe80::123:45:67:1
192.168.0.0/16	192.168.5.13	fe80::192:168:5:13
172.16.0.0/12	172.31.25.32	fe80::172:31:25:32
10.0.0.0/8	10.30.1.43	fe80::10:30:1:43

As with global unicast addresses, you could use just the interface identifier part of each IPv4 address, which would result in the following manually assigned IPv6 link-local addresses:

Subnet	IPv4 Address	Corresponding IPv6 Address
123.45.67.00/24	123.45.67.1	fe80::1
192.168.0.0/16	192.168.5.13	fe80::5:13
172.16.0.0/12	172.31.25.32	fe80::15:25:32
10.0.0.0/8	10.30.1.43	fe80::30:1:43

Note that the addresses 123.45.67.1/24 and 192.168.0.1/16 would both produce fe80::1 as the equivalent IPv6 address, but this would not produce a conflict since they are in different subnets, and link-local addresses are valid only within a single subnet.

Obviously, no addresses generated with Stateless Address Autoconfiguration will use this convention, although you should be careful to make sure there are no conflicts between addresses you create and automatically generated addresses. Duplicate Address Detection during automated address creation should detect such conflicts. On the other hand, you can easily create DHCPv6 address pools that will be consistent with these schemes.

Warning: There is a perfectly valid (but not often used) textual representation of IPv6 addresses that would allow you to use the exact same bits as a 32-bit IPv4 interface identifier and even specify those 32 bits in dotted decimal. However, it mixes hexadecimal and decimal numbers, plus colons and dots in a single address representation, which to me is extremely confusing and inelegant. It represents the first 96 bits of an address in *coloned hex* notation and the last 32 bits of that address in *dotted decimal* notation. When you use this *mixed* notation, you must always specify all four dotted decimal fields, and they must be the least significant 32 bits. It is possible that some software applications will not accept this representation. Also, many things that report addresses (e.g., ipconfig) have no way to display some addresses in mixed notation and others in regular coloned hex notation, so they just display all addresses in coloned hex notation. This can lead to confusion. As examples of addresses with this mixed notation, the preceding IPv4 addresses would have corresponding IPv6 addresses that look like this:

```
IPv4 Address     IPv6 Address in "Mixed" Notation
    Same Address in Coloned Hex
123.45.67.1      2001:df8:5403:3000::123.45.67.1
    2001:df8:5403:3000::7b2d:4301
192.168.5.13     2001:df8:5403:3100::192.168.5.13
    2001:df8:5403:3000::c0a8:50d
172.16.25.3      2001:df8:5403:3200::172.16.25.3
    2001:df8:5403:3000::ac10:1903
10.30.1.43       2001:df8:5403:3300::10.30.1.43
    2001:df8:5403:3000::a1e:12b
```

I recommend that you avoid use of this mixed notation altogether. If you use the Simple IPv6 Address Assignment scheme, be very careful to use colons (not dots) between all fields, as software that understands the mixed address syntax will interpret addresses with dots in the last four groups as perfectly valid "mixed" notation. This will result in some odd problems. The mixed notation was really intended for use with IPv4-mapped IPv6 addresses, but it works anywhere. You should never create addresses using it, but you need to know about it in case you see addresses written in it by someone else.

Multiple IPv6 Subnet Numbers on a Single Network Link

A single network link can actually have addresses with more than one 16-bit subnet number at any given time. For example, the prefix 2001:df8:5403:1600::/64 may be used with stateless autoconfiguration, while the prefix 2001:df8:5403:1601::/64 could be used with stateful autoconfiguration using DHCPv6 on the same network link. You could also have manually assigned addresses using a *third* prefix (e.g., 2001:df8:5403:1602::/64) on the same network link. Addresses with different subnet numbers, but the same interface identifier, *are not in conflict.* Normally, you only broadcast one 64-bit prefix with Router Advertisement messages onto a given network link, so all addresses created with stateless autoconfiguration in a given subnet will have only that one 64-bit prefix. It is possible in some implementations to advertise many prefixes on each network link. If multiple prefixes are advertised, there will still be only one default gateway, which is the link-local address of the gateway that is sending Router Advertisement messages. Another alternative is to define a subnet size *greater than /64* on a single network link that includes all the desired subnet numbers. With a "/60" subnet, you can actually have 16 sequential /64 subnet numbers in a single network link (the first subnet number has to be an integral multiple of 16). This is called *supernetting.* Do this only if you really understand what you are doing.

Multiple IPv6 Addresses on a Single Node

Unlike with IPv4, it is completely normal for IPv6 nodes to have *multiple* valid addresses. They don't even all have to have the same subnet number (if you are running multiple subnet numbers on a single link). A single node could have addresses with each of the preceding 64-bit prefixes (or even multiple manually assigned addresses) at any given time. It could also have various multicast addresses. One of the unicast addresses (chosen at random) will be used as the source address of packets sent by that node, but incoming packets addressed to any of the addresses owned by the node will be accepted.

A host is required to recognize any of the following addresses as referring to itself. Any node has most of these *by default* without anyone having to assign them. The default link-local address is created with Stateless Address Autoconfiguration even if there are no Router Advertisement messages. *Solicited node multicast* addresses are created and assigned automatically when unicast or anycast addresses are assigned:

- *The loopback address* (::1): Always present.

- The all-nodes multicast addresses (ff01::1, ff02::1, etc.): Only the "on node" and "on link" scoped multicast addresses are created automatically – ones with larger scope must be specifically assigned to each node that you wish to accept such addresses.

- The automatically generated link-local unicast address.

- Any additional unicast and anycast addresses that have been assigned to any of the node's interfaces, manually or automatically.

- The solicited node multicast address for each of its unicast and anycast addresses (created automatically for you when the corresponding unicast or anycast address is assigned).

- Multicast addresses for all other groups to which the node has subscribed.

A router (gateway) is required to recognize all addresses that a host is required to recognize, plus the following special addresses for routers, as identifying itself:

- The subnet-router anycast address for all interfaces for which it is configured to act as a router

- All other anycast addresses with which the router has been configured

- The all-routers multicast addresses (ff01::2, ff02::2, ff05::2)

Automatically Generated Interface Identifiers Based on EUI-64

By default, every IPv6 interface will create a unique link-local address (fe80::w:x:y:z). If there is a Router Advertisement Daemon configured and running on the link, the node will *also* automatically create a global unicast address by using the 64-bit subnet prefix from the Router Advertisement message. It either can generate the interface identifier (low 64 bits) from the node's MAC address (using EUI-64) or can use a random 64-bit value. This is described in RFC 4291, "IP Version 6 Addressing Architecture," and RFC 2464, "Transmission of IPv6 Packets over Ethernet Networks."

An EUI-64 address is created by taking the *first 24 bits* of a MAC address (the *Organizationally Unique Identifier* part of the MAC address), setting the seventh bit of this to 1 (counting rightward from the most significant bit), appending the 16-bit value FFFE, and then appending the last 24 bits of the MAC address (the *device identifier*). Hence, the 48-bit MAC address

```
00-18-8B-78-DA-1A
```

produces an EUI-64 identifier of

```
0218:8BFF:FE78:DA1A
```

This is a reversible mapping, so given an EUI-64 identifier, it is trivial to determine the MAC address of the node (discard the FFFE in the middle 16 bits and invert the seventh bit of the remaining 48-bit value). Note: The seventh bit in the first byte of all valid Organizationally Unique Identifiers, hence of all MAC addresses, will always be 0.

One of the security advantages of IPv6 is supposed to be that the number of possible addresses in a subnet (2^{64}) is so large that it is impractical to scan all of them to discover all the nodes on a subnet (this is called *mapping* a subnet). If EUI-64 interface identifiers are used, there are so few of these (in comparison with the total possible number of interface identifiers) that it *is* possible to scan for them (especially with the knowledge of which Organizationally Unique Identifiers are actually in use, which is not difficult to determine).

Randomized Interface Identifiers

There are several privacy concerns related to using addresses with EUI-64 interface identifiers. One is the ability for a hacker to create a map of all nodes on the subnets via scanning. It would also be possible to identify any person's traffic at any point through which the traffic flows, if you know the MAC address of their network interface. You could certainly associate various traffic flows that all have the same MAC address as coming from a single node. In IPv4, MAC addresses never leave your LAN. With EUI-64-based IPv6 unicast addresses, MAC addresses can go anywhere in the world. Fortunately, there is a way to generate a random interface identifier instead of using the EUI-64 identifier. This is defined in RFC 4941,[8] "Privacy Extensions for Stateless Address Autoconfiguration in IPv6," September 2007. The randomized identifier even changes automatically over time. I may have had *that* address yesterday, but today I've got a completely different one! Interface identifier randomization is enabled by default in Windows 7, but it can be enabled or disabled with the following commands:

```
netsh interface IPv6 set global randomizeidentifiers=enabled
netsh interface IPv6 set global randomizeidentifiers=disabled
```

The reason you might want to disable randomization is that some servers will only accept a connection from nodes for which they can perform a reverse DNS lookup. This often will fail with randomized identifiers. Note that use of randomized interface identifiers can make it very difficult to determine to whom specific traffic in a log belongs, unless a record is kept of randomized interface identifiers used by each node.

---

[8] https://tools.ietf.org/html/rfc4941

When a randomized address changes, the old address is kept around for some time, but marked as *deprecated*, which means your node will not use it for further outgoing connections. It will accept incoming replies addressed to a deprecated address until that address becomes *invalid*, which it eventually will be. Since you aren't making new outgoing connections with it, replies to it will cease fairly quickly. Addresses with randomized interface identifiers are used primarily for outgoing connections (and replies thereto). A node that can accept incoming connections from anyone should have (possibly in addition to other addresses) a static (unchanging) unicast address, which is published in DNS. This would be used by other nodes that want to connect to it. A node that only ever makes outgoing connections need not have such a static address assigned to it, and there is no need to publish its name and IPv6 address in DNS (at least not in your *external* DNS). Remember in IPv6, it is *much* more likely that other nodes will be connecting to your node (for VoIP, VPNs, P2P, etc.). The age of NAT (and one-way connectivity) is over.

IPv6 Address Allocation

The standard allocation block to be given to organizations is a "/48," which is 65,536 subnets, each of which is a "/64" block consisting of 2^{64} or about 18 billion, billion addresses (about 4 billion times the total number of addresses in the Second Internet). Some ISPs may choose to allocate only a single "/64" block to individuals or home users, who have no need for multiple subnets. It is not practical to allocate only a single IPv6 address (a "/128" block) to a user, due to the fact that nodes often create new addresses. One "/48" block will supply 65,536 individuals or homes with "/64" blocks. Perhaps I'm a bit unusual, but I already have two subnets in my home today (one dual stack, one IPv6-only). Who knows, I might have a bunch someday! My company has a "/48" (2001:df8:5403::/48), which we divided into 16 "/52" sub-blocks, each of which has 4096 subnets. I have one of these "/52" sub-blocks (subnets 3000–3fff) routed to my house. That should just about take care of me for some time to come. A single "/64" block should work for most home users.

ISPs are allocated *really* big "/32" blocks of addresses, which are enough to allocate "/48" blocks for up to 65,536 customers. Should they use up an entire "/32" block, there are *plenty* more "/32" blocks where that one came from (about 536 million of them just in the 2000::/3 block marked for allocation). The RIRs (ARIN, RIPE, APNIC, LACNIC, and AfriNIC) will be happy to give an ISP all they can use. If you assume there are 7 billion people alive, there are over 5000 "/48" blocks for every human alive, just out of the 2000::/3 range currently marked for allocation. It is extremely unlikely that *any* single human will ever be able to use any appreciable percentage of their "fair share" of addresses, let alone have the IANA run out. The folks in Taiwan say they want to connect 3 billion devices to the Internet in the next couple of years. This would take three-fourths of the entire Second Internet's address space but could be handled with a tiny fraction (less than 1 billionth) of a single "/64" block with IPv6, should they want to have them all in one block for some bizarre reason. It will be quite a while before anyone worries about IPv6 address space exhaustion (famous last words?).

The People's Republic of China believes that they were cheated out of sufficient IPv4 addresses to participate fully in the Second Internet. By the time China started deploying IPv4, if they had taken *all* the remaining addresses, over 90% of the people there would not have gotten one. The Second Internet recently passed an interesting threshold. There are now more Chinese-speaking users on it than English-speaking users. If you recall the chart of allocated addresses earlier in this book, the United States has over 43% (28% ARIN + 15% legacy, both of which are mostly US users) of the total IPv4 address space for less than 5% of the world's population. In comparison, APNIC, which includes China, India, and several other populous countries (all together about 50% of the world's population), has only 16% of the IPv4 address space. When the IPv4 addresses were all gone in September 2011, APNIC would probably still have less than 20% of the IPv4 address space (about .28 addresses per person), while the United States would probably have about 45% (about 6.4 addresses per person). However, note that about one-third of that 45% are held by fewer than 50 organizations (like MIT, Apple, HP, etc.). The distribution of addresses in the Second Internet was (and remains) anything but equitable. It's really pretty much impossible to do anything about that now. We're doing it right on the *Third Internet*. The Second Internet was really an American thing that they shared (to some extent) with the rest of the world. The *Third Internet* is the first truly global Internet. Every country can have as many public addresses as they can conceivably use.

Should We Reserve Some IPv6 Addresses for Developing Nations?

There has been talk from the ITU (International Telecommunication Union) about reserving some IPv6 address space for developing nations to make *absolutely* certain that nobody ever gets left out again, as has happened in the Second Internet. There are *so many* IPv6 addresses that there is essentially no chance of this ever happening. The ITU might as well try to reserve a few trillion grains of sand (maybe a dump truck's worth) to make sure that every country can be assured of getting their fair share of grains of sand. The total number of IPv6 addresses is on the same general scale as the number of grains of sand on earth.

Note that block 2000::/3 (which you can also think of as blocks 2000::/16 through 3fff::/16) is currently the only part of the overall space marked for unicast address allocation. This is only one-eighth of the total IPv6 address space. Even so, this is still 2^{125}, or about 4.15 e+37, addresses. You can also view this as 2^{45} (about 35.2 trillion) "/48" blocks or just over 5000 "/48" blocks *per human alive* in 2010 (using the worldwide population as 7 billion). Should we ever use this up, there are still at least 5.5 times that much space currently not used for anything (from 4000::/16 to efff::/16) that we could repurpose for additional allocation.

I personally don't think there is any reason to reserve a special block of addresses for anyone, including developing nations. Unlike with IPv4, there are plenty of addresses for everyone this time around.

The People's Republic of China (and every other country) will have *plenty* of addresses in the Third Internet, and this is one reason they are investing so heavily in it. India is now determined to deploy IPv6 nationwide and had quite a bit deployed by the end of 2010. By some measure they were at 60% deployment in 2019. The inequitable distribution of addresses in the Second Internet may also account for some of the lack of urgency to migrate to the *Third Internet* in the United States. Unfortunately, it is not simply a matter of still having enough IPv4 addresses. Imagine if the United States stayed with Standard-Definition NTSC TV, while the entire rest of the world went with globally standard High-Definition TV. The United States would not be able to export their programming to anyone else nor import programming from the rest of the world. If they choose to stay with IPv4, they will be isolating themselves in some very serious ways. It's not completely ridiculous to think that the United States might decide not to deploy the *Third Internet*. Look what happened with the metric system. If IPv4 is "riding horses" and IPv6 is "driving cars," you don't need to wait until the last horse dies before you get a car. The "cars" (IPv6) are ready and widely available today. Those who adopt cars first will leave those still riding horses *way behind*. I'd suggest you migrate to IPv6 *as soon as possible*. Countries that master it and start creating products and applications based on it will have a giant head start in the twenty-first century over those who wait until the last possible minute.

How Is the Entire IPv6 Address Space Divided Up?

Here are the official allocations of the IPv6 address space as of May 13, 2008 (from IANA), along with the RFCs that allocated the blocks listed:

```
IPv6 Prefix          Allocation            Reference     Note
-----------          ----------            ---------     ----
0000::/8             Reserved by IETF      [RFC4291]     [1] [5]
0100::/8             Reserved by IETF      [RFC4291]
0200::/7             Reserved by IETF      [RFC4048]     [2]
0400::/6             Reserved by IETF      [RFC4291]
0800::/5             Reserved by IETF      [RFC4291]
1000::/4             Reserved by IETF      [RFC4291]
2000::/3             Global Unicast        [RFC4291]     [3]
4000::/3             Reserved by IETF      [RFC4291]
6000::/3             Reserved by IETF      [RFC4291]
8000::/3             Reserved by IETF      [RFC4291]
A000::/3             Reserved by IETF      [RFC4291]
C000::/3             Reserved by IETF      [RFC4291]
E000::/4             Reserved by IETF      [RFC4291]
F000::/5             Reserved by IETF      [RFC4291]
F800::/6             Reserved by IETF      [RFC4291]
FC00::/7             Unique Local Unicast  [RFC4193]
FE00::/9             Reserved by IETF      [RFC4291]
FE80::/10            Link Local Unicast    [RFC4291]
FEC0::/10            Reserved by IETF      [RFC3879]     [4]
FF00::/8             Multicast             [RFC4291]
```

Notes:

[0] The IPv6 address management function was formally delegated to IANA in December 1995 [RFC1881].

[1] The "unspecified address", the "loopback address", and the IPv6 Addresses with Embedded IPv4 Addresses are assigned out of the 0000::/8 address block.

[2] 0200::/7 was previously defined as an OSI NSAP-mapped prefix set [RFC4548]. This definition has been deprecated as of December 2004 [RFC4048].

[3]   The IPv6 Unicast space encompasses the entire IPv6 address range
      with the exception of FF00::/8. [RFC4291] IANA unicast address
      assignments are currently limited to the IPv6 unicast address
      range of 2000::/3. IANA assignments from this block are registered
      in the IANA registry: iana-IPv6-unicast-address-assignments.

[4]   FEC0::/10 was previously defined as a Site-Local scoped address
      prefix. This definition has been deprecated as of September 2004
      [RFC3879].

[5]   0000::/96 was previously defined as the "IPv4-compatible IPv6
      address" prefix. This definition has been deprecated by [RFC4291].

The referenced RFCs are

RFC 1881, "IPv6 Address Allocation Management,"
December 1995

RFC 3879, "Deprecating Site Local Addresses," September 2004
(affects FEC0::/10)

RFC 4048, "RFC 1888 Is Obsolete," April 2005 (dropping mapping
of OSI addresses)

RFC 4193, "Unique Local IPv6 Unicast Addresses," October 2005

RFC 4291, "IP Version 6 Addressing Architecture," February 2006

The 6bone was an early worldwide IPv6 testbed. It used addresses from 3ffe::/16 (as per RFC 2471,[9] "IPv6 Testing Address Allocation," December 1998). These have since been returned to the overall allocation pool as per RFC 3701,[10] "6bone (IPv6 Testing Address Allocation) Phase-Out," March 2004, once the 6bone had served its purpose and was shut down. Interestingly, some addresses from this block still show up on the IPv6 backbone. Among other places, they are still used in IPv6-ready tests, so if an IPv6-ready test network is connected to the main Internet, those addresses could be accidentally routed. Even though they are just more IPv6 unicast addresses now, I would recommend against using them in production systems, just in case. It's not like there aren't plenty of other IPv6 unicast addresses to use.

---

[9] https://tools.ietf.org/html/rfc2471'
[10] https://tools.ietf.org/html/rfc3701

As of January 2010, the RIRs have the following number of IPv6 prefixes that actually have traffic on the backbone:

```
RIPE      1998
ARIN      1207
APNIC     852
LACNIC    267
AfriNIC   82
```

Here are the top ten countries plus a few from Asia (from SixXS, January 24, 2010) ranked by the number of IPv6 prefixes allocated. 'V' means visible (actual traffic detected), 'A' means allocated (obtained from an ISP or RIR), and 'VP' is the percentage of all allocated blocks that are visible (total for the world would be 100%):

Rank	Country	V	A	VP
1	United States	422	1143	9.30%
2	Germany	179	324	3.24%
3	United Kingdom	100	225	2.20%
4	Netherlands	102	176	2.25%
5	Japan	93	176	2.05%
6	Australia	41	152	0.90%
7	Russia	54	117	1.19%
8	France	49	111	1.08%
9	Brazil	29	106	0.64%
10	Switzerland	56	102	1.23%
19	Korea	15	58	0.33%
20	China	21	54	0.46%
24	India	7	36	0.15%
31	Taiwan	19	33	0.42%
33	Vietnam	4	28	0.09%
34	Philippines	8	27	0.18%
35	Thailand	12	27	0.26%

Note that this data does not reflect the actual number of addresses or the volume of traffic, just the number of distinct 48-bit prefixes, which is a rough indication of the *number of organizations* investigating IPv6. Much of this in the United States is probably research or academic. As percentages of the gigantic total number of "/48" blocks available for allocation, all these are essentially zero (pretty much *all* the 2000::/3 IPv6 address space is still available for allocation). This tiny percentage is more an indication of the colossal size of the IPv6 address space than of any lack of interest or activity.

Here is a graph of the percentage of traffic that is IPv6, for the world and the five RIRs, as of late 2018. You can see how much things changed since 2010.

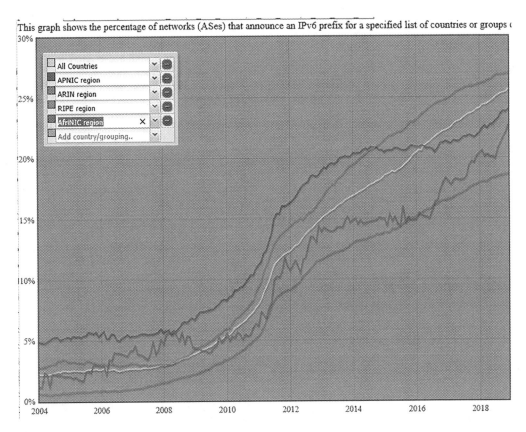

Figure 6-5. IPv6 deployment by region

Classless Inter-Domain Routing (CIDR)

There is no reason to implement CIDR for IPv6. It was done in IPv4 *only* to extend the lifetime of the IPv4 address space long enough for IPv6 to be fully developed, which has now happened. There is no need to extend the lifetime of the IPv6 address space. If IPv6 had been ready and we had migrated to it in the mid-1990s, we would never have had to suffer through the complexities brought about by CIDR and NAT. The reason we are having to deal with these issues today is that we have already stayed with IPv4 *far too long*. Imagine trying to do serious work today with an 8-bit processor and 64K bytes of RAM.

Network Ports

Network ports work exactly the same way under IPv6 as they do in IPv4. There are still 65,536 of them associated with every IPv6 address. They could have gone to 32-bit port numbers (yielding 4.3 billion ports for each address), but this would have required even more changes in packet headers and other places, so this was not done. 65,536 is plenty for almost any need, especially since you can assign any number of global unicast addresses to a single interface (each of which has 65,536 ports). The same *well-known port* numbers are used in IPv6 as in IPv4. The only difference is that you will never see port numbers on IPv6 addresses being shifted by a NAPT gateway, since there is no NAT for IPv6 to IPv6. Note that a given port being used over IPv4 does not prevent it from being used by the same or even a different application, over IPv6 (and vice versa).

Subnetting in IPv6

There is no CIDR in IPv6 (although the CIDR "slash notation" is still used). As a result, subnetting is much simpler in IPv6. All subnets are "/64." The only exception is if you do *supernetting* (e.g., a "/60" subnet) to allow multiple "/64" blocks to be used on a single network link. This will likely only be done in large, advanced corporate networks, so most network engineers will never see anything but "/64" subnets.

The only reason for doing this might be to use different "/64" subnets for specific purposes, such as 1000 for SLAAC, 1001 for DHCPv6-assigned addresses, and 1002 for manually assigned addresses. If you use EUI-64 interface identifiers for SLAAC, it is not difficult to partition a single "/64" so there will be no overlap between SLAAC, DHCPv6, and manual assignments. If you use random interface identifiers, they may fall anywhere in a "/64" address space. However, the probability of one colliding with an address assigned manually or via DHCPv6 stateful mode is incredibly low, and Duplicate Address Detection should prevent the odd collision. Having at least two "/64" subnets in a single network (one for SLAAC, one for manually and DHCPv6-assigned addresses) removes *all* possibility of an address collision.

Each subnet needs to be *at least* a "/64," since EUI-64 can generate "node within subnet" values that are 64 bits long. Randomized interface identifiers are also 64 bits in length. But a "/64" subnet is *already larger* than any organization could conceivably use (18 billion, billion addresses). There are so many "/64" blocks in a single "/48" (65,536) that we can use them even for subnets between a border router and a firewall, which have only two addresses. There is never an excuse to use any subnet smaller than a "/64," although I have seen some old-school IPv4-trained administrators allocate "/124" IPv6 subnets for the link between a border gateway and firewall case (in IPv4, tiny subnets like /30 would be used in such a case). Old habits die hard. After living with increasing scarcity with IPv4 addresses, it is hard for some of us to realize that there are PLENTY of addresses this time around.

Link Layer Addresses

The software in the *Application Layer*, the *Transport Layer*, and the *Internet Layer* of the IPv6 stack think in terms of IP addresses. But the *Link Layer* (and the hardware) thinks in terms of MAC addresses. In IPv6 the mapping from IPv6 address to Link Layer (MAC) address is done with the Neighbor Discovery protocol. Note that in this book, I often use the terms *Link Layer address* and *MAC address* interchangeably.

NOTE   A Link Layer address is a "MAC address" only for Ethernet-based network hardware (and a few others), so when I use the term MAC address, think "physical layer address for the actual network hardware in use." The term Link Layer address is more accurate (a MAC address is just a special case of *Link Layer address*), but it is easy to confuse it with the similar-sounding term *link-local address*. Just realize that if the actual network in use is not Ethernet, there may be some other name for the physical layer addresses that IP addresses have to be mapped onto, and it may not look anything like the 48-bit MAC address.

IPv6 addresses are not actually used at the lowest layer of the IPv6 network stack (the *Link Layer*). The 48-bit MAC addresses covered in Chapter 3 still exist and are used the same way at the *Link Layer* (at least for Ethernet networks).

Neighbor Discovery (ND) Protocol

There is no ARP (Address Resolution Protocol) in IPv6. The new ND (Neighbor Discovery) protocol, which is defined in RFC 4861,[11] "Neighbor Discovery for IP version 6 (IPv6)," September 2007, accomplishes the same thing and many other functions as well, including the following:

---

[11] https://tools.ietf.org/html/rfc4861

- *Router discovery*: A host can locate router(s) residing on any link to which it is attached.

- *Prefix discovery*: A host can discover the correct 64-bit prefix for any link to which it is attached.

- *Parameter discovery*: A host can determine the correct IPv6 parameters, for any link to which it is attached, such as MTU.

- *Stateless Address Autoconfiguration (SLAAC)*: A host can automatically obtain a *link-local* address and, if a Router Advertisement Daemon exists, also a *global unicast* address.

- *Address resolution*: Mapping IPv6 addresses to MAC addresses (as the replacement for ARP).

- *Next-hop determination*: Hosts can determine the next-hop router for a given destination address.

- *Neighbor Unreachability Detection (NUD)*: Determine that a given neighbor is no longer reachable on any attached link (there is no corresponding IPv4 functionality).

- *Duplicate Address Detection (DAD)*: Hosts can determine if a proposed address is already in use.

- *Redirect*: A router can inform a host about a better (or working) first hop.

There are five ICMPv6 messages that ND uses to accomplish these things:

- *Router Solicitation:* Request a Router Advertisement message.

- *Router Advertisement*: Router advertises the 64-bit prefix and parameters for a link, usually sent by a Router Advertisement Daemon living in a gateway router or firewall. The Router Advertisement Daemon can send different information into each attached link, if there are multiple links. This also tells nodes whether or not there is a DHCPv6 server available.

- *Neighbor Solicitation:* Any node can say "Howdy, neighbor" to another node to see if it responds.

- *Neighbor Advertisement:* Response to a "Howdy, neighbor" message from someone else.

- *Redirect*: A router can inform any node that there is a better first hop available than one it has just tried ("there's a bridge out along *that* road; try going down *this* road"), based on its discovered knowledge of the surrounding network.

By the way, some people use "NDP" as the initialism for the Neighbor Discovery protocol (see Wikipedia). If you read the RFCs, the creators of the protocol use just "ND," so we will use that convention in this book. The initialisms of some protocols include the "P" (for Protocol) (e.g., TCP), while others don't (like MLD). I follow the conventions used in the RFCs.

IPv6 Router Advertisement messages carry link-layer (MAC) addresses, so no additional packet exchange is required to resolve the router's Link Layer address. They also carry prefixes, so no separate mechanism is needed to configure a netmask.

By using link-local addresses to uniquely identify routers, hosts can maintain router associations. This capability is necessary for Router Advertisements and for redirects. Hosts need to maintain router associations if the site switches to a new global prefix.

ND is immune to spoofing attacks that originate from off-link nodes. In IPv4, off-link nodes can send ICMPv4 Redirect messages and IPv4 Router Advertisement messages.

In the following, DAD refers to *Duplicate Address Detection*, which is one of the functions performed by ND. Addresses may be in any one of the following states at any given time:

- TENTATIVE: Generated, but not yet determined by DAD to be unique – attempts to bind() to the address fail with EADDRNOTAVAIL, as if the address doesn't exist (this can cause race conditions)

- DUPLICATED: Generated and determined by DAD to be duplicated (hence unusable)

- PREFERRED: Generated and determined by DAD to be unique (hence valid)

- DEPRECATED: A preferred address that has passed its *preferred lifetime* (still valid, and incoming packets addressed to it will be accepted, but no further outgoing packets will be sent using it)

- INVALID: A deprecated address that has passed its *valid lifetime* (may no longer be used for sending *or* receiving packets)

Here are the details of the various functions that ND can perform.

Router Discovery

At any time (but typically at power on), any node can determine the link-local address of the router(s) on the local link.

> *Step 1*: The node sends a Router Solicitation message to the "all routers on link" multicast group (*ff02::2*). If the node's link-local address has already been created, then that will be used as the source address; else, the unspecified address ("::") will be used as the source address.

> *Step 2*: All routers on the link will respond with Router Advertisement messages, usually to the "all nodes on link" multicast group (*ff02::1*), but if the source address of the Router Solicitation message was a link-local address, the router can choose to send the Router Advertisement message directly to that address. The source address of each received Router Advertisement message is added to a default gateway table (from which the preferred link-local default gateway will be chosen). The Prefix Information option in all the responses should be the same, so the subnet prefix from the last received Router Advertisement message will be used.

IPv6 router discovery corresponds roughly to IPv4 router discovery (which was defined in RFC 1256, "ICMP Router Discovery Messages," September 1991), but in IPv6 it is a part of the base protocol. There is no need for hosts to *snoop* the routing protocols to discover a router. IPv4 router discovery contains a preference field, which is not needed in IPv6 router discovery because of Neighbor Unreachability Detection. IPv4 Router Advertisements and Solicitations (ICMP type 9) work only with multicast-capable IPv4 routers and are not commonly used. All IPv6 nodes support multicast, and Router Advertisements are a fundamental part of almost every nontrivial network.

Address Resolution (Mapping IPv6 Addresses to MAC Addresses)

Say Alice (one IPv6 node) is trying to send a packet to Bob (another IPv6 node). Address resolution is done as follows:

> *Step 1*: Alice checks her *Neighbor Cache* (similar to the ARP table in IPv4) to see if it already has an entry with Bob's IPv6 address. If it does, then Alice sends the packet immediately to Bob using Bob's MAC address from her Neighbor Cache, and she is finished. If Alice's Neighbor Cache doesn't have an entry for Bob's IPv6 address, the process continues.

> *Step 2*: Alice adds a new Neighbor Cache entry for Bob, in the INCOMPLETE state. Alice then sends a Neighbor Solicitation message to Bob, using Bob's *solicited node multicast address* as the destination address. Any of the addresses assigned to Alice's interface can be used as the source address of this packet, but if possible, it should match the source address of the original packet Alice wanted to send. Alice includes her MAC address as the Source Link Layer Address option in this packet. This ensures Bob will have Alice's MAC address when it's time for him to reply.

> *Step 3*: Bob receives the Neighbor Solicitation message and responds with a Neighbor Advertisement message, sent to Alice's MAC address.

> *Step 4*: Alice receives the Neighbor Advertisement message from Bob and then updates Bob's entry in her Neighbor Cache.

> *Step 5*: Alice can now send the original packet she wanted to send to Bob using his MAC address.

Prefix Discovery

At any time, a node can discover the default network prefix. A Router Advertisement message can contain up to three "options":

- The Source Link Layer Address (the sending router's MAC address.

- The MTU (the maximum packet size supported on this link)

- The Prefix Information (the preferred address prefix for this subnet).

When a router sends an unsolicited Router Advertisement message, it includes all three options. In a solicited Router Advertisement message, at least the Prefix Information and MTU options will be included, so in either case, the node will obtain the preferred prefix for the link.

> *Step 1*: The node wanting to discover the subnet prefix sends a Router Solicitation message, using its own link-local address as the source and the "all routers in local link" multicast group (*ff02::02*) as the destination address.

> *Step 2*: All routers on the local link respond with Router Advertisement messages, with their own link-local address as source and the "all nodes on local link" multicast group (*ff02::1*) as the destination. The Router Advertisement message includes at least the subnet prefix option. This prefix is extracted from the prefix option and stored as the subnet prefix. All routers will respond with the same prefix, but the last Router Advertisement message received will have the subnet prefix that is used.

Duplicate Address Detection (DAD)

DAD is used to determine if a proposed (tentative) address is a duplicate of any address on the local link. Both hosts and routers perform DAD on *all* unicast and anycast addresses regardless of how they are obtained (Stateless Address Autoconfiguration, DHCPv6, or even manual assignment). DAD is accomplished using Neighbor Solicitation and Neighbor Advertisement messages.

> *Step 1*: The node owning the tentative address sends a number of Neighbor Solicitation messages using the unspecified address (::) as the *source address*, the solicited node multicast address as the *destination address*, and the TENTATIVE address as the *target address*.

Step 2: If any node on the link is already using the TENTATIVE address, it will respond by sending a Neighbor Advertisement to the "all nodes on local link" multicast group (*ff02::1*). If no such response is seen during a short interval (configurable), then the TENTATIVE address is considered to be unique.

Stateless Address Autoconfiguration (SLAAC)

This is one of the most important new aspects of IPv6. It is specified in RFC 4682,[12] "IPv6 Stateless Address Autoconfiguration," September 2007. It is primarily used to allow IPv6-capable hosts (as opposed to routers) to automatically obtain address information (link-local and global unicast node addresses and link-local default gateway). Routers use it to generate and validate their link-local addresses (but not their global addresses, which must be statically configured). The process makes strong use of link-local and multicast addresses, and all network communication is done with ICMPv6 messages that are part of ND. If a source of Router Advertisement messages (e.g., a router or firewall) is available, then at least one global unicast IPv6 address will also be generated. The acronym for Stateless Address Autoconfiguration is "SLAAC."

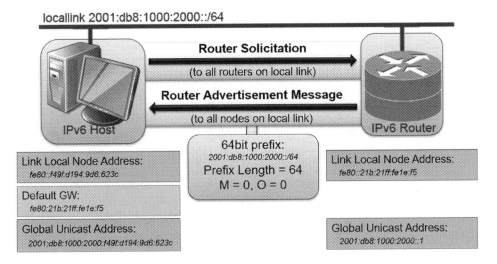

Figure 6-6. *SLAAC operation, M=0, O=0*

---

[12] https://tools.ietf.org/html/rfc4862

There are four steps involved in Stateless Address Autoconfiguration:

Step 1: The node creates a 64-bit interface identifier. This can be created using the MAC address and the EUI-64 algorithm or can be a randomly generated value ("randomized interface identifier").

Step 2: The host creates a TENTATIVE link-local address. This is done by appending the chosen interface identifier to the prefix *fe80://10*. DAD is performed to determine if the link-local address is unique. If so, that address goes to the PREFERRED state, its lifetime starts counting, and the process continues. If the address is duplicated, the address goes to the DUPLICATED state, the interface is disabled, and the SLAAC process fails without having generated any addresses.

Step 3: The host sends a Router Solicitation message to the "all routers on link" multicast group (*ff02::2*). If the node's link-local address has already been created, then that will be used as the source address; else, the unspecified address ("::") will be used as the source address. All routers on the link will respond with Router Advertisement messages, usually to the "all nodes on link" multicast group (*ff02::1*), but if the source address of the Router Solicitation message was a link-local address, the router can choose to send the Router Advertisement message via unicast to just that address. The source address of each received Router Advertisement message is added to a default gateway table (from which the preferred link-local default gateway will be chosen). The Prefix Information option in all of the responses should be the same, so the subnet prefix from the last received Router Advertisement message will be used.

If no router responds to the Router Solicitation message within a certain time, then the SLAAC process terminates, having created a valid link-local node address, but no link-local default gateway and no global unicast address.

Step 4: If we reach this step, a valid Router Advertisement was received with a subnet prefix, so the host combines the discovered subnet prefix with the created interface identifier, to create a TENTATIVE global unicast address for the node. DAD is performed on the tentative global unicast address, and if the address is unique, it goes to the PREFERRED state, and its lifetime starts counting. If not, the address goes to the DUPLICATED state, the interface is disabled, and the SLAAC process terminates, again having created a valid link-local address and a link-local default gateway address (but no global unicast address).

Anytime a link-local or global address lifetime expires (enters the INVALID state), address regeneration is done. If using randomized interface identifiers, a different random interface identifier is created for each address regeneration. If using EUI-64 interface identifiers, the regeneration process basically just confirms that the addresses are still valid – they don't actually change. If something has changed since the last validation (e.g., gateway down, link broken, etc.), the SLAAC process may fail, and the address is marked INVALID.

Next-Hop Determination

When one node needs to send a packet to another node, the sending node must determine whether the destination address is *on-link* or *off-link*. To be considered *on-link*, the address must match at least one of the following criteria:

- The prefix of the address must match one of the prefixes assigned to the link.

- The address is the target of a Redirect message sent by a router.

- The address is the target address of a Neighbor Advertisement message.

- The address is the source address of any Neighbor Discovery message received by the node.

If the address is *on-link,* then the next-hop address is the same as the destination address. If the address is *off-link,* then the next-hop address is selected from the default router list.

Neighbor Unreachability Detection (NUD)

Each entry in the Neighbor Cache contains the IP address, the link-layer (MAC) address, and the *reachability status* for that node. There are five possible values for that status, and the state transition rules are as follows:

> INCOMPLETE: Cache entry is newly created, and address resolution is in progress. Any transmitted packets are queued. When the address resolution completes, the link-layer address is added into the Neighbor Cache, and the state changes to REACHABLE.

> REACHABLE: Any queued packets are immediately sent. Any newly transmitted packets are sent normally. If more than a certain time passes without any traffic to or from the address, the state changes to STALE.

> STALE: The reachability of the node is *UNKNOWN*. The address remains in this state until traffic to that node is generated. At that point, the traffic is queued, and the state changes to DELAY.

> DELAY: The address remains in the DELAY state for a short period. The status is still *UNKNOWN*. Once the delay expires, the probe packet is sent, and the state changes to PROBE.

> PROBE: A probe packet has been sent to determine reachability (after the delay), but the result has not yet been obtained. The status is still *UNKNOWN*. When the result is seen, REACHABILITY is confirmed, and the state changes to REACHABLE. If a certain amount of time elapses without any response, then the node is considered unreachable, any queued traffic is discarded, and an error is generated to the sender.

Note that there is nothing comparable to Neighbor Unreachability Detection in IPv4. IPv6 NUD improves packet delivery in the presence of failing routers and over partially failing or partitioned links. It improves delivery to nodes that change their link-layer (MAC) addresses. For example, mobile nodes can move off-link without losing any connectivity due to stale ARP caches. NUD detects dead routers and dead switches that block access to working routers.

Redirect

A router can send a Redirect message to a packet sender, if there is a better first-hop router or if the destination is an on-link neighbor. In the first case, the Target Address field contains the link-local address of the better first-hop router. In the second case, the Target Address field contains a copy of the destination address. The Destination Address field contains the address of the ultimate packet destination. The router uses its knowledge of the larger environment to generate this information. You might think of a Redirect message as saying something like "There is a bridge out down *that* road – try going down *this* road, instead."

IPv6 redirects contain the link-layer (MAC) address of the new first hop, which eliminates the need for an additional packet exchange to resolve the IP address. Unlike with IPv4 redirects, the recipient of an IPv6 redirect assumes that the new next hop is on-link. The IPv6 redirect is useful on non-broadcast and shared media links. On such links, nodes should not check for all prefixes for on-link destinations.

Viewing the Neighbor Cache

To view the Neighbor Cache in Windows 7 or later:

1. Start a command prompt (cmd) and enter the following commands in it.

2. Enter the command netsh –c "interface ipv6".

3. At the netsh prompt, enter the command show interface.

4. In the resulting list, find the interface index for "Local Area Connection" (say it is 11).

5. At the netsh prompt, enter the command show neighbors 11 (or whatever interface index).

6. You should see global unicast addresses, link-local addresses, and a lot of multicast addresses:

```
C:\>netsh -c "interface IPv6"
netsh interface IPv6>show interface
```

```
Idx       Met        MTU        State           Name
---    ----------  ----------  ------------   ----------------------------
  1          50    4294967295  connected      Loopback Pseudo-Interface 1
 12          50          1280  disconnected   isatap.infoweapons.com
 13          50          1280  connected      Local Area Connection* 11
 11          10          1500  connected      Local Area Connection

netsh interface IPv6>show neighbors 11

Interface 11: Local Area Connection

Internet Address                              Physical Address   Type
-------------------------------------------   -----------------  ----------
2001:df8:5403:2410::fff2                      00-15-17-30-b8-ec  Reachable
                                                                 (Router)
2001:df8:5403:2410::10:11                     00-e0-81-48-62-7a  Stale
fe80::215:17ff:fe30:b8ec                      00-15-17-30-b8-ec  Reachable
                                                                 (Router)
fe80::230:48ff:fe61:d6be                      00-30-48-61-d6-be  Stale
ff02::2                                       33-33-00-00-00-02  Permanent
ff02::c                                       33-33-00-00-00-0c  Permanent
ff02::16                                      33-33-00-00-00-16  Permanent
ff02::1:2                                     33-33-00-01-00-02  Permanent
ff02::1:3                                     33-33-00-01-00-03  Permanent
ff02::1:ff00:69                               33-33-ff-00-00-69  Permanent
ff02::1:ff00:fff2                             33-33-ff-00-ff-f2  Permanent
ff02::1:ff03:186                              33-33-ff-03-01-86  Permanent
ff02::1:ff10:11                               33-33-ff-10-00-11  Permanent
ff02::1:ff10:14                               33-33-ff-10-00-14  Permanent
ff02::1:ff10:26                               33-33-ff-10-00-26  Permanent
ff02::1:ff13:f5                               33-33-ff-13-00-f5  Permanent
ff02::1:ff2b:6589                             33-33-ff-2b-65-89  Permanent
ff02::1:ff30:b8ec                             33-33-ff-30-b8-ec  Permanent
ff02::1:ff3f:58e5                             33-33-ff-3f-58-e5  Permanent
ff02::1:ff61:d6be                             33-33-ff-61-d6-be  Permanent
ff02::1:ff62:62                               33-33-ff-62-00-62  Permanent
ff02::1:ffc6:ed59                             33-33-ff-c6-ed-59  Permanent
ff02::1:ffce:5d59                             33-33-ff-ce-5d-59  Permanent
ff05::1:3                                     33-33-00-01-00-03  Permanent
```

SEcure Network Discovery (SEND)

Note that there are some potentially exploitable vulnerabilities in ND. ARP in IPv4 has several well-known and easily exploited vulnerabilities, used in many hacking attacks. For details of these, search for "ARP Vulnerabilities Black Hat." You should find an excellent PowerPoint presentation that was presented by Mike Beekey at a Black Hat Briefing security conference. It shows exactly how ARP is vulnerable and how this is exploited by hackers. ARP does not exist in IPv6, so its vulnerabilities do not affect IPv6 networks. However, ND (which replaces ARP) has some new vulnerabilities that do not affect IPv4 networks.

A secure version of ND is defined in RFC 3971,[13] "SEcure Neighbor Discovery (SEND)," March 2005. This is still a *Proposed Standard*. SEND uses cryptographically generated addresses, which are defined in RFC 2972,[14] "Cryptographically Generated Addresses (CGA)," March 2005 (this is also a *Proposed Standard* and has already been updated by RFCs 4581 and 4982). SEND does not depend on IPsec. It is still very much in experimental status even in 2019.

Note that SEND only *digitally signs* ND packets; it does not *encrypt* them.

Types of IPv6 Packet Transmission

Unicast, anycast, multicast, and broadcast have already been covered in section 5.3.2.2, because in IPv6, this is considered to be part of the addressing model.

IPv6 Broadcast

Most things that you would use broadcast for in IPv4, you would use some form of multicast, with a more restricted scope, in IPv6. A multicast transmission to the address ff01::2 would go to the same nodes (all nodes on local link) as an IPv4 broadcast. However, there are other scopes, such as site, organization, and global for multicast, that (unlike IPv4 broadcast) will cross routers, but other than "all nodes in local link," multicast to the wider scopes requires that all recipients intentionally add the necessary multicast address to their node.

---

[13] https://tools.ietf.org/html/rfc3971
[14] https://tools.ietf.org/html/rfc2972

IPv6 Multicast

The basic multicast address type has been covered, but there is a lot more to a full multicast system, as you saw in the section "IPv4 Multicast." For an in-depth discussion of all aspects of IPv6 multicast, I recommend Chapter 6, "Providing IPv6 Multicast Services," from the book *Deploying IPv6 Networks*,[15] by Ciprian Popoviciu, Eric Levy-Abegnoli, and Patrick Grossetete, Cisco Press, 2006.

Multicast exists in IPv4, but there are some serious problems with it, which are resolved in IPv6.

Not all IPv4 routers support multicast. In general, it is difficult to deploy except in a "walled garden," such as the customers of a single ISP like Comcast. In IPv6, support for multicast is mandatory – all compliant routers support it, and it works across ISPs, even worldwide.

The *Internet Group Management Protocol* (IGMP) is not part of IPv4, and not all IPv4 routers include it. In IPv6, the *Multicast Listener Discovery* (MLD) protocol is standardized and is actually just a subset of the ICMPv6 messages. Because of this, all IPv6-compliant routers include it.

Multicast in IPv4 was an afterthought, grafted on long after the original protocol was designed. In IPv6, multicast was incorporated from the beginning and is present in all address scopes. Multicast link-local addresses are used extensively in SLAAC and other places.

For IPTV applications, IPv6 networks will be the first time that really global Internet TV services can be deployed and work reliably. This is as exciting as when Ted Turner first relayed the signal from his small UHF TV station via a satellite. That breakthrough resulted in WTBS, CNN, CNN Headline News, TNT, Cartoon Network, and, indirectly, the entire multibillion-dollar satellite/cable television network industry.

There are many other areas in which working, scalable multicast can be used to improve applications. You could build chat, VoIP, or even video conferencing clients that could build fully meshed networks, with each new participant subscribing to all existing clients' multicast "channels" and all existing clients subscribing to the new participant's multicast "channel." Even if the initial participant left, all remaining participants would

---

[15] www.amazon.com/Deploying-IPv6-Networks-Author-Popoviciu/dp/B010BALRUK/ref=sr_1_1?
keywords=deploying+ipv6+networks&qid=1554605574&s=gateway&sr=8-1

still have a fully functional mesh network. This also eliminates the need for any central exchange point (other than perhaps a search or directory facility to help in setting up the conference and allowing participants to locate each other).

The following standards are relevant to multicast in IPv6:

RFC 2375, "IPv6 Multicast Address Assignments," July 1998 (Informational)

RFC 2710, "Multicast Listener Discovery (MLD) for IPv6," October 1999 (Standards Track)

RFC 3306, "Unicast-Prefix-based IPv6 Multicast Addresses," August 2002 (Standards Track)

RFC 3307, "Allocation Guidelines for IPv6 Multicast Addresses," August 2002 (Standards Track)

RFC 3590, "Source Address Selection for the Multicast Listener Discover (MLD) Protocol," September 2003 (Standards Track)

RFC 3810, "Multicast Listener Discovery Version 2 (MLDv2) for IPv6," June 2004 (Standards Track)

RFC 3956, "Embedding the Rendezvous Point (RP) Address in an IPv6 Multicast Address," November 2004 (Standards Track)

RFC 4489, "A Method for Generating Link-Scoped IPv6 Multicast Addresses," April 2006 (Standards Track)

RFC 4607, "Source-Specific Multicast for IP," August 2006 (Standards Track)

Multicast Listener Discovery (MLD) Protocol

MLD is used by IPv6 routers to discover the presence of multicast listeners (nodes that wish to receive multicast packets) and the specific multicast addresses to which they want to subscribe. MLD (defined in RFC 2710) is commonly referred to as MLDv1. It is the IPv6 equivalent to IPv4's IGMPv2 (defined in RFC 2236). MLDv1 and IGMPv2 multicast protocols are used to set up *any-source multicast* (ASM), which allows multiple sources in a group (*,G) or "channel." This is also known as *traditional multicast.*

MLDv2 extends the definition of MLDv1 by adding support for "source filtering." It includes all the functionality of MLDv1, so there is no need to deploy both on a given node. This allows a node to indicate interest *only* in packets from specific source addresses (INCLUDE mode) or in packets from all multicast addresses *except* for specific source addresses (EXCLUDE mode). MLDv2 is the IPv6 equivalent of IPv4's IGMPv3. MLDv2 and IGMPv3 multicast protocols are used to set up *source-specific multicast* (SSM), which allows a specific source (S) in a group (G) to deliver packets to all members that join (S,G) known as a "channel." This is described in RFC 4604,[16] "Using Internet Group Management Protocol Version 3 (IGMPv3) and Multicast Listener Discovery Protocol Version 2 (MLDv2) for Source-Specific Multicast," and in RFC 4607,[17] "Source-Specific Multicast (SSM) for IP."

There is another RFC that defines MLD proxying: RFC 4605,[18] "Internet Group Management Protocol (IGMP)/Multicast Listener Discovery (MLD)-Based Multicast Forwarding ("IGMP/MLD Proxying")." A *proxy* would exist on a forwarding gateway that links together multiple subnets and relay messages across that gateway between an MLD Querier on *one* subnet and MLD listeners on *a different* subnet.

MLDv1 and MLDv2 are *sub-protocol*s of ICMPv6. All MLDv2 messages are just additional ICMPv6 messages. All IPv6-compliant devices should include support for MLD. MLD messages must be sent with a link-local IPv6 source address, a Hop Limit of 1, and an IPv6 Router Alert Option in the Hop-by-Hop Options extension packet header. When used in Neighbor Discovery protocol's *Stateless Address Autoconfiguration*, the source address can be the unspecified address (::). IGMP is *not* a sub-protocol of ICMPv4. It does not use ICMPv4 messages, but an entirely new protocol. IGMP is not mandatory on all IPv4 routers.

MLD can co-exist with IGMPv3 in a dual-stack network, as MLD (v1 or v2) will only involve IPv6 messages and IGMP (v1, v2, or v3) will only involve IPv4 messages. However, in general, multicast will work far better on IPv6 than on IPv4.

With MLD, there is a "router role" (performed by at most one router in a subnet) and a "listener role" (performed by any number of listener nodes in that subnet) in the protocol.

For the router role, only one router on a subnet can be *the Querier* at any given time. If there is more than one router on a subnet, there is an election mechanism that selects one of them to be the Querier. Should that router fail at some point, all other routers on that subnet have been listening in and maintaining state, so another election will select one of the surviving routers on that subnet to become the Querier. Only the Querier sends periodic or triggered Query messages on its subnet.

---

[16] https://tools.ietf.org/html/rfc4604

[17] https://tools.ietf.org/html/rfc4607

[18] https://tools.ietf.org/html/rfc4605

There are three types of MLDv2 Query messages sent by the Querier to the "all nodes on local link" multicast address (*ff02::1*). They should be sent with a valid IPv6 link-local source address. Any Query message received with the source address being the unspecified address (::), or any other address that is not a valid IPv6 link-local address, should be silently discarded.

- General queries

- Multicast address–specific queries

- Multicast address– and source-specific queries

There are two types of reports sent by listeners to the Querier, to a special multicast address (*ff02::16*) to which all MLDv2-compliant multicast routers listen. If a single Report message is not large enough to hold all of the state information, multiple Report messages can be sent.

- Current State Report (sent in response to a query)

- State Change Report (sent unsolicited in response to some change on the listener)

General queries are sent from the Querier to all listeners on the subnet periodically to learn multicast address listener information, to build and refresh state inside all multicast routers on the subnet. Even though only the Querier sends out periodic queries, all routers listen to the responses and update their state.

When a listener node gets a General Query message, it responds by sending a *Current State Report*, with its per-interface state information. It is also possible for a listener node to immediately report a state change (such as someone "unsubscribing" to a multicast channel) through an unsolicited *State Change Report*. Current State Reports are sent only once (if one is lost, it will probably be received in response to the next periodic query). State Change Reports are sent multiple times for robustness (to increase the probability of all routers getting the message).

When the Querier gets a State Change Report from a listener, it sends a *multicast address–specific query* to see if there are still any *other* listeners to that multicast address. If not, the Querier will delete that multicast address from its multicast address listener state table, which stops relaying the corresponding traffic. If there are source-specific listeners, the Querier will send a *multicast address– and source-specific query* instead.

There must be a *service interface* (API routines) available, which allows an application to cause a *State Change Report* to be sent to the Querier. A sample API is documented in RFC 3678,[19] "Socket Interface Extensions for Multicast Source Filters," January 2004. The full API includes the ability to JOIN or LEAVE a multicast group ("subscribe to a multicast channel") and to BLOCK and UNBLOCK specific source addresses, as well as to set and retrieve source filter sets.

For details on the syntax of the various MLDv2 messages, see RFC 3810.[20]

Protocol Independent Multicast (PIM) for IPv6

PIM is a multicast protocol, which deals with router-to-router communications. IPv6 PIM is similar to IPv4 PIM, has the same variants (Dense Mode, Sparse Mode, and Bidirectional Mode), and is defined in the same RFCs (in the sections relevant to IPv6). The IPv6 implementation uses the Neighbor Discovery protocol, Multicast Listener Discovery protocol, Path MTU Discovery, and IPv6 multicast, rather than the corresponding IPv4 mechanisms. As with TCP, the PIM message checksum factors in the source and destination IP addresses, so the pseudo header used in the calculation of the checksum (which includes IPv6 addresses) is different from the one used in IPv4. The following items are IP version specific in all variants:

Item	IPv4	IPv6
Source-specific multicast	232.0.0.0/8	ff3x:/32
Wildcard Group set	224.0.0.0/3	ff00::/8
ALL-PIM-ROUTERS group	224.0.0.13	ff02::d

PIM for IPv6 does not include routing, but provides multicast forwarding by using static IPv6 routes or routing tables created by IPv6 unicast routing protocols, such as RIPng, OSPFv3, IS-ISv6, or BGP4+.

PIM Dense Mode is defined in RFC 3973,[21] "Protocol Independent Multicast – Dense Mode (PIM-DM)," January 2005 (for both IPv4 and IPv6). This uses dense multicast routing, which builds shortest-path trees by flooding multicast traffic domain-wide and then pruning branches where no receivers are present. It does not scale well.

---

[19] https://tools.ietf.org/html/rfc3678
[20] https://tools.ietf.org/html/rfc3810
[21] https://tools.ietf.org/html/rfc3973

PIM Sparse Mode is defined in RFC 4601,[22] "Protocol Independent Multicast – Sparse Mode (PIM-SM): Protocol Specification (Revised)," August 2006 (for both IPv4 and IPv6). As in IPv4, PIM-SM builds unidirectional shared trees routed at a rendezvous point per group and can create shortest-path trees per source. It scales fairly well for wide-area use.

Bidirectional PIM is defined in RFC 5015,[23] "Bidirectional Protocol Independent Multicast (BIDIR-PIM)," October 2007 (for both IPv4 and IPv6). It builds shared bidirectional trees. It never builds a shortest-path tree, so there may be longer end-to-end delays, but it scales very well.

There is one new standard specific to IPv6 PIM, RFC 3956,[24] "Embedding the rendezvous point (RP) Address in an IPv6 Multicast Address," November 2004. This defines an address allocation policy in which the address of the Rendezvous Point (RP) is encoded in an IPv6 multicast group address. For PIM-SM, this can be seen as a specification of a group-to-RP mapping mechanism. This supports easy deployment of scalable inter-domain multicast and simplifies configuration as well.

> *Example 1*: An ISP manages 2001:db8::/32 and wants an RP for the network and all its customers, on an existing subnet, for example, 2001:db8:beef:feed::/64. The group address would be something like ff7x:y40:2001:db8:beef:feed::/96, and the RP address would be 2001:db8:beef:feed::y (y can be any value from 1 to F, but not 0).

> *Example 2*: An organization wants to have its own PIM-SM domain. It should pick multicast addresses such as ff7x:y30:2001:db8:beef::/80. The RP address would be 2001:db8:beef::y (y can be any value from 1 to F, but not 0).

---

[22] https://tools.ietf.org/html/rfc4601
[23] https://tools.ietf.org/html/rfc5015
[24] https://tools.ietf.org/html/rfc3956

ICMPv6: Internet Control Message Protocol for IPv6

ICMPv6 is a key protocol in the *Internet Layer* that complements version 6 of the Internet Protocol (IPv6). It was originally defined in RFC 1885 (December 1995) and then enhanced in RFC 2463 (December 1998). It is currently defined in RFC 4443,[25] "Internet Control Message Protocol (ICMPv6) for the Internet Protocol Version 6 (IPv6) Specification," March 2006.

There are many more ICMPv6 messages defined than there are ICMPv4 messages (in fact, Neighbor Discovery and Multicast Listener Discovery protocols are just subsets of the ICMPv6 messages). ICMPv6 messages have a much greater range of functionality than ICMPv4 messages. Even if you block all ICMPv4 messages (common practice by some IPv4 network administrators), normal network operation will usually occur. *This is not true with ICMPv6.* ICMPv6 messages are used in normal operation of IPv6.

There are two classes of ICMPv6 messages:

- Error messages, with message type ranging from 0 to 127

- Informational messages, with message type ranging from 128 to 255

ICMPv6 Error Messages

> 1 Destination Unreachable (ICMPv6, RFC 4443)

> 2 Packet Too Big (ICMPv6, RFC 4443)

> 3 Time Exceeded (ICMPv6, RFC 4443)

> 4 Parameter Problem (ICMPv6, RFC 4443)

ICMPv6 Informational Messages

> 128 Echo Request (ICMPv6, RFC 4443)

> 129 Echo Reply (ICMPv6, RFC 4443)

> 130 Multicast Listener Query message (MLDv2, RFC 3810)

> 131 Multicast Listener Report (MLDv1, RFC 2710)

> 132 Multicast Listener Done (MLDv1, RFC 2710)

> 133 Router Solicitation message (ND, RFC 2461)

---

[25] https://tools.ietf.org/html/rfc4443

134 Router Advertisement message (ND, RFC 2461)

135 Neighbor Solicitation message (ND, RFC 2461)

136 Neighbor Advertisement message (ND, RFC 2461)

137 Redirect message (ND, RFC 2461)

138 Router Renumbering (RR, RFC 2894)

139 ICMP Node Information Query (NIQ, RFC 4620)

140 ICMP Node Information Response (NIQ, RFC 4620)

141 Inverse Neighbor Discovery Solicitation message (IND, RFC 3122)

142 Inverse Neighbor Discovery Advertisement message (IND, RFC 3122)

143 Multicast Listener Report message (MLDv2, RFC 3810)

144 Home Agent Address Discovery Request message (MIPv6, RFC 3775)

145 Home Agent Address Discovery Reply message (MIPv6, RFC 3775)

146 Mobile Prefix Solicitation (MIPv6, RFC 3775)

147 Mobile Prefix Advertisement (MIPv6, RFC 3775)

148 Certification Path Solicitation (SEND, RFC 3971)

149 Certification Path Advertisement (SEND, RFC 3971)

151 Multicast Router Advertisement (MRD, RFC 4286)

152 Multicast Router Solicitation (MRD, RFC 4286)

153 Multicast Router Termination (MRD, RFC 4286)

154 FMIPv6 messages (MIPv6, RFC 5568)

IND Inverse Neighbor Discovery

MIPv6 Mobile IPv6

MLDv1 Multicast Listener Discovery, version 1

MLDv2 Multicast Listener Discovery, version 2

MRD Multicast Router Discovery

ND Neighbor Discovery

NIQ Node Information Query

RR Router Renumbering

SEND SEcure Neighbor Discovery

Note that there is no equivalent ICMPv6 message corresponding to the following ICMPv4 messages (or else its function is now contained in another message).

4 Source Quench

5 Redirect

13 Timestamp

14 Timestamp Reply

15 Information Request

16 Information Reply

Destination Unreachable Error

```
 0                   1                   2                   3
 0 1 2 3 4 5 6 7 8 9 0 1 2 3 4 5 6 7 8 9 0 1 2 3 4 5 6 7 8 9 0 1
+-+-+-+-+-+-+-+-+-+-+-+-+-+-+-+-+-+-+-+-+-+-+-+-+-+-+-+-+-+-+-+-+
|     Type      |     Code      |          Checksum             |
+-+-+-+-+-+-+-+-+-+-+-+-+-+-+-+-+-+-+-+-+-+-+-+-+-+-+-+-+-+-+-+-+
|                            Unused                             |
+-+-+-+-+-+-+-+-+-+-+-+-+-+-+-+-+-+-+-+-+-+-+-+-+-+-+-+-+-+-+-+-+
|                 As much of invoking packet                   |
+                 as possible without the ICMPv6 packet        +
|                 exceeding the minimum IPv6 MTU               |
```

IPv6 Fields:

Destination Address

Copied from the Source Address field of the invoking packet.

ICMPv6 Fields:

Type   1

Code   0 - No route to destination
                1 - Communication with destination
                    administratively prohibited
                2 - Beyond scope of source address
                3 - Address unreachable
                4 - Port unreachable
                5 - Source address failed ingress/egress policy
                6 - Reject route to destination

Unused   This field is unused for all code values.
                It must be initialized to zero by the originator
                and ignored by the receiver.

Description

A Destination Unreachable message SHOULD be generated by a router, or
by the IPv6 layer in the originating node, in response to a packet
that cannot be delivered to its destination address for reasons other
than congestion. (An ICMPv6 message MUST NOT be generated if a
packet is dropped due to congestion.)

Packet Too Big Message

```
     0                   1                   2                   3
     0 1 2 3 4 5 6 7 8 9 0 1 2 3 4 5 6 7 8 9 0 1 2 3 4 5 6 7 8 9 0 1
    +-+-+-+-+-+-+-+-+-+-+-+-+-+-+-+-+-+-+-+-+-+-+-+-+-+-+-+-+-+-+-+-+
    |     Type      |     Code      |           Checksum            |
    +-+-+-+-+-+-+-+-+-+-+-+-+-+-+-+-+-+-+-+-+-+-+-+-+-+-+-+-+-+-+-+-+
    |                             MTU                               |
    +-+-+-+-+-+-+-+-+-+-+-+-+-+-+-+-+-+-+-+-+-+-+-+-+-+-+-+-+-+-+-+-+
    |                 As much of invoking packet                   |
    +                 as possible without the ICMPv6 packet        +
    |                 exceeding the minimum IPv6 MTU               |
```

IPv6 Fields:

Destination Address

> Copied from the Source Address field of the invoking
> packet.

ICMPv6 Fields:

Type   2

Code   Set to 0 (zero) by the originator and ignored by the
              receiver.

MTU   The Maximum Transmission Unit of the next-hop link.

Description

A Packet Too Big MUST be sent by a router in response to a packet
that it cannot forward because the packet is larger than the MTU of
the outgoing link. The information in this message is used as part
of the Path MTU Discovery process.

Time Exceeded Message

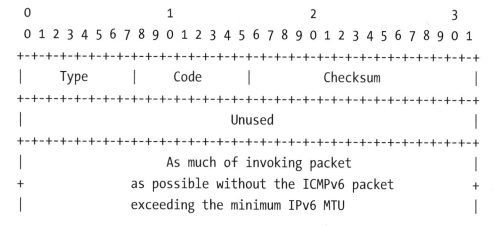

IPv6 Fields:

Destination Address

> Copied from the Source Address field of the invoking
> packet.

ICMPv6 Fields:

Type   3

Code   0 - Hop limit exceeded in transit
                1 - Fragment reassembly time exceeded

Unused   This field is unused for all code values.
                It must be initialized to zero by the originator
                and ignored by the receiver.

Description

If a router receives a packet with a Hop Limit of zero, or if a
router decrements a packet's Hop Limit to zero, it MUST discard the
packet and originate an ICMPv6 Time Exceeded message with Code 0 to
the source of the packet. This indicates either a routing loop or
too small an initial Hop Limit value.

Parameter Problem Message

```
 0                   1                   2                   3
 0 1 2 3 4 5 6 7 8 9 0 1 2 3 4 5 6 7 8 9 0 1 2 3 4 5 6 7 8 9 0 1
+-+-+-+-+-+-+-+-+-+-+-+-+-+-+-+-+-+-+-+-+-+-+-+-+-+-+-+-+-+-+-+-+
|     Type      |     Code      |          Checksum             |
+-+-+-+-+-+-+-+-+-+-+-+-+-+-+-+-+-+-+-+-+-+-+-+-+-+-+-+-+-+-+-+-+
|                           Pointer                             |
+-+-+-+-+-+-+-+-+-+-+-+-+-+-+-+-+-+-+-+-+-+-+-+-+-+-+-+-+-+-+-+-+
|                    As much of invoking packet                 |
+                  as possible without the ICMPv6 packet        +
|                    exceeding the minimum IPv6 MTU             |
```

IPv6 Fields:

Destination Address

                Copied from the Source Address field of the invoking
                packet.

ICMPv6 Fields:

Type   4

Code   0 - Erroneous header field encountered
                1 - Unrecognized Next Header type encountered
                2 - Unrecognized IPv6 option encountered

Pointer   Identifies the octet offset within the
                invoking packet where the error was detected.

                The pointer will point beyond the end of the ICMPv6
                packet if the field in error is beyond what can fit
                in the maximum size of an ICMPv6 error message.

Description

If an IPv6 node processing a packet finds a problem with a field in
the IPv6 header or extension headers such that it cannot complete
processing the packet, it MUST discard the packet and SHOULD
originate an ICMPv6 Parameter Problem message to the packet's source,
indicating the type and location of the problem.

Echo Request Message

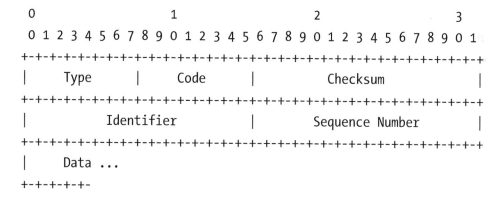

IPv6 Fields:

Destination Address

                Any legal IPv6 address.

ICMPv6 Fields:

Type   128

Code   0

Identifier   An identifier to aid in matching Echo Replies
                to this Echo Request. May be zero.

Sequence Number

                A sequence number to aid in matching Echo Replies
                to this Echo Request. May be zero.

Data   Zero or more octets of arbitrary data.

Description

Every node MUST implement an ICMPv6 Echo responder function that
receives Echo Requests and originates corresponding Echo Replies.
A node SHOULD also implement an application-layer interface for
originating Echo Requests and receiving Echo Replies, for diagnostic
purposes.

Echo Reply Message

```
 0                   1                   2                   3
 0 1 2 3 4 5 6 7 8 9 0 1 2 3 4 5 6 7 8 9 0 1 2 3 4 5 6 7 8 9 0 1
+-+-+-+-+-+-+-+-+-+-+-+-+-+-+-+-+-+-+-+-+-+-+-+-+-+-+-+-+-+-+-+-+
|     Type      |     Code      |          Checksum             |
+-+-+-+-+-+-+-+-+-+-+-+-+-+-+-+-+-+-+-+-+-+-+-+-+-+-+-+-+-+-+-+-+
|           Identifier          |        Sequence Number        |
+-+-+-+-+-+-+-+-+-+-+-+-+-+-+-+-+-+-+-+-+-+-+-+-+-+-+-+-+-+-+-+-+
|     Data ...
+-+-+-+-+-
```

IPv6 Fields:

Destination Address

Copied from the Source Address field of the invoking
Echo Request packet.

ICMPv6 Fields:

Type   129

Code   0

Identifier   The identifier from the invoking Echo Request message.

Sequence Number

                The sequence number from the invoking Echo Request
                message.

Data   The data from the invoking Echo Request message.

Description

Every node MUST implement an ICMPv6 Echo responder function that
receives Echo Requests and originates corresponding Echo Replies.
A node SHOULD also implement an application-layer interface for
originating Echo Requests and receiving Echo Replies, for diagnostic
purposes.

The source address of an Echo Reply sent in response to a unicast
Echo Request message MUST be the same as the destination address of
that Echo Request message.

An Echo Reply SHOULD be sent in response to an Echo Request message
sent to an IPv6 multicast or anycast address. In this case, the
source address of the reply MUST be a unicast address belonging to
the interface on which the Echo Request message was received.

The data received in the ICMPv6 Echo Request message MUST be returned
entirely and unmodified in the ICMPv6 Echo Reply message.

IPv6 Routing

IPv6 has to solve the same problems as IPv4 in terms of how to get packets from one point to another through a packet-switched network. However, the differences in IP address length and addressing model mean that the existing routing protocols for IPv4 do not work. All the popular routing protocols have been extended to support IPv6. These include RIPng, EIGRP, IS-ISv6, OSPF for IPv6, and BGP4 with Multiprotocol Extensions (BGP4+).

The following standards are relevant to routing in IPv6:

RFC 2080, "RIPng for IPv6," January 1997 (Standards Track)

RFC 2185, "Routing Aspects of IPv6 Transition," September 1997 (Informational)

RFC 2545, "Use of BGP-4 Multiprotocol Extensions for IPv6 Inter-Domain Routing," March 1999 (Standards Track)

RFC 5308, "Routing IPv6 with IS-IS," October 2008 (Standards Track)

RFC 5340, "OSPF for IPv6," July 2008 (Standards Track)

RIPng: RIP Next Generation. Defined in RFC 2080, "RIPng for IPv6," January 1997. This IETF standard specifies extensions to the RIP (as defined in RFCs 1058 and 1723), to support IPv6. Like RIP for IPv4, RIPng also uses the distance vector algorithm. Unlike RIP for IPv4, RIPng is implemented only in routers. IPv6 itself provides mechanisms for router discovery (part of ND). RIPng is a UDP-based protocol, using port 521 (compare with port 520 for RIP). It supports 128-bit IPv6 addresses instead of 32-bit IPv4 addresses. It has the same limitations as RIP, such as being useful only in small networks, with less than 15 hops. It does have some of the extensions of RIPv2. When a response is sent to all neighbors, the multicast group *ff02::9* (all-rip-routers) is used. RIPng only routes IPv6. On a dual-stack network, you would need both RIP (for IPv4) and RIPng. Since RIPng runs over IPv6, it can use the IPsec Authentication header (AH) and Encapsulating Security Payload (ESP) mechanisms to ensure integrity and authentication/confidentiality of routing exchanges.

EIGRP: Enhanced Interior Gateway Routing Protocol (proprietary Cisco routing protocol). This already includes extensions to allow it to route IPv4 and/or IPv6 packets. For details, see Cisco documentation.

IS-ISv6: Extension of IS-IS to support IPv6. Based on two levels, L2 = Backbone, L1 = Stub, L2L1 = Interconnected L2 and L1. It runs over CLNS (Connectionless Network Service, an OSI Network Layer protocol, similar to IP). Each IS node still sends out Link State Packets and sends information via Tag/Length/Values. There are two new TLVs, IPv6 Reachability and IPv6 Interface Address, and a new Network Layer Identifier, IPv6 NLPID (Network Layer Protocol IDentifier). Other than that, IS-ISv6 is pretty much the same as the original IS-IS. It is still suitable mainly for large ISPs.

OSPF for IPv6: Open Shortest Path First for IPv6 (also known as *OSPFv3*). Defined in RFC 5340, "OSPF for IPv6," July 2008. This is still an Interior Gateway Routing Protocol and is suitable for use within organizations, but not between autonomous systems (BGP4+ is needed for this).

The basic OSPF for IPv4 mechanisms (flooding, designated router election, area support, Short Path First calculations) are unchanged. Some changes are required because of new protocol semantics or larger address size. Most fields and packet-size limitations in OSPF for IPv4 have been relaxed, and option handling is more flexible. The protocol processing is now per link, instead of per subnet. There is now a *flooding scope* to reflect the scopes of IPv6 addresses. It uses IPv6 link-local addresses. The addressing semantics have been removed (with a few exceptions), leaving a mostly network protocol–independent core. OSPF Router IDs, Area IDs, and Link State IDs are still 32 bits, so those can no longer be IP addresses (which in IPv6 are 128 bits).

The new *flooding scope* allows control over how widely to flood information: link local, area wide, or AS wide (the entire routing domain). It is now possible to run multiple instances of the OSPF protocol on a single link (every message now includes an *Instance ID* value). Link-local addresses are used where they are meaningful (for transactions completely within a link), but global-scope IPv6 addresses must still be used in some places (e.g., source address for OSPF protocol packets). The AuType and Authentication fields have been removed from OSPF for IPv4, as IPsec AH and ESP are available and superior. As with TCP, the header checksum covers the entire OSPF packet and a prepended IPv6 pseudo header. All support for MOSPF (Multicast OSPF) has been removed.

OSPF for IPv6 runs only over IPv6 and only routes IPv6. On a dual-stack network, you would need both OSPF for IPv4 (OSPFv2) and OSPF for IPv6 (OSPFv3) deployed, similar to RIP and RIPng. It is possible that a future version of OSPF will support both IPv4 and IPv6 routing.

BGP4 with Multiprotocol Extensions (also known informally as BGP4+): Defined in RFC 4760, "Multiprotocol Extensions for BGP-4," January 2007. BGP4 is currently defined in RFC 4271, "A Border Gateway Protocol 4 (BGP-4)," January 2006. BGP4 supports only IPv4. The multiprotocol extensions have been around since RFC 2283, February 1998, but have been updated with each new version of BGP4.

These extensions allow BGP4+ to carry routing information for multiple Network Layer protocols,(e.g., IPv6, IPX, L3VPN, etc.). L3VPN is a "layer 3 Virtual Private Network." BGP4+ is designed to be backward compatible, such that a BGP4+-compliant router can exchange IPv4 routing information with a router that does not support the multiprotocol extensions (basic BGP4).

Currently BGP4+ is the primary protocol used for routing IPv6 packets between autonomous systems (very large networks under the control of a single entity, such as ISPs or major corporations). Most IPv6 engineers will never work with it, unless they work for an ISP or a really large company.

One of the issues that ISPs face when supporting IPv6 is to migrate their BGP4 gateways to BGP4+ gateways. They typically must also upgrade many routers to dual stack. At the ISP level, many routers have hardware acceleration, so this can be expensive. These may involve "forklift" upgrades, where entirely new high-end routers must be purchased, and there may be relatively little resale value for legacy IPv4-only equipment (hint to ISPs: migrate to IPv6 now and sell your old gear while it still has SOME value!).

Looking at Local Routing Information

In Windows, you can view all currently known routes with the "route print" command. If you have enabled IPv6 and are connected to an IPv6 network, you might see something like the following (the "-6" tells it to print only IPv6 route information):

```
C:\Users\lhughes>route print -6
===========================================================================
Interface List
 11...00 18 8b 78 da 1a ......Broadcom NetXtreme 57xx Gigabit Controller
  1.........................Software Loopback Interface 1
 12...00 00 00 00 00 00 00 e0 Microsoft ISATAP Adapter
 13...00 00 00 00 00 00 00 e0 Teredo Tunneling Pseudo-Interface
===========================================================================
```

IPv6 Route Table

```
===============================================================================
Active Routes:
 If Metric Network Destination        Gateway
 11     26 ::/0                        fe80::21b:21ff:fe1e:f4
  1    306 ::1/128                     On-link
 13     58 2001::/32                   On-link
 13    306 2001:0:cf2e:3096:30c3:380d:f5fd:fa12/128
                                       On-link
 11     18 2001:df8:5403:2410::/64     On-link
 11    266 2001:df8:5403:2410:3c79:b2ca:90ce:5d59/128
                                       On-link
 11    266 2001:df8:5403:2410:a446:d5ef:d313:f5/128
                                       On-link
 11    266 fe80::/64                   On-link
 13    306 fe80::/64                   On-link
 13    306 fe80::30c3:380d:f5fd:fa12/128
                                       On-link
 11    266 fe80::3c79:b2ca:90ce:5d59/128
                                       On-link
  1    306 ff00::/8                    On-link
 13    306 ff00::/8                    On-link
 11    266 ff00::/8                    On-link
===============================================================================
Persistent Routes:
  None
```

Network Address Translation

NAT (Network Address Translation) was introduced to extend the lifetime of the IPv4 address space long enough for its replacement, IPv6, to be defined and refined and compliant infrastructure products and applications to be developed. IPv6 is now fully developed and ready for prime time. NAT has served its purpose. It is time to put it out to pasture.

There is a common belief that the practice of hiding nodes behind a single routable IPv4 address ("hide-mode NAT") adds security. It really doesn't.

First, anytime you make an outgoing connection, either directly or via NAT, the connection you make is a two-way path, and the node you connect to can easily attack you right through your packet filtering firewall and Network Address Translation. You should have "defense in depth" and protect your node with a host-based firewall whether or not you are behind a firewall and NAT gateway.

Second, if a hacker manages to breach your firewall by installing a Trojan horse onto any node in your network, they can attack you from that compromised node. Hackers have a term for networks that have a strong perimeter defense but limited internal defenses. It is "hard crunchy outside, soft chewy inside." Again, host-based firewalls on all nodes are a good idea.

Third, if you are using almost any peer-to-peer software, VoIP (e.g., Skype), or IPsec VPN, it probably includes a mechanism called NAT traversal (e.g., STUN, TURN, SOCKS, etc.). NAT traversal basically bores a hole right through your NAT protection (required for any of the preceding applications). Anything that includes NAT traversal can easily be used to attack you. Many people think Skype is a productivity tool. Network security people think it is a security vulnerability.

Fourth, any time you open a document from outside (Word document, Excel spreadsheet, JPEG image, etc.), it may contain malware that infects your node right through firewalls and NAT.

It is better to allow direct connections to your node over IPv6, through various layers of firewalls, including a host-based firewall, together with good active anti-malware software, than to have NAT giving you a false sense of security.

On the other hand, NAT causes problems with any connectivity model other than simple client server outgoing connections, such as web browser to web server. This was covered in some detail in Chapter 3, section "Network Address translation (NAT)."

The real kicker is that NAT is the hacker's friend! It is easy for a hacker to hide behind a NAT gateway and do all kinds of mischief, sending of malware, etc. It is quite difficult for the authorities to figure out which of the nodes hidden behind the common address is doing the bad stuff. To do this, the ISP must log EVERY connection, including source address, destination address, timestamp, and port. This mounts up to several TERABYTES for each ISP customer over a year, which is not a trivial amount of storage. With a flat address space (as in IPv6), it is far easier to figure out where the attack is coming from.

Because of these issues, *there is no IPv6-to-IPv6 NAT* defined in any IETF standard. There is no need for it to extend the IPv6 address space lifetime, it has no other real benefit, it causes many problems, and it is greatly impeding innovation. Other than those minor things, I guess it's okay (sarcasm warning!).

On the other hand, there is a real need for IPv4-to-IPv6 (and IPv6-to-IPv4) Network Address Translation, and there are about eight proposed methods in the IETF now. All of them have various problems and tradeoffs (that is the nature of NAT). One of the more promising schemes is NAT64 in combination with DNS64. These will be discussed in more detail in Chapter 8 on migration to IPv6.

TCP: The Transmission Control Protocol in IPv6

There is very little difference between TCP over IPv4 and TCP over IPv6. The main difference is that more storage must be provided in the implementations to hold the four times larger addresses (16 bytes vs. 4 bytes, for each address). The other aspect involves the TCP header checksum, which uses a *pseudo header* to allow inclusion of the IP addresses in the calculation of the checksum (in addition to the contents of the payload). Of course, there is a different pseudo header format for IPv4 and IPv6, given the difference in address size. There are no new RFCs for TCP over IPv6.

There is one new feature for both TCP and UDP over IPv6 called "Jumbograms." This is defined in RFC 2675,[26] "IPv6 Jumbograms," August 1999. Jumbograms are very large packets, with a payload containing more than 65,535 bytes. The standard *Payload Length* field is only 16 bits, so the maximum payload size is 65,535 bytes. RFC 2675 defines a new Hop-by-Hop Option that includes a 32-bit Payload Length field, allowing packet lengths of up to 4.3 billion bytes. Of course, such packets require paths with very large MTUs. The simple 16-bit checksum becomes a less reliable error detection scheme as the payload length increases significantly. Of course, even a 1-bit error would require retransmission of an entire packet, so this should be used only on extremely reliable links.

---

[26] https://tools.ietf.org/html/rfc2675

TCP Packet Header

No changes are required to the TCP packet header, as port numbers are still 16 bits in length. The only differences are in how the header checksum is calculated (using the IPv6 pseudo header) and the availability of Jumbograms.

UDP: The User Datagram Protocol in IPv6

UDP over IPv6 has basically the same differences from UDP over IPv4 as was described for TCP.

DHCPv6: Dynamic Host Configuration Protocol for IPv6

Unlike with DNS, it was not possible to add new functionality into DHCPv4 to support IPv6 (let alone a single server that could handle both IPv4 and IPv6). DHCPv6 is pretty much a new design from the ground up. DHCPv4 was built from an earlier protocol called BOOTP and contains many now unnecessary features from that. DHCPv6 was cleaned up considerably and contains none of the things left over from BOOTP.

DHCPv4 runs over IPv4 and supplies only 32-bit IPv4 information (assigned IPv4 addresses, IPv4 addresses of DNS servers, etc.). DHCPv6 runs only over IPv6 and supplies only 128-bit IPv6 information (assigned IPv6 addresses, IPv6 addresses of DNS servers, etc.). There is no conflict between DHCPv4 and DHCPv6 in terms of functionality or ports used, so it is possible to run both on a single, dual-stack node.

Hosts communicate only with DHCPv6 servers or relay agents on their local link, using link-local addresses (typically *ff02::1:2,* "all DHCPv6 relay agents and servers"). DHCPv6 uses UDP ports 546 and 547 (compare with DHCPv4, which uses UDP ports 67 and 68). As with DHCPv4, relay agents are used to allow hosts to communicate with remote DHCPv6 servers (ones not on the local link). This is still done via UDP but using a site-scope address (*ff05::1:3* "all DHCPv6 servers, but not relay agents"), which is used only by relay agents.

In some simple networks, there is no need for DHCPv6 because of Stateless Address Autoconfiguration. Currently, however, DHCPv6 is the only way for IPv6-capable nodes to automatically learn the IPv6 addresses of DNS servers. This is particularly important for IPv6-only ("pure IPv6") networks, of which there are not many yet. For dual-stack networks, there is no conflict between DHCPv4 and DHCPv6, and both can exist even on a single node. In this case, the IPv4 side of a node would get its IPv4 configuration from the DHCPv4 server, and the IPv6 side of a node would get its IPv6 configuration from the DHCPv6 server.

DHCPv6 allows the administrator far better control over distribution of interface identifiers (low 64 bits of each address) than with Stateless Address Autoconfiguration. With SLAAC, interface identifiers can either make use of only a tiny percentage of the possible 2^{64} address space (when using EUI-64-generated interface identifiers) or have interface identifiers scattered randomly all over the possible 2^{64} address space (when using cryptographically generated addresses). Either of these can lead to problems with network access control (NAC) or firewall rules. In general, administrators like to cluster IP addresses by department (or other groupings), so that a single firewall or NAC rule can be used for an entire group, by specifying an *address range* (e.g., all addresses that fall between *2001:df8:5403:3000::1000* and *2001:df8:5403:3000::1fff*, inclusive).

IPv6-capable nodes can be informed that there is a DHCPv6 server available via 2 bits in the Router Advertisement message. The Router Advertisement message and the relevant bits are described in RFC 4861, "Neighbor Discovery for IP version 6 (IPv6)," September 2007. In the Router Advertisement message, there are 2 bits, M and O (first and second bits of the sixth byte of the Router Advertisement message), with the following semantics:

> *M*: "Managed address configuration" flag. When set it indicates that addresses are available via DHCPv6. If set, then the O flag can be ignored. This enables *stateful DHCPv6*, where both the stateless information (IPv6 addresses of DNS and other servers) *and* global unicast addresses can be obtained from DHCPv6.

> *O*: "Other configuration" flag. When set, it indicates that other configuration information is available via DHCPv6. This includes things such as IPv6 addresses of DNS or other servers. This is called *stateless DHCPv6* and is used in conjunction with Stateless Address Autoconfiguration (for obtaining global unicast addresses).

If both M and O bits are clear, then SLAAC is the only way to get addresses, and there is no source of IPv6 addresses for any servers, including DNS.

Relevant RFCs for DHCPv6

There are several RFCs that define DHCPv6. The most important ones are

RFC 3319, "Dynamic Host Configuration Protocol (DHCPv6) Options for Session Initiation Protocol (SIP) Servers," July 2003

RFC 3646, "DNS Configuration Options for Dynamic Host Configuration Protocol for IPv6 (DHCPv6)," December 2003 (Standards Track)

RFC 3898, "Network Information Service (NIS) Configuration Options for Dynamic Host Configuration Protocol for IPv6 (DHCPv6)," October 2004 (Standards Track)

RFC 4075, "Simple Network Time Protocol (SNTP) Configuration Option for DHCPv6," May 2005

RFC 4076, "Renumbering Requirements for Stateless Dynamic Host Configuration Protocol for IPv6 (DHCPv6)," May 2005 (Informational)

RFC 4339, "IPv6 Host Configuration of DNS Server Information Approaches," February 2006 (Informational)

RFC 4477, "Dynamic Host Configuration Protocol (DHCP): IPv4 and IPv6 Dual-Stack Issues," May 2006 (Informational)

RFC 4580, "Dynamic Host Configuration Protocol for IPv6 (DHCPv6) Relay Agent Subscriber-ID Option," June 2006 (Standards Track)

RFC 4649, "Dynamic Host Configuration Protocol for IPv6 (DHCPv6) Relay Agent Remote-ID Option," August 2006 (Standards Track)

RFC 4704, "The Dynamic Host Configuration Protocol for IPv6 (DHCPv6) Client Fully Qualified Domain Name (FQDN) Option," October 2006 (Standards Track)

RFC 4994, "DHCPv6 Relay Agent Echo Request Option," September 2007 (Proposed Standard)

RFC 5007, "DHCPv6 Leasequery," September 2007 (Standards Track)

RFC 5460, "DHCPv6 Bulk Leasequery," February 2009 (Standards Track)

RFC 5908, "Network Time Protocol (NTP) Server Option for DHCPv6," June 2010 (Proposed Standard)

RFC 5970, "DHCPv6 Options for Network Boot," September 2010 (Proposed Standard)

RFC 6011, "Session Initiation Protocol (SIP) User Agent Configuration," October 2010 (Informational)

RFC 6221, "Lightweight DHCPv6 Relay Agent," February 2011 (Proposed Standard)

RFC 6334, "Dynamic Host Configuration Protocol for IPv6 (DHCPv6) Option for Dual-Stack Lite," August 2011 (Proposed Standard)

RFC 6355, "Definition of the UUID-Based DHCPv6 Unique Identifier (DUID-UUID)," August 2011 (Proposed Standard)

RFC 6422, "Relay Supplied DHCP Options," December 2011 (Proposed Standard)

RFC 6603, "Prefix Exclude Option for DHCPv6-Based Prefix Delegation," May 2012 (Proposed Standard)

RFC 6607, "Virtual Subnet Selection Options for DHCPv4 and DHCPv6," April 2012 (Proposed Standard)

RFC 6644, "Rebind Capability in DHCPv6 Reconfigure Messages," July 2012 (Proposed Standard)

RFC 6653, "DHCPv6 Prefix Delegation in Long-Term Evolution (LTE) Networks," July 2012 (Informational)

RFC 6784, "Kerberos Options for DHCPv6," November 2012 (Proposed Standard)

RFC 6853, "DHCPv6 Redundancy Deployment Considerations," February 2013 (Best Current Practice)

RFC 6939, "Client Link-Layer Address Option in DHCPv6," May 2013 (Proposed Standard)

RFC 6977, "Triggering DHCPv6 Reconfiguration from Relay Agents," July 2013 (Proposed Standard)

RFC 7031, "DHCPv6 Failover Requirements," September 2013 (Informational)

RFC 7037, "RADIUS Option for the DHCPv6 Relay Agent," October 2013 (Proposed Standard)

RFC 7078, "Distributing Address Selection Policy Using DHCPv6," January 2014 (Proposed Standard)

RFC 7227, "Guidelines for Creating New DHCPv6 Options," May 2014 (Best Current Practice)

RFC 7341, "DHCPv4-over-DHCPv6 (DHCP 4o6) Transport," August 2014 (Proposed Standard)

RFC 7598, "DHCPv6 Options for Configuration of Softwire Address and Port-Mapped Clients," July 2015 (Proposed Standard)

RFC 7610, "DHCPv6-Shield: Protecting Against Rogue DHCPv6 Servers," August 2015 (Best Current Practice)

RFC 7653, "DHCPv6 Active Leasequery," October 2015 (Proposed Standard)

RFC 7774, "Multicast Protocol for Low-Power and Lossy Networks (MPL) Parameter Configuration Option for DHCPv6," March 2016 (Proposed Standard)

RFC 7824, "Privacy Considerations for DHCPv6," May 2016 (Informational)

RFC 7839, "Access-Network-Identifier Option in DHCP," June 2016 (Proposed Standard)

RFC 7844, "Anonymity Profiles for DHCP Clients," May 2016, (Proposed Standard)

RFC 7934, "Host Address Availability Recommendations," July 2016 (Best Current Practice)

RFC 7943, "A Method for Generating Semantically Opaque Interface Identifiers (IISs) with the Dynamic Host Configuration Protocol for IPv6 (DHCPv6)," September 2016 (Informational)

RFC 7969, "Customizing DHCP Configuration on the Basis of Network Topology," October 2016 (Informational)

RFC 8026, "Unified IPv4-in-IPv6 Softwire Customer Premises Equipment (CPE): A DHCPv6-Based Prioritization Mechanism," November 2016 (Proposed Standard)

RFC 8115, "DHCPv6 Option for IPv4-Embedded Multicast and Unicast IPv6 Prefixes," March 2017 (Proposed Standard)

RFC 8156, "DHCPv6 Failover Protocol," June 2017 (Proposed Standard)

RFC 8168, "DHCPv6 Prefix-Length Hint Issues," May 2017 (Proposed Standard)

RFC 8415, "Dynamic Host Configuration Protocol for IPv6 (DHCPv6)," November 2018 (Standards Track)

RFC 8539, "Softwire Provisioning Using DHCPv4 over DHCPv6," March 2019 (Proposed Standard)

DHCPv6 has a failover mechanism. Two servers can manage a single pool of addresses for redundancy (in case of failure of one of the servers). This also can be used for load balancing.

All IPv6 hosts have automatically generated *link-local* addresses that can be used to exchange packets with any other node on the local link. DHCPv4 requires some complex hacks to allow hosts to communicate before they get an address. All IPv6 hosts support link-local multicast. All DHCPv6 servers listen to DHCPv6 multicast groups. With DHCPv4, clients have to do a general broadcast to all nodes on the link, which generates significant broadcast traffic on the link and unnecessary traffic handling on all nodes.

With DHCPv6, a single request can configure all interfaces on a node. The server can offer multiple addresses, one for each interface, and each interface can even have different options. With DHCPv4, each interface would require a separate DHCP operation.

Some of the *stateless* information (i.e., other than assigned IPv6 addresses for each node) includes

- IPv6 prefix

- Vendor-specific options

- Addresses of SIP servers

- Addresses of DNS servers and search options

- NIS configuration

- SNTP servers

There are several implementations of DHCPv6 already on the market. Windows Server 2008 contains a very complete implementation, in addition to its DHCPv4 server. You can view the IPv6-ready phase 2 products list for other options, including my own company's *Sixscape DNS* appliance. Some implementations only support *stateless* mode, which means they can supply stateless information (like DNS addresses) but not actually allocate addresses. Be sure the DHCPv6 server you select includes full support for stateful mode as well (where it can supply addresses to each node, in addition to stateless information). You should also be sure that the gateway router or firewall you select has the ability to inform nodes that DHCPv6 servers are available on the subnet.

Address Reservations with DHCPv6

In the case of DHCPv4, it is possible to make an *address reservation*, linked to the MAC address of a node. Anytime the node with a MAC address for which an address reservation has been made asks DHCPv4 for an address, it will get the specific address that was reserved for that MAC address. In the case of DHCPv6, the same concept applies, except that you use two identifiers called the IAID (Interface Association ID) and DUID (DHCP Unique IDentifier).

A DUID consists of a 2-byte type code represented in network byte order, followed by a variable number of bytes that make up the actual identifier. A DUID can be no more than 128 bytes long (not including the type code). The following types are currently defined:

Link Layer Address plus Time (DUID-LLT): This type is recommended for all general-purpose computing devices, such as desktop computers, printers, routers, etc. They must contain some form of writable non-volatile storage. Note that the device should configure the time on the node before this DUID is generated, if possible. The only purpose of the timestamp is to lower the chance of an identifier conflict. The Link Layer address is typically the MAC address for Ethernet media. The DUID is defined as follows:

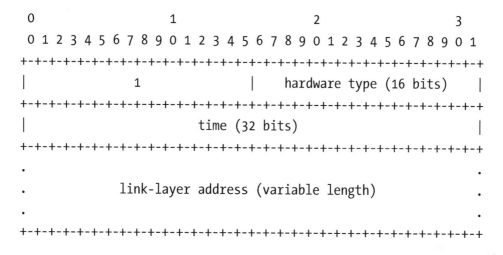

Vendor-Assigned Based on Enterprise Number (DUID-EN): This type is assigned to the device by the vendor. This type of DUID is for devices that have some form of non-volatile storage (e.g., EEPROM). The enterprise number is the IANA 32-bit assigned number for the vendor. The identifier can be anything the vendor chooses but must be unique within that vendor for each device.

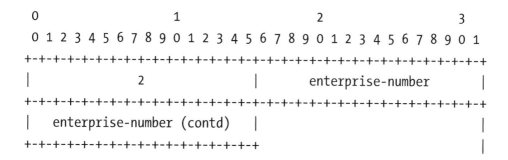

```
.                                identifier                         .
.                             (variable length)                     .
.                                                                    .
+-+-+-+-+-+-+-+-+-+-+-+-+-+-+-+-+-+-+-+-+-+-+-+-+-+-+-+-+-+-+-+-+
```

> *Link Layer Address (DUID-LL)*: This type is just like the DUID-
> LLT, without the timestamp. It is recommended for permanently
> connected devices that have a Link Layer address, but no non-
> volatile, writeable stable storage.

```
 0                   1                   2                   3
 0 1 2 3 4 5 6 7 8 9 0 1 2 3 4 5 6 7 8 9 0 1 2 3 4 5 6 7 8 9 0 1
+-+-+-+-+-+-+-+-+-+-+-+-+-+-+-+-+-+-+-+-+-+-+-+-+-+-+-+-+-+-+-+-+
|               3               |        hardware type (16 bits)    |
+-+-+-+-+-+-+-+-+-+-+-+-+-+-+-+-+-+-+-+-+-+-+-+-+-+-+-+-+-+-+-+-+
.                                                                    .
.              link-layer address (variable length)                 .
.                                                                    .
+-+-+-+-+-+-+-+-+-+-+-+-+-+-+-+-+-+-+-+-+-+-+-+-+-+-+-+-+-+-+-+-+
```

Viewing Your Node's DUID

In Windows 7, using a command prompt, type the command ipconfig /all. In the section related to the interface you are interested in, look for the field *DHCPv6 Client DUID*. Note that this is a DUID-LLT (type code 00-01). The next six hex digit pairs (00-01-12-D6-97-E5) are the timestamp. The last six hex digit pairs (00-18-8B-78-DA-1A) are the same as the *physical address* (MAC address).

```
Ethernet adapter Local Area Connection:

   Connection-specific DNS Suffix  . : infoweapons.com
   Description . . . . . . . . . . . : Broadcom NetXtreme 57xx Gigabit
                                       Controller
   Physical Address. . . . . . . . . : 00-18-8B-78-DA-1A
   DHCP Enabled. . . . . . . . . . . : Yes
   Autoconfiguration Enabled . . . . : Yes
   IPv6 Address. . . . . . . . . . . : 2001:df8:5403:2410:3c79:b2ca:90ce:5d
                                       59(Preferred)
```

```
Temporary IPv6 Address. . . . . . : 2001:df8:5403:2410:882:cf5c:e810:363
                                    d(Preferred)
Link-local IPv6 Address . . . . . : fe80::3c79:b2ca:90ce:5d59%11(
                                    Preferred)
IPv4 Address. . . . . . . . . . . : 10.2.5.237(Preferred)
Subnet Mask . . . . . . . . . . . : 255.240.0.0
Lease Obtained. . . . . . . . . . : Friday, March 12, 2010 10:52:52 AM
Lease Expires . . . . . . . . . . : Friday, March 12, 2010 4:11:44 PM
Default Gateway . . . . . . . . . : fe80::21b:21ff:fe1e:f4%11
                                    10.0.0.10
DHCP Server . . . . . . . . . . . : 10.1.0.14
DHCPv6 IAID . . . . . . . . . . . : 234887307
DHCPv6 Client DUID. . . . . . . . : 00-01-00-01-12-D6-97-
                                    E5-00-18-8B-78-DA-1A

DNS Servers . . . . . . . . . . . : 2001:df8:5403:2400::14
                                    2001:df8:5403:2400::13
                                    10.1.0.14
                                    10.1.0.13
NetBIOS over Tcpip. . . . . . . . : Enabled
Connection-specific DNS Suffix Search List :
                                    cebu.infoweapons.com
```

You can also see a *DHCPv6 IAID* value (in this case 234887307). This identifies a particular *Identity Association*, which allows a server and a client to identify, group, and manage a set of related IPv6 addresses. Each IA consists of an IAID, one or more IPv6 addresses, and the time T1 and T2 for that IA. Each IA is associated with exactly one interface. For further details, see RFC 3315, section 11.

DHCPv6 Ports and Messages

Clients and servers exchange DHCPv6 messages using UDP over IPv6. The client uses a link-local address, or addresses obtained via other mechanisms, as the source address for transmitting and receiving DHCPv6 messages. Servers receive messages from clients using a reserved link-scoped multicast address, so that clients don't need to be configured with the addresses of DHCPv6 servers. To allow hosts to communicate with servers on other links, DHCPv6 relay agents are used. Clients listen for DHCPv6 messages on UDP port 546. Servers and relay agents listen for DHCPv6 messages on UDP port 547.

The link-scoped multicast address used by a client to communicate with an on-link relay agent or server is *ff02::1:2*. All DHCPv6 servers and relay agents are members of this multicast group.

The site-scoped multicast address used by a relay agent to communicate with servers is *ff05::1:3*, if it wants to send a message to all DHCPv6 servers or does not know the unicast address of the servers. All DHCPv6 servers in a given site are members of this multicast group.

There are several DHCPv6 messages:

>   The **SOLICIT** message (1) is sent (multicast) by a client to locate servers.

>   The **ADVERTISE** message (2) is sent (multicast) by a server to indicate that it is available to provide DHCPv6 service, in response to a Solicit message from a client.

>   The **REQUEST** message (3) is sent (unicast) by a client to request configuration parameters, including IP addresses, from a specific server.

>   The **CONFIRM** message (4) is sent (multicast) by a client to any available server to determine whether the addresses it was assigned are still appropriate on the link to which the client is connected.

>   The **RENEW** message (5) is sent (unicast) by a client to the server that originally provided the client's address and configuration parameters, to extend the lifetime on the addresses assigned to the client and update other configuration parameters.

>   The **REBIND** message (6) is sent (multicast) by a client to any available server to extend the lifetimes on the addresses assigned to the client and to update other configuration parameters. This message is sent after a client receives no response to a RENEW message.

>   The **REPLY** message (7) is sent (unicast) by a server to a client in response to a SOLICIT, REQUEST, RENEW, or REBIND message received from a client. A server sends a REPLY message containing

configuration parameters in response to an INFORMATION-REQUEST message. It sends a REPLY message in response to a CONFIRM message confirming or denying that the addresses assigned to the client are appropriate on the link to which the client is connected. A server sends a REPLY message to acknowledge receipt or a RELEASE or DECLINE message.

The *RELEASE* message (8) is sent (unicast) by a client to the server that assigned addresses to the client to indicate that the client will no longer use one or more of the assigned addresses.

The *DECLINE* message (9) is sent (unicast) to a server to indicate that the client has determined that one or more addresses assigned by the server are already in use on the link to which the client is connected.

The *RECONFIGURE* message (10) is sent (unicast) by a server to a client to inform the client that the server has new or updated configuration parameters and that the client should initiate a RENEW/REPLY or INFORMATION-REQUEST/REPLY transaction with the server in order to obtain the updated information.

The *INFORMATION-REQUEST* message (11) is sent (unicast) by a client to a server to request configuration parameters, without the assignment of any IP addresses to the client.

The *RELAY-FORW* message (12) is sent (multicast) by a relay agent to forward messages to servers, either directly or through another relay agent. The received message, either a client message or a RELAY-FORW message from another relay agent, is encapsulated in an option in the RELAY-FORW message.

The *RELAY-REPL* message (13) is sent (unicast) by the server to a relay agent containing a message that the relay agent should then deliver to a client. The RELAY-REPL message may be relayed by other relay agents for delivery to the destination relay agent. The server encapsulates the client message as an option in the RELAY-REPL message, which the relay agent extracts and then relays to the next relay agent or directly to the client.

DHCPv6 Status Codes

The following codes are used to communicate the success or failure of operations requested in messages from clients and servers and additional information about the specific cause in the event of a failure to perform the operation.

```
Name           Code Description
----------     ---- -----------
Success           0 Success.
UnspecFail        1 Failure, reason unspecified; this
                    status code is sent by either a client
                    or a server to indicate a failure
                    not explicitly specified in this
                    document.
NoAddrsAvail      2 Server has no addresses available to assign to
                    the IA(s).
NoBinding         3 Client record (binding) unavailable.
NotOnLink         4 The prefix for the address is not appropriate for
                    the link to which the client is attached.
UseMulticast      5 Sent by a server to a client to force the
                    client to send messages to the server.
                    using the All_DHCP_Relay_Agents_and_Servers
                    address.
```

DHCPv6 Message Syntax

All messages sent between clients and servers share the following syntax:

```
 0                   1                   2                   3
 0 1 2 3 4 5 6 7 8 9 0 1 2 3 4 5 6 7 8 9 0 1 2 3 4 5 6 7 8 9 0 1
+-+-+-+-+-+-+-+-+-+-+-+-+-+-+-+-+-+-+-+-+-+-+-+-+-+-+-+-+-+-+-+-+
|    msg-type   |               transaction-id                  |
+-+-+-+-+-+-+-+-+-+-+-+-+-+-+-+-+-+-+-+-+-+-+-+-+-+-+-+-+-+-+-+-+
|                                                               |
.                            options                            .
.                           (variable)                          .
|                                                               |
+-+-+-+-+-+-+-+-+-+-+-+-+-+-+-+-+-+-+-+-+-+-+-+-+-+-+-+-+-+-+-+-+
```

`msg-type`	Identifies the DHCP message type
`transaction-id`	The transaction ID for this message exchange.
`options`	Options carried in this message.

The DHCPv6

DHCPv6 works in somewhat the same way as DHCPv4, except that different messages are used and communication between client and server takes place using link-local scoped multicast and unicast addresses.

When it first comes up, before any DHCPv6 operation, an IPv6-capable client node obtains a link-local unicast address through ND (and possibly a global unicast address as well, using information from a Router Advertisement message). If a Router Advertisement message is seen, then the client can check the M and O bits in it to determine if there is stateful DHCPv6, stateless DHCPv6, or no DHCPv6 available. If no Router Advertisement is available, a client can still attempt DHCPv6 server discovery, as follows.

The client sends a *SOLICIT* message to multicast group *ff02::1:2*. This address specifies *all DHCPv6 servers or relay agents on the local link*. The included options are

> ClientID
>
> Option Request Option (IA-NA, DNS-Servers, Domain-List)

One or more DHCPv6 servers on the link (or servers on remote links, via DHCPv6 relay agents) will reply with an *ADVERTISE* message to the client that sent the *SOLICIT* message (via unicast). The included options are

> ServerID, ClientID
>
> DNS-Servers, IA-NA (IAID, IAPREFIX).

The client will select one responding DHCPv6 server and send a *REQUEST* message to it (via unicast). This will actually ask for an address lease. The included options are

> ServerID, ClientID

Option Request Option (IA-NA, DNS-Servers, Domain-List)

The selected server will send a *REPLY* message to the client that sent the *REQUEST* message (via unicast). This will confirm the address lease. The included options are

ServerID, ClientID

DNS-Servers: 2001:xxx:yyy:zzz::a, 2001:xxx:yyy:zzz::b

IA-NA: IAID: 1,

IAPREFIX: Preferred lifetime: nnnnnn,

Valid lifetime: nnnnnn,

Prefix: 2001:xxx:yyy:zzz::c/64

For Further Information on DHCPv6

For details on how clients send and respond to DHCPv6 messages, see RFC 3315, section 17.

For details on DHCP Client-initiated Configuration Exchanges, see RFC 3315, section 18.

For details on DHCP Server-initiated Configuration Exchanges, see RFC 3315, section 19.

For details on relay agent behavior, see RFC 3315, section 20.

For details on the optional authentication mechanism, for use of DHCPv6 in unsecured environments, such as wireless networks, see RFC 3315, section 21.

For available DHCPv6 message options and their syntax, see RFC 3315, section 22.

Stateless DHCPv6 assumes that assigned IPv6 addresses are obtained some other way, such as Stateless Address Autoconfiguration, and that only stateless information (IPv6 addresses of DNS servers, SIP servers, etc.) will be obtained from DHCPv6. RFC 3736, "Stateless Dynamic Host Configuration Protocol (DHCP) Server for IPv6," April 2004, defines the subset of messages and options from the full (stateful) DHCPv6 functionality that are required to provide stateless DHCPv6 service.

For details on publishing the address of SIP servers with DHCPv6, see RFC 3633, "IPv6 Prefix Options for Dynamic Host Configuration Protocol (DHCP) version 6," December 2003.

For details on publishing the address of DNS servers with DHCPv6, see RFC 3646, "DNS Configuration Options for Dynamic Host Configuration Protocol for IPv6 (DHCPv6)," December 2003.

For details on publishing the address of NIS (Network Information Service) servers with DHCPv6, see RFC 3898, "Network Information Service (NIS) Configuration Options for Dynamic Host Configuration Protocol for IPv6 (DHCPv6)," October 2004.

For details on publishing the address of SNTP (Simple Network Time Protocol) servers with DHCPv6, see RFC 4075, "Simple Network Time Protocol (SNTP) Configuration Option for DHCPv6," May 2005.

Useful Commands Related to DHCPv6

In Windows 7, there are some commands available in a command prompt box related to DHCPv6:

> ipconfig /release6: Release assigned IPv6 address (es) and de-configure network.
>
> ipconfig/renew 6: Do a new configuration request for IPv6.
>
> ipconfig/all: View all network configuration settings (IPv4 and IPv6).

This is an example of the output from "ipconfig /all":

```
...
Ethernet adapter Local Area Connection:

   Connection-specific DNS Suffix  . : hughesnet.local
   Description . . . . . . . . . . . : Realtek PCIe GBE Family Controller
   Physical Address. . . . . . . . . : 00-22-15-24-32-9C
   DHCP Enabled. . . . . . . . . . . : Yes
   Autoconfiguration Enabled . . . . : Yes
   IPv6 Address. . . . . . . . . . . : 2001:df8:5403:3000::2:1(Preferred)
   Lease Obtained. . . . . . . . . . : Friday, March 12, 2010 9:43:06 PM
   Lease Expires . . . . . . . . . . : Wednesday, March 24, 2010 9:43:09 PM
   IPv6 Address. . . . . . . . . . . :
                                       2001:df8:5403:3000:b5ea:976d:679f:30
                                       f5(Preferred)
   Temporary IPv6 Address. . . . . . :
                                       2001:df8:5403:3000:218a:4956:7d8c:7c
                                       2c(Preferred)
```

```
Link-local IPv6 Address . . . . . : fe80::b5ea:976d:679f:30f5%11(
                                     Preferred)
IPv4 Address. . . . . . . . . . . : 172.20.2.1(Preferred)
Subnet Mask . . . . . . . . . . . : 255.255.0.0
Lease Obtained. . . . . . . . . . : Friday, March 12, 2010 9:42:57 PM
Lease Expires . . . . . . . . . . : Thursday, March 18, 2010 9:43:00 PM
Default Gateway . . . . . . . . . : fe80::21b:21ff:fe1d:c159%11
                                     172.20.0.1
DHCP Server . . . . . . . . . . . : 172.20.0.11
DHCPv6 IAID . . . . . . . . . . . : 218112533
DHCPv6 Client DUID. . . . . . . . : 00-01-00-01-11-99-BD-28-00-22-15-24
                                     -32-9C
DNS Servers . . . . . . . . . . . : 2001:df8:5403:3000::c
                                     2001:df8:5403:3000::b
                                     172.20.0.11
                                     172.20.0.12
NetBIOS over Tcpip. . . . . . . . : Enabled
Connection-specific DNS Suffix Search List : hughesnet.local
```

...

In the preceding, notice the following:

- The MAC address ("physical address") of the interface is *00-22-15-23-32-9C*.

- A 64-bit interface identifier (*b5ea:976d:679f:30f5*) was created, which is a cryptographically generated value (not from EUI-64). A link-local unicast address was generated from this (by prepending *fe80::/64*). The link-local address of the default gateway (*fe80::21b:21ff:fe1d:c159*) was then obtained using ND router discovery.

- A Router Advertisement message supplied the subnet prefix (*2001:df8:5403:3000::/64*), so the node used it to create two global unicast addresses, one of which (*2001:df8:5403:3000:b5ea:976d:679 f:30f5*) used the 64-bit random interface identifier from ND and the other (*2001:df8:5403:3000:218a:4956:7d8c:7c2c*) used yet another random interface identifier.

- *Obtain an IP address automatically* and *Obtain DNS server address automatically* were selected in the IPv4 GUI configuration ("DHCP Enabled"), and a working DHCPv4 server was found ("Autoconfiguration Enabled"). So an IPv4 address (172.20.2.1), the subnet mask (255.255.0.0), the default gateway (172.20.0.1), and the IPv4 addresses of two DNS servers were obtained from the DHCPv4 server. The lease for this was obtained on March 12, 2010, 9:42 p.m., and would expire on March 18, 2010, at 9:43 p.m. The MAC address (00-22-15-23-32-9C) was used to make a DHCPv4 reservation for this node, so this node will always get that IPv4 address.

- *Obtain an IPv6 address automatically* and *Obtain DNS server address automatically* were selected in the IPv6 GUI configuration, and both the M and O bits were set in the Router Advertisement message (stateful and stateless DHCPv6 available), so another global unicast IPv6 address (*2001:df8:5403:3000::2:1*) was obtained from DHCPv6, plus the IPv6 addresses of two DNS servers. The lease for this was obtained on March 12, 2010, at 9:43 p.m., and would expire on March 24, 2010, at 9:43 p.m.

- The DUID of the node is *00-01-00-01-11-99-BD-28-00-22-15-24-32-9C*. The first two hex digit pairs contain 00-01. That means this is a type 1 DUID (DUID-LLT), "Link Layer plus Timestamp." The next six hex digit pairs (*00-01-11-99-BD-28*) are the timestamp, and the last six hex digit pairs (*00-22-15-23-32-9C*) contain the interface MAC address. This DUID, along with the IAID (*218112533*), was used to make a DHCPv6 address reservation for this node. So this node will always get that IPv6 address.

IPv6 Network Configuration

Let's assume our LAN has the following configuration:

> *Network prefix*: 2001:df8:5403:3000::/64
>
> *Default gateway*: 2001:df8:5403:3000::1
>
> *DHCPv6 server address*: 2001:df8:5403:3000::11

> *DNS server addresses*: 2001:df8:5403:3000::11
>
> 2001:df8:5403:3000::12
>
> *Domain name*: redwar.org

Furthermore, assume the DHCPv6 server is correctly configured with this information and is managing the address range 2001:df8:5403:3100::1000 to 2001:df8:5403:3100::1fff (and that some leases have already been granted).

Any node connected to a network with IPv6 (that will access IPv6 nodes on the Internet) must have certain items configured, including

- IPv6 link-local node address (obtained automatically)

- All nodes on link multicast address (*ff01::1*), there by default

- IPv6 global unicast address

- IPv6 address of default gateway (link-local address of gateway obtained automatically)

- IPv6 addresses of DNS servers (manually configured or from DHCPv6)

- Nodename

- DNS domain name

Manual Network Configuration for IPv6-Only

It is possible to perform IPv6 configuration manually, either by editing ASCII configuration files, as in FreeBSD or Linux, or via GUI configuration tools, as in Windows. If you have understood the material in this chapter, it should be fairly easy to configure your node(s). In most cases, if you have ISP service, the ISP will give you all the information necessary to configure your node(s). In the coverage of dual-stack networks, we will show configuration of *both* IPv4 and IPv6 on a single node.

Auto Network Configuration Using Stateless Address Autoconfiguration

It is easy for a FreeBSD node to be automatically configured using Stateless Address Autoconfiguration. Note that the global unicast address will be created with the EUI-64 algorithm from your MAC address.

Let's configure a FreeBSD 7.2 node automatically with SLAAC. Assign it the following configuration:

FreeBSD interface name: vr0

Nodename: us1.redwar.org

Node IP address (whatever SLAAC comes up with)

Default gateway (whatever SLAAC comes up with)

DNS domain name: redwar.org

DNS Server 1: 2001:df8:5403:3000::11

DNS Server 2: 2001:df8:5403:3000::12

You need to edit the following files (you will need root privilege to do this):
/etc/rc.conf

```
...
hostname="us1.redwar.org"
IPv6_enable="YES"
...
```

/etc/resolv.conf

```
Domain        redwar.org
nameserver    2001:df8:5403:3000::11
nameserver    2001:df8:5403:3000::12
```

If you make these changes, then reboot. You can check the configuration as shown:

```
$ ifconfig vr0
vr0: flags=8843<UP,BROADCAST,RUNNING,SIMPLEX,MULTICAST> metric 0 mtu 1500
        options=2808<VLAN_MTU,WOL_UCAST,WOL_MAGIC>
        ether 00:15:f2:2e:b4:1c
        inet6 2001:df8:5403:3000::215:f2ff:fe2e:b41c prefixlen 64
        inet6 fe80::215:f2ff:fe2e:b41c%vr0 prefixlen 64 scopeid 0x1
        media: Ethernet autoselect (100baseTX <full-duplex>)
        status: active
$ uname -n
us1.redwar.org
```

```
$ nslookup
> server

> exit

$ netstat -finet6 -rn
```

Auto Network Configuration Using Manually Specified (Static) IPv6 Address

Let's configure a FreeBSD 7.2 node manually with a static node address. Assign it the following configuration:

> *FreeBSD interface name*: vr0
>
> *Nodename*: us1.redwar.org
>
> *Node IP address*: 2001:df8:5403:3000::13
>
> *Default gateway*: 2001:df8:5403::1
>
> *DNS domain name*: redwar.org
>
> *DNS Server 1*: 2001:df8:5403:3000::11
>
> *DNS Server 2*: 2001:df8:5403:3000::12

You need to edit the following files (you will need root privilege to do this):

/etc/rc.conf

```
...
hostname='us1.redwar.org'
IPv6_enable='YES'
IPv6_ifconfig_vr0='2001:df8:5403:3000::13 prefixlen 64'
IPv6_defaultrouter='2001:df8:5403:3000::1'
...
```

/etc/resolv.conf

```
Domain      redwar.org
nameserver  2001:df8:5403:3000::11
nameserver  2001:df8:5403:3000::12
```

If you make these changes, then reboot. You can check the configuration as shown:

```
$ ifconfig vr0
vr0: flags=8843<UP,BROADCAST,RUNNING,SIMPLEX,MULTICAST> metric 0 mtu 1500
        options=2808<VLAN_MTU,WOL_UCAST,WOL_MAGIC>
        ether 00:15:f2:2e:b4:1c
        inet6 2001:df8:5403:3000::13 prefixlen 64
        inet6 fe80::215:f2ff:fe2e:b41c%vr0 prefixlen 64 scopeid 0x1
        media: Ethernet autoselect (100baseTX <full-duplex>)
        status: active
$ uname -n
us1.hughesnet.local
$ nslookup
> server

> exit

$ netstat -finet6 -rn
```

Note   If you specify a static IPv6 address in FreeBSD 7.x ("IPv6-config_vr0="..."),
the node will not obtain a link-local default gateway address automatically.
Therefore, in this case it is essential that you also manually specify a default
gateway address (which can be global unicast or link local), using the
"IPv6_defaultrouter=..." option in /etc/rc.conf. If no default gateway is defined,
communication with other on-link nodes will work okay, but communication with
off-link nodes will fail.

This is different from the behavior of Windows 7 and Linux, where the addition of a manually configured global unicast address does not stop the node from obtaining the link-local default gateway automatically.

Summary

In this chapter, we covered the "core" protocols related to IPv6:

- IPv6 itself

- IPCMPv6 (the helper protocol)

- ND (Neighbor Discovery, which is really just a subset of ICMPv6)

- Stateless Address Autoconfiguration (SLAAC) (which is one of the ND mechanisms)

We also covered the new IPv6 packet header, as well as the all-new packet header extensions. We compared that with the older IPv4 packet header, noting that the basic header is twice as long, but simpler. The advanced features have now been moved to the header extensions.

We covered the IPv6 addressing model, which is more complex than the IPv4 model. For the first time, *address scopes* have been fully implemented. IPv4 private addresses were kind of an address scope, but in IPv6 that concept is fully realized.

We covered DHCPv6, although technically that does not live in the Internet Layer (like DHCPv4, it lives in the Application Layer). With the new SLAAC, there is not much need for DHCPv6, especially now that SLAAC advertises the DNS addresses for the network. The only real need for DHCPv6 in current networks is for Prefix Delegation (how ISPs advertise the network prefix to subscribers).

Finally, we covered how you actually configure IPv6 addresses in FreeBSD.

CHAPTER 7

IPsec and iKEv2

This chapter covers two advanced protocols for TCP/IP called IPsec and IKEv2. IPsec is for "Internet Protocol Security" and adds authentication and encryption at the Internet Layer. IKEv2 is the *Internet Key Exchange* protocol for use with IPsec, and the current version is 2. You can use IPsec without IKEv2 with manual key management, but this is not scalable or particularly secure. Both IPsec and IKEv2 are available for IPv4 *and* IPv6, but NAT breaks both IPsec itself and IKEv2, so IPsec works far better over IPv6 (where there is no NAT to break them). IPsec was created for both IPv4 and IPv6, in RFC 1825,[1] "Security Architecture for the Internet Protocol," August 1995:

> *This memo describes the security mechanisms for IP version 4 (IPv4) and IP version 6 (IPv6) and the services that they provide.*

You *can* deploy IPsec on IPv4 today, but if the path crosses a NAT gateway, you have to also deploy NAT traversal, which introduces more security issues than IPsec solves. Because of this, SSL-VPN[2] is easier to work with and more widely used than IPsec *on IPv4*. Unfortunately, SSL-VPN is a very badly designed scheme misusing SSL in the wrong part of the network stack. There is also no IETF standard for SSL-VPN, because the IETF doesn't consider it a viable protocol. So there is no guarantee that two vendors' implementations will actually interoperate. IPsec is the only IETF-approved scheme for implementing VPNs, and as IPv6 becomes more widely deployed, it will finally become widely used. The problems people run into today are more problems with IPv4 and NAT than with IPsec. It works great on IPv6.

---

[1] https://tools.ietf.org/html/rfc1825
[2] https://openvpn.net/faq/why-ssl-vpn/

L. E. Hughes, *Third Generation Internet Revealed*, https://doi.org/10.1007/978-1-4842-8603-6_7

Originally IPsec was *mandatory* for IPv6 implementations (but of course did not have to be *enabled* on every connection), leading to the myth that IPv6 was more secure than IPv4. RFC 6434,[3] "IPv6 Node Requirements," December 2011, changed MUST to SHOULD (IPsec became *optional* in implementations but is still *strongly recommended*). Most IPv6 implementations I use (on real computers, not sensors) include IPsec.

Note that you can still build a subnet-to-subnet VPN with IPsec over IPv6 even if both subnets are dual stack or even IPv4-only, if IPv6 is available on a path connecting them. You can tunnel IPv4 through an IPv6 path.

These facts add strong incentives for organizations to begin supporting IPv6 sooner rather than later.

Internet Protocol Layer Security (IPsec)

The official name for this technology (as used by the IETF) is the *Security Architecture for the Internet Protocol*. Since it takes place in the Internet Layer, protocols at the Transport and Application Layers need not even be aware of it and do not need to be modified to use it. In effect, you build secure tunnels at the IP Layer that the higher-layer protocols go through unmodified. Compare with SSL/TLS where the Application Layer is heavily impacted, both in the design and implementation stage (many changes to source code and application design) and in the deployment stage (obtaining and installing server digital certificates into the server application). You could say that all applications get a "free ride" over IPsec, gaining security features (without having to include any support for security themselves) simply by running over Internet Layer links secured with IPsec.

By *security* we are here referring to three specific aspects of security:

> *Privacy* (keeping others from being able to view the content of your transmissions), accomplished using the Encapsulating Security Payload (ESP) feature

> *Authentication* (knowing for sure whom the packets came from), accomplished with the Authentication header (AH) feature

---

[3] https://tools.ietf.org/html/rfc6434

> ***Message integrity*** (knowing if someone has made any changes
> to the Data field or certain fields of the header, including the
> source and destination addresses), also accomplished with the
> Authentication header (AH) feature

The AH and ESP features are mutually independent. You can make use of neither, either, or both, depending on your requirements. If you need only authentication and message integrity, you can use only AH. If you need *only* privacy, you can use only ESP. If you require both, you can use both AH and ESP (there is no conflict between them).

Note, however, that since AH protects the source and destination addresses in the IP header and the source and destination ports in the TCP or UDP header, any changes at all to these fields will be detected as tampering (to AH there is no way to distinguish malicious tampering from changes your own network makes to these fields). This means AH will report the changes to IP addresses and ports in the header by NAT to be a *hacking attack*. It is possible to combine NAT traversal with IPsec (as discussed in RFC 3715,[4] "IPsec-Network Address Translation (NAT) Compatibility Requirements"), but this greatly complicates the creation and deployment of IPsec applications and introduces new security issues, which may outweigh the benefits of using IPsec in the first place. IPsec is not very suitable for use on existing IPv4 networks since NAT is so widely deployed. Many network professionals have gotten a bad impression of it, but this is because of NAT in IPv4 networks, not any shortcoming of IPsec. There is no other technology for building VPNs supported by the IETF. IPsec works *great* in IPv6 networks. This is because of better support in IPv6 headers to some extent, but primarily because there is no NAT.

Note that IPsec AH does not use a "heavyweight" scheme like PKI digital signatures for authentication. It uses the much lighter-weight HMAC[5] scheme (a hash algorithm that uses a key). If it used real digital signatures (based on asymmetric key cryptography), it would reduce throughput dramatically.

Any security features at this level (which must be performed once per packet) must be very lightweight (they cannot use mechanisms that require a lot of CPU power). This rules out the use of asymmetric key cryptography (as used in digital signatures and digital envelopes), at least on a per-packet basis. Use of asymmetric key cryptography would cause severe degradation of network throughput. Fortunately, there is a

---

[4] https://tools.ietf.org/html/rfc3715
[5] https://en.wikipedia.org/wiki/HMAC

lightweight alternative to digital signatures, which is *Hash-Based Message Authentication Codes* (essentially a key-driven message digest), which is used in AH. Encryption is handled using only symmetric key encryption and decryption with the same symmetric key for many packets. The key can be manually distributed (*shared secret deployment*) or securely distributed via the Internet Key Exchange (IKE) protocol, which *does* use asymmetric key cryptography (but IKE is used infrequently – the exchanged key is used to encrypt or decrypt a large number of packets before another key is exchanged). Even with these lightweight algorithms, there can still be an impact on network throughput (especially on systems with lower-performance CPUs). Network Interface Cards (NICs) are available that include hardware acceleration of the IPsec algorithms (HMAC generation and checking, as well as symmetric key encryption/decryption). These allow wire-speed network throughput even on systems with low-performance CPUs.

Relevant Standards for IPsec

The following standards are relevant to IPsec and IKE:

> RFC 2410, "The NULL Encryption Algorithm and Its Use with IPsec," November 1998 (Standards Track)

> RFC 2412, "The Oakley Key Determination Protocol," November 1998 (Informational)

> RFC 2709, "Security Model with Tunnel-Mode IPsec for NAT Domains," October 1999 (Informational)

> RFC 3193, "Securing L2TP Using IPsec," November 2001 (Standards Track)

> RFC 3554, "On the Use of Stream Control Transmission Protocol (SCTP) with IPsec," July 2003 (Standards Track)

> RFC 3456, "Dynamic Host Configuration Protocol (DHCPv4) Configuration of IPsec Tunnel Mode," January 2003 (Standards Track)

> RFC 3457, "Requirements for IPsec Remote Access Scenarios," January 2003 (Informational)

RFC 3554, "On the Use of Stream Control Transmission Protocol (SCTP) with IPsec," July 2003 (Standards Track)

RFC 3566, "The AES-XCBC-MAC-96 Algorithm and Its Use with IPsec," September 2003 (Standards Track)

RFC 3585, "IPsec Configuration Policy Information Model," August 2003 (Standards Track)

RFC 3602, "The AES-CBC Cipher Algorithm and Its Use with IPsec," September 2003 (Standards Track)

RFC 3715, "IPsec-Network Address Translation (NAT) Compatibility Requirements," March 2004 (Informational)

RFC 3884, "Use of IPsec Transport Mode for Dynamic Routing," September 2004 (Informational)

RFC 3947, "Negotiation of NAT-Traversal in the IKE," January 2005 (Standards Track)

RFC 3948, "UDP Encapsulation of IPsec ESP Packets," January 2005 (Standards Track)

RFC 4025, "A Method for Storing IPsec Keying Material in DNS," March 2005 (Standards Track)

RFC 4106, "The Use of Galois/Counter Mode (GCM) in IPsec Encapsulating Security Payload (ESP)," June 2005 (Standards Track)

RFC 4109, "Algorithms for Internet Key Exchange version 1 (IKEv1)," May 2005 (Standards Track)

RFC 4196, "The SEED Cipher Algorithm and Its Use with IPsec," January 2006 (Standards Track)

RFC 4301, "Security Architecture for the Internet Protocol," December 2005 (Standards Track)

RFC 4302, "IP Authentication Header," December 2005 (Standards Track)

**RFC 4303, "IP Encapsulating Security Payload (ESP),"
December 2005 (Standards Track)**

RFC 4304, "Extended Sequence Number (ESN) Addendum
to IPsec Domain of Interpretation (DOI) for Internet Security
Association and Key Management Protocol (ISAKMP)," December
2005 (Standards Track)

**RFC 4308, "Cryptographic Suites for IPsec," December 2005
(Standards Track)**

**RFC 4309, "Using Advanced Encryption Standard (AES) CCM
Mode with IPsec Encapsulating Security Payload (ESP),"
December 2005 (Standards Track)**

RFC 4312, "The Camellia Cipher Algorithm and Its Use with
IPsec," December 2005 (Standards Track)

RFC 4322, "Opportunistic Encryption Using the Internet Key
Exchange (IKE)," December 2005 (Informational)

RFC 4430, "Kerberized Internet Negotiation of Keys (KINK),"
March 2006 (Standards Track)

RFC 4434, "The AES-XCBC-PRF-128 Algorithm for the Internet
Exchange Protocol (IKE)," February 2006 (Standards Track)

RFC 4494, "The AES-CMAC-96 Algorithm and Its Use with IPsec,"
June 2006 (Standards Track)

RFC 4543, "The Use of Galois Message Authentication Code
(GMAC) in IPsec ERP and AH," May 2006 (Standards Track)

RFC 4555, "IKEv2 Mobility and Multihoming Protocol (MOBIKE),"
June 2006 (Standards Track)

**RFC 4615, "The Advanced Encryption Standard-Cipher-based
Message Authentication Code-Pseudo-Random Function-128
(AES-CMAC-PRF-128) Algorithm for the Internet Key
Exchange Protocol (IKE)," August 2006 (Standards Track)**

RFC 4807, "IPsec Security Policy Database Configuration MIB,"
March 2007 (Standards Track)

RFC 4809, "Requirements for an IPsec Certificate Management Profile," February 2007 (Informational)

RFC 4868, "Using HMAC-SHA-256, HMAC-SHA-384, and HMAC-SHA-512 with IPsec," May 2007 (Standards Track)

RFC 4877, "Mobile IPv6 Operation with IKEv2 and the Revised IPsec Architecture," April 2007 (Standards Track)

RFC 4891, "Using IPsec to Secure IPv6-in-IPv4 Tunnels," May 2007 (Informational)

RFC 4894, "Use of Hash Algorithms in Internet Key Exchange (IKE) and IPsec," May 2007 (Informational)

RFC 4945, "The Internet IP Security PKI Profile of IKEv1/ ISAKMP, IKEv2, and PKIX," August 2007 (Standards track)

RFC 5265, "Mobile IPv4 Traversal Across IPsec-Based VPN Gateways," June 2008 (Standards Track)

RFC 5374, "Multicast Extensions to the Security Architecture for the Internet Protocol," November 2008 (Standards Track)

RFC 5386, "Better-Than-Nothing Security: An Unauthenticated Mode of IPsec," November 2008 (Standards Track)

RFC 5387, "Problem and Applicability Statement for Better-Than-Nothing Security (BTNS)," November 2008 (Informational)

RFC 5406, "Guidelines for Specifying the Use of IPsec Version 2," February 2009 (Best Current Practice)

RFC 5529, "Modes of Operation for Camellia for Use with IPsec," April 2009 (Standards Track)

RFC 5566, "BGP IPsec Tunnel Encapsulation Attribute," June 2009 (Standards Track)

RFC 5660, "IPsec Channels: Connection Latching," October 2009 (Standards Track)

RFC 5755, "An Internet Attribute Certificate Profile for Authorization," January 2010 (Standards Track)

RFC 5856, "Integration of Robust Header Compression over IPsec Security Associations," May 2010 (Informational)

RFC 5857, "IKEv2 Extensions to Support Robust Header Compression over IPsec," May 2010 (Standard)

RFC 5858, "IPsec Proposed Extensions to Support Robust Header Compression over IPsec," May 2010 (Proposed Standard)

RFC 5879, "Heuristics for Detecting ESP-NULL Packets," May 2010 (Informational)

RFC 6027, "IPsec Cluster Problem Statement," October 2010 (Informational)

RFC 6071, "IP Security (IPsec) and Internet Key Exchange (IKE) Document Roadmap," February 2011 (Informational)

RFC 6040, "Tunneling of Explicit Congestion Notification," November 2010 (Proposed Standard)

RFC 6071, "IP Security (IPsec) and Internet Key Exchange (IKE) Document Roadmap," February 2011 (Informational)

RFC 6151, "Updated Security Considerations for the MD5 Message-Digest and the HMAC-MD5 Algorithms," March 2011 (Informational)

RFC 6193, "Media Description for the Internet Key Exchange Protocol (IKE) in the Session Description Protocol (SDP)," April 2011 (Informational)

RFC 6311, "Protocol Support for High Availability of IKEv2/IPsec," (Standards Track)

RFC 6379, "Suite B Cryptographic Suites for IPsec," October 2011 (Historic)

RFC 6380, "Suite B Profile for Internet Protocol Security (IPsec)," October 2011 (Historic)

RFC 6467, "Secure Password Framework for Internet Key Exchange Version 2 (IKEV2)," December 2011 (Informational)

RFC 6479, "IPsec Anti-Reply Algorithm Without Bit Shifting," January 2012 (Informational)

RFC 6538, "The Host Identity Protocol (HIP) Experiment Report," March 2012 (Informational)

RFC 7018, "Auto-Discovery VN Problem Statement and Requirements," September 2012 (Informational)

RFC 7146, "Securing Block Storage Protocols over IP: RFC 3723 Requirements Update for IPsec v3," April 2014 (Proposed Standard)

RFC 7296, "Internet Key Exchange (IKEv2) Protocol," December 2005 (Standards Track)

RFC 7427, "Signature Authentication in the Internet Key Exchange Version 2 (IKEv2)," October 2014 (Proposed Standard)

RFC 7634, "ChaCha20, Poly1305 and Their Use in the Internet Key Exchange Version 2 (IKEv2)," January 2015 (Proposed Standard)

RFC 7815, "Minimal Internet Key Exchange Version 2 (IKEV2) Initiator Implementation," March 2016 (Informational)

RFC 7236, "Guidelines on the Cryptographic Algorithms to Accompany the Usage of Standards GOST R 34.10-2012 and GOST R 34.11-2012," October 2017 (Proposed Standard)

RFC 8221, "Cryptographic Algorithm Implementation Requirements and Usage Guidance for Encapsulating Security Payload (ESP) and Authentication Header (AH)," October 2017 (Proposed Standard)

RFC 8229, "TCP Encapsulation of IKE and IPsec Packets," August 2017 (Proposed Standard)

RFC 8247, "Algorithm Implementation Requirements and Usage Guidance for the Internet Key Exchange Version 2 (IKEv2)," September 2017 (Standards Track)

Security Association, Security Association Database, and Security Parameter Index

A *Security Association* (SA) is a collection of which protocols and algorithms to use for authentication and encryption, together with the keys used, for communication in one direction between two IPsec-enabled nodes (e.g., from Alice to Bob). A different SA is created for communication in the other direction, between the same two nodes (e.g., from Bob to Alice). Each pair of communicating IPsec nodes requires an SA for each direction that data will be sent between them (normally both directions). Each node stores both SAs for a connection in its Security Association Database (SADB). It refers to the SA for outgoing traffic when it sends packets and the SA for incoming traffic when it receives packets. When a secure connection is set up the first time (or if anything is changed), then the relevant Security Associations are negotiated and stored.

A *Security Association Database* (SADB) is a collection of Security Associations, on a given node. Each IPsec-enabled node has its own SADB (this is not stored in a central DBMS and in fact does not resemble what most people would call a database – it's really more of a simple table). As the node negotiates Security Associations with other nodes, it stores each new one into its SADB.

The *Security Parameter Index* (SPI) is an index into the SADB. An SPI together with a destination IP address uniquely identifies a particular Security Association.

IPsec Transport Mode and IPsec Tunnel Mode

There are two *modes* in which a given IPsec connection can operate: *Transport Mode* and *Tunnel Mode*.

In *Transport Mode*, with AH, only the packet payload and certain fields in the packet headers (including source and destination IP addresses and source and destination port numbers) are authenticated. This means the contents of the payload and those header fields are included in the calculation of the AH cryptographic checksum using HMAC. None of those fields are modified in any way by IPsec. Therefore, the original addresses are used for routing of the packet. However, if NAT modifies any of these header fields as the packet goes through a NAT gateway (which is the normal behavior of NAT), then the cryptographic checksum will fail when the packet is received (after all, someone has tampered with the packet contents).

In IPv4 Transport Mode, the AH packet header is inserted after the IP header, but before the TCP (or UDP) header, as follows:

```
                    BEFORE APPLYING AH

        ----------------------------
IPv4   |orig IP hdr  |     |        |
       |(any options)| TCP | Data   |
        ----------------------------

                    AFTER APPLYING AH

        -----------------------------------------------------
IPv4   |original IP hdr (any options) | AH | TCP |   Data    |
        -----------------------------------------------------
       |<- mutable field processing ->|<- immutable fields ->|
       |<----- authenticated except for mutable fields ----->|
```

In IPv6 Transport Mode, the original IP packet header comes first, followed by one or more extension headers, one of which is the new AH, followed by the TCP (or UDP) header and then the Data field (payload).

```
                    BEFORE APPLYING AH

        -----------------------------------------
IPv6   |             | ext hdrs |     |       |
       | orig IP hdr |if present| TCP | Data  |
        -----------------------------------------

                    AFTER APPLYING AH

        ---------------------------------------------------------
IPv6   |             |hop-by-hop, dest*, |   | dest |     |     |
       |orig IP hdr  |routing, fragment. | AH| opt* | TCP | Data|
        ---------------------------------------------------------
       |<--- mutable field processing -->|<-- immutable fields -->|
       |<---- authenticated except for mutable fields ---------->|

          * = if present, could be before AH, after AH, or both
```

In *Transport Mode* with ESP, only the TCP (or UDP) header and the Data field (payload) are encrypted. No other header fields are encrypted, or the packet could not be delivered. Transport mode is used only for host-to-host communications. No IPsec gateway is required. Any node involved in a Transport Mode IPsec connection must have support for IPsec Transport Mode. If automated key exchange is to be used, those nodes must also support a common key exchange protocol (IKEv1, IKEv2, or KINK). If available, IKEv2 is preferred. If IKE is used, each node requires an appropriate IPsec digital certificate that binds the public key to its IP address(es). If KINK is used, then a Kerberos Key Distribution Center (KDC) must be available to all nodes using it.

In IPv4 Transport Mode with ESP, the original IP header comes first, followed by the ESP header, followed by the TCP (or UDP) header, followed by the data. With ESP, after the Data field, there is an *ESP trailer* and an *Integrity Check Value* (ICV). Encryption is done on the TCP (or UDP) header, the Data field (payload), and the ESP trailer. Integrity (for the ICV) covers those fields plus the ESP header.

```
             BEFORE APPLYING ESP
         ----------------------------
 IPv4   |orig IP hdr  |     |      |
        |(any options)| TCP | Data |
         ----------------------------

             AFTER APPLYING ESP
         ---------------------------------------------------
 IPv4   |orig IP hdr  | ESP |     |      |   ESP   | ESP|
        |(any options)| Hdr | TCP | Data | Trailer | ICV|
         ---------------------------------------------------
                             |<---- encryption ---->|
                       |<-------- integrity ------->|
```

In IPv6 Transport Mode with ESP, the original IP header comes first, followed by one or more extension headers, one of which is the ESP extension header, followed by the original TCP (or UDP) header and then the Data field (the packet payload). As with IPv4, there is an *ESP trailer* and an *ESP ICV*. Encryption is done on any extension headers after the ESP extension header, the TCP (or UDP) header, the Data field (payload), and the ESP trailer. Integrity (for the ICV) covers those fields plus the ESP header.

```
               BEFORE APPLYING ESP
          ----------------------------------------
 IPv6   |                | ext hdrs |     |      |
        | orig IP hdr  |if present| TCP | Data |
```

```
-----------------------------------------
            AFTER APPLYING ESP
         -------------------------------------------------------
IPv6  | orig |hop-by-hop,dest*,|    |dest|   |   | ESP  | ESP|
      |IP hdr|routing,fragment.|ESP|opt*|TCP|Data|Trailer| ICV|
      -------------------------------------------------------
                                  |<--- encryption ---->|
                                |<------ integrity ------>|
```

               * = if present, could be before ESP, after ESP, or both

In *Tunnel Mode* (with AH and/or ESP), the entire original IP packet (all headers plus the payload) is encrypted and/or authenticated. The result is encapsulated into a new IP packet with new headers. This encapsulation is added to packets on the way out by an IPsec tunnel gateway and removed on the way at the other end of the network path by another IPsec tunnel gateway. In between, it looks like a normal IPv4 (or IPv6) packet that has an odd-looking payload and is routed like any other packet. The IP version of the inner packet does not have to be the same as the IP version of the outer packet. You can tunnel IPv6 packets over IPv4 or IPv4 packets over IPv6. The node you connect to must support the version(s) of IP you send to it, though. After the encapsulation is removed and the authentication and/or encryption is removed, the resulting packet is forwarded to the inside of the tunnel gateway, where it continues on to its destination.

In IPv4 Tunnel Mode with AH, the outer IP header comes first (possibly with options), followed by the new AH header and followed by the original entire packet.

In IPv6 Tunnel Mode with AH, the outer IP header comes first, followed by one or more extension headers, including the AH extension header, followed by the original entire packet (which itself may contain extension packet headers, but they won't be processed until after the packet is de-tunneled).

```
         -------------------------------------------------------
IPv4 |                           |   | orig IP hdr* |  |      |
     |new IP header * (any options) | AH | (any options) |TCP| Data |
     -------------------------------------------------------
       |<- mutable field processing ->|<------ immutable fields ----->|
       |<- authenticated except for mutable fields in the new IP hdr->|

     -------------------------------------------------------
```

```
IPv6 |                 | ext hdrs*|    |                | ext hdrs*|    |    |
     |new IP hdr*|if present| AH |orig IP hdr*|if present|TCP|Data|
     -----------------------------------------------------------------
     |<--- mutable field -->|<--------- immutable fields -------->|
     |         processing        |
     |<-- authenticated except for mutable fields in new IP hdr ->|

       * = if present, construction of outer IP hdr/extensions and
           modification of inner IP hdr/extensions is discussed in
           the Security Architecture document.
```

In IPv4 Tunnel Mode with ESP, the outer IP header comes first, followed by the ESP header, followed by any options from the original packet header, followed by the TCP (or UDP) header, followed by the Data field (payload). Again, the data is followed by an *ESP trailer* and *ESP ICV*. Encryption covers everything after the ESP header, up to and including the ESP trailer. The integrity value covers all that plus the ESP header.

In IPv6 Tunnel Mode with ESP, the outer IP packet header comes first followed by any new packet header extensions, followed by the ESP header and then the entire original packet. As before, the Data field (payload) is now followed by the *ESP trailer* and the *ESP ICV*. Encryption covers everything after the ESP header, up to and including the ESP trailer. The integrity value covers all that plus the ESP header.

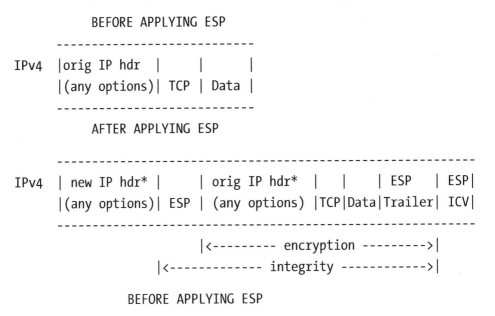

```
                    BEFORE APPLYING ESP
          ----------------------------
IPv4  |orig IP hdr |     |      |
      |(any options)| TCP | Data |
          ----------------------------
                    AFTER APPLYING ESP

          -------------------------------------------------------------
IPv4  | new IP hdr* |     | orig IP hdr* |    |    | ESP   | ESP|
      |(any options)| ESP | (any options) |TCP|Data|Trailer| ICV|
          -------------------------------------------------------------
                              |<--------- encryption --------->|
                              |<------------ integrity ----------->|

                    BEFORE APPLYING ESP
```

274

```
              ----------------------------------------
IPv6   |                    | ext hdrs |     |      |
       |  orig IP hdr |if present|  TCP  | Data |
              ----------------------------------------

                      AFTER APPLYING ESP

              ----------------------------------------------------------
IPv6   |  new*  |new ext |    |  orig*|orig ext |    |    | ESP   | ESP|
       |IP hdr|  hdrs*   |ESP|IP hdr|  hdrs *  |TCP|Data|Trailer|  ICV|
              ----------------------------------------------------------
                            |<--------- encryption ---------->|
                            |<------------ integrity ------------>|
```

  * = if present, construction of outer IP hdr/extensions and
      modification of inner IP hdr/extensions is discussed in
      the Security Architecture document.

The IPsec tunnel processing can be physically located inside a gateway firewall or router, or it can be in a node that does *just* IPsec tunneling, on the inside of an existing router or firewall. All IPsec-related processing (generating or validating the HMAC cryptographic checksum and/or packet encryption/decryption) takes place in the tunneling IPsec node. IPsec Tunnel Mode is primarily used for network-to-network tunnels, but it *can* be used for host-to-network communications (e.g., a road warrior connecting into the home network securely) or even host-to-host communications (e.g., for private chat or VoIP). If any hosts (as opposed to gateways) are involved in an IPsec Tunnel Mode connection, those hosts would need to support IPsec Tunnel Mode. If the tunnel is built between two gateway nodes (network-to-network tunnel), then any other node in either network can send things through that tunnel to nodes in the other network without having to know anything about IPsec. If automated key exchange is to be used, the participating nodes must also support a common key exchange protocol (IKEv1, IKEv2, or KINK). If available, IKEv2 is preferred. If IKE is used, each node requires an appropriate IPsec digital certificate that binds the public key to its IP address(es), for mutual authentication. If KINK is used, then a Kerberos Key Distribution Center must be available to all nodes using it.

An IPsec tunnel must be set up in the sending and receiving nodes so that the sending node knows what address to put in the *outer packet header*. This configuration would specify things like "All traffic destined for 123.45.56.00/24 is to be tunneled and sent to the gateway located at 123.45.67.1" or "Accept IPsec-tunneled traffic from the node that is located at 87.65.34.21, de-tunnel it, and route the inside packet onto the LAN." Regardless of what transport the original traffic used (TCP or UDP), the tunneled traffic will be over UDP. This introduces additional overhead and can complicate the built-in error recovery mechanism in TCP. The IP addresses in the outer packet header are not authenticated.

If IPv4-tunneled traffic crosses any router, the addresses of all nodes must be valid global unicast IP addresses (although this could be simulated using BINAT with a node that has a private IP address behind a NAT gateway). It would be possible to create an IPsec path over IPv4 entirely within a routing domain using private IP addresses. Basically, there must be a flat address space (or a reasonable facsimile thereof) over the entire path of the connection. A classic problem with VPNs connecting nodes in disjoint private networks is that the private address spaces in the two networks must not *overlap*. If two companies are both using 10.0.0.0/8 as their private addresses, either VPNs will not work between them or at least one network must renumber (e.g., to 172.16.0.0/12). If you have an HQ network and several branch offices, you might want to use 10.0.0.0/16 for HQ, 10.1.0.0/16 for the first branch office, 10.2.0.0/16 for the second branch office, etc. This would allow up to 256 branch offices, each with up to 65,535 nodes. This way, if you do want to build VPNs between them someday, there will not be any overlap.

IPsec over IPv6

In IPv6, none of the issues related to NAT arise, since all nodes can easily obtain global unicast IPv6 addresses, and the entire world is a single flat address space. IPsec works *beautifully* over IPv6, and this is one of the strongest arguments for migrating to dual stack sooner rather than later. Use the IPv6 side for protocols that require a flat address space, such as IPsec, SIP, P2P, etc. Use the IPv4 side for legacy applications like web surfing, email, etc. You can gradually move those over to IPv6 as well.

The IPv6 packet header design supports IPsec very well. There are two packet header extensions defined, one for AH and one for ESP. The AH extension header will be inserted if and only if AH is used on that packet. For details, see RFC 4302, "IP Authentication Header." The ESP extension header will be inserted if and only if ESP is used on that packet. For details, see RFC 4303, "IP Encapsulating Security Payload (ESP)."

IPsec in Multicast Networks

It is possible to deploy IPsec in a multicast network. Security Association negotiation is rather more complicated in a one-to-many connection than in a one-to-one connection. RFC 5374, "Multicast Extensions to the Security Architecture for the Internet Protocol," covers the details. IPsec in multicast networks could allow content providers to control access to multicast content using ESP and some clever key management. For example, each valid subscriber could be issued a unique IPsec digital certificate (tied to their set top box) that would allow decrypting the symmetric session key used to encrypt the content. If they don't pay their bill, their certificate could be revoked.

Using IPsec to Secure L2TP Connections

L2TP (Layer 2 Tunneling Protocol) itself does not provide either privacy or authentication. RFC 3193, "Securing L2TP Using IPsec," specifies how to use IPsec in conjunction with it to add privacy (with ESP) and/or authentication (with AH) to an L2TP-based system.

Internet Key Exchange (IKE)

The Internet Key Exchange (IKEv1 and IKEv2) is based on ISAKMP (Internet Security Association and Key Management Protocol), which is a *framework* for key exchange. It uses parts of the Oakley and SKEME (*Secure Key Exchange MEchanism for Internet*) protocols within this framework. Oakley describes a series of key exchanges, known as modes, and specifies things such as *perfect forward secrecy* for keys, identity protection, and authentication. It is discussed in RFC 2412, "The Oakley Key Determination Protocol." SKEME is a key exchange technique that provides anonymity and quick key refreshment. There is no RFC that covers SKEME, but there are some papers available online. There is coverage of the parts of Oakley and SKEME used in IKE in RFCs 2408 and 2409.

IKE uses the Diffie-Hellman Key Agreement protocol to securely exchange a shared secret, from which symmetric session keys for AH and ESP are derived. IKE is also used to mutually authenticate nodes to each other. Authentication can be accomplished with a pre-shared secret (manually distributed to each node) or by use of IPsec digital certificates (ones that bind IPv4 and/or IPv6 addresses to the public key). Each pair of nodes that use IPsec use IKE to do key exchange and mutual authentication, which results in setting up a Security Association (SA) at each end for the other node in the pair.

IKE is usually implemented as a daemon process (a software application that starts running when the computer boots and stays running until shutdown) on each IPsec-enabled node, in *user space* (the part of memory where user applications run). IKE communicates via UDP over port 500. There is no client or server role; the communication is between peers. Either node of a pair can initiate an IKE connection. The other node of the pair will accept it. The AH and ESP packet processing is embedded in the TCP/IP stack (specifically in the IP Layer), which usually runs in *Kernel Space* (the part of memory where the operating system kernel runs, typically protected from access by user applications).

Note in the following that Oakley defines *modes*: Main Mode, Aggressive Mode, and Quick Mode. ISAKMP defines *phases*: phase 1 and phase 2. Main Mode and Aggressive Mode take place during ISAKMP's phase 1, while Quick Mode takes place during ISAKMP's phase 2.

IKE phase 1 establishes an encrypted, authenticated communication channel between the two parties. It first uses the Diffie-Hellman Key Agreement protocol, which produces a shared secret. A symmetric session key is derived from the shared secret by both parties, which is used to encrypt further IKE exchanges. The goal of phase 1 is to create a *bidirectional* ISAKMP Security Association (SA). Mutual authentication can be accomplished either using a pre-shared secret or via cryptographic challenge/response using IPsec public key digital certificates. Either a pre-shared secret or an IPsec certificate must be installed on each IPsec-enabled node when the system is deployed. Phase 1 can operate in either *Main Mode* or *Aggressive Mode*. Main Mode protects the identities of the nodes but takes longer. Aggressive Mode is faster but does not protect the identities of the nodes.

IKE phase 2 uses the secure channel established in phase 1 to negotiate additional Security Associations (SAs) for services such as IPsec. The output of phase 2 is a pair of *unidirectional* SAs (one for traffic from Alice to Bob and one for traffic from Bob to Alice). On each node, one of these SAs is used for inbound traffic and the other one for outbound traffic. Phase 2 operates in *Quick Mode*.

Cryptographic challenge/response is used by each end to authenticate itself to the other. This is based on public/private key pairs (asymmetric cryptography) and digital certificates. Each direction works as follows (flipping roles A and B when the second direction is done):

Step 1: Node A sends its public key digital certificate to Node B.

Step 2: Node B verifies Node A's digital certificate by checking its digital signature, its expiration date, and its revocation status. It also climbs the chain of trust to a trusted root key. If the identity in the certificate is an IP address, it must match the source IP address of the IKE connection.

Step 3: Node B generates a random string and encrypts it using Node A's public key (from its digital certificate) and sends it as a *challenge* to Node A.

Step 4: Node A decrypts the *challenge* using its own private key and returns the result to Node B as its *response*.

Step 5: Node B compares the *response* with the original string. If they match, that is proof that Node A possesses the private key associated with the public key in Node A's digital certificate (without Node A revealing its private key to anyone). This authenticates Node A to Node B.

An IPsec digital certificate is like a server (SSL) digital certificate or a client digital certificate. The primary difference is what identity information the public key is bound to. In a *server* cert, the identity information is an FQDN (e.g., `www.example.com`) and an organization name (in the distinguished name). In a *client* cert, the identity information is a person's name and email address (in the distinguished name). It is also possible for an IPsec certificate to specifically include IKE as a valid key usage (id_kp_ipsecIKE attribute). In an IPsec cert, there are several possibilities for the identity:

- Individual IPv4 and/or IPv6 address (ID_IPV4_ADDR and ID_IPV6_ADDR). These do not work well if the connection traverses NAT. There is no problem with using IPv6 addresses.

- IPv4 or IPv6 subnet (ID_IPv4_ADDR_SUBNET and ID_IPv6_ADDR_SUBNET). The same issues are involved if the connection traverses NAT. For example, you could identify your node as being in the subnet *2001:418:5403:3000::/64*.

- IPv4 or IPv6 address range (ID_IPv4_ADDR_RANGE and ID_IPV6_ADDR_RANGE). This is used for specifying a block of addresses that doesn't happen to fall on (power of 2) subnet boundaries (e.g., all addresses from *2001:418:5403:3000::100* to *2001:418:5403:3000::200*).

- FQDN (ID_FQDN). This depends on trusting the mapping from FQDN to IP address; hence, DNS should only be used if DNSSEC is deployed. Otherwise, the resolution from FQDN to IP address must be handled by some other means, which *is* trusted. Again, NAT can cause problems with this.

If IP addresses are specified, during the authentication process, the source IP address of the IKE connection must match the IP address (byte for byte) in the IPsec digital certificate. This is similar to SSL/TLS, where the nodename you are connecting to must match the FQDN in the server certificate. In SSL, if you connect to an alias name or a numeric IP address, you will get an error.

Internet Key Exchange Version 2 (IKEv2)

IKEv1 was defined in November 1998. There are still some IPsec implementations that support only IKEv1, so some IKEv2 implementations also support IKEv1 for backward compatibility (e.g., RACOON2 in BSD). There were many issues with IKEv1, which led to the creation of IKEv2 in December 2005. The issues with IKEv1 include the following:

- The specification of IKEv1 was spread over three basic RFCs (2407, 2408, and 2409), plus others for NAT traversal (3715) and so on. In comparison, almost all of IKEv2 was specified in RFC 4306. The list of supported cryptographic algorithms was split off into RFC 4307 for ease of updating the algorithm suite in the future.

- IKEv1 had no support for SCTP or Mobile IP. Both are supported in IKEv2. For SCTP, see RFC 3554, "On the Use of Stream Control Transmission Protocol (SCTP) with IPsec." For Mobile IP, see RFC 4877, "Mobile IPv6 Operation with IKEv2 and the Revised IPsec Architecture" (section 7.3), and RFC 4555, "IKEv2 Mobility and Multihoming Protocol (MOBIKE)."

- IKEv1 had *eight* distinct initial exchange mechanisms, each of which had advantages and disadvantages, vs. *one* four-message initial exchange in IKEv2.

- IKEv1 included an excessive number of cryptographic algorithms, which resulted in complex implementation and long (and costly) certification processes (e.g., Common Criteria and FIPS 140-2). For details, see RFC 2409, updated by RFC 4109, "Algorithms for Internet Key Exchange version 1 (IKEv1)." In comparison, IKEv2 reduced the number of supported cryptographic algorithms. For details, see RFC 4306, "Cryptographic Algorithms for Use in the Internet Key Exchange Version 2 (IKEv2)."

- IKEv1 had reliability issues due to poor state management. This could result in a hung node, requiring Dead Peer Detection (which was never standardized, leading to interoperability issues). IKEv2 added sequence numbers and acknowledgements to greatly improve state management.

- IKEv1 had issues with Denial of Service attacks, where connections from spoofed addresses could cause it to do *expensive* asymmetric key processing. IKEv2 does such processing only *after* it verifies the validity and existence of a client.

- The IPv6-ready test centers have no certification tests for IKEv1, but they do for IKEv2. They also have certification tests for IPsec over IPv6.

Some IPv4 routers (especially ones for home or small office) include an *IPsec Helper* function. This routes all IPsec traffic to the first node that negotiates a Security Association (SA). The assumption is that there is only one IPsec endpoint inside the home network. This is yet another attempt to make IPsec work over NAT. In IPv6 routers, there is no need for any IPsec Helper function.

Another issue with NAT involves fragmentation. During IKE, if an IPsec digital certificate is sent, this often is larger than a single packet, which leads to packet fragmentation. Many NAT gateways will simply drop fragmented packets, as these are usually part of a hacking attack. This is never a problem in IPv6.

It should be obvious by now why there are so many problems using IPsec in the legacy Second Internet. NAT helped keep the Internet going while IPv6 was being created, but now that IPv6 is complete and available, we should switch to it *at least* for certain protocols (especially IPsec and SIP) rather than create ever more complex workarounds to fix the problems caused by NAT.

Kerberized Internet Negotiation of Keys: KINK

There is an alternative to IKE for securely exchanging keys among IPsec-enabled nodes, called *KINK*. It is defined in RFC 4430,[6] "Kerberized Internet Negotiation of Keys (KINK)." With IKE, each node needs an IPsec digital certificate to authenticate itself to other nodes, which is done on a peer-to-peer basis. A full PKI must be deployed to support the issuance and maintenance of these certificates. With KINK, nodes need only mutually authenticate with the authentication server of the Key Distribution Center (KDC) of a Kerberos facility. No IPsec digital certificates are required for each node, and no PKI is required. However, there must be a Kerberos KDC that can do the necessary authentication, and all participating IPsec-enabled nodes need client-side support for Kerberos and KINK. Deploying Kerberos securely can be just as big a challenge as deploying a PKI. If your nodes will need to connect over IPsec to nodes in other organizations, KINK is probably not the best way to go. IKE is clearly the preferred key exchange technology.

Summary

In this chapter we covered IPsec (IP Layer security) and IKE (Internet Key Exchange). We compared it with SSL-VPN (which is not an IETF standard and uses HTTPS in a nonstandard way). With IPv4, IPsec and IKE are blocked by NAT, so SSL-VPN has become widely used there. In IPv6, there is no NAT to block IPsec and IKE, so it will quickly replace SSL-VPN.

IKEv1 is now deprecated, so everyone should be using IKEv2 at this point.

It is possible to do mutual authentication with IPsec, which requires use of IPsec VPN digital certificates at both ends. These are normally tied to the devices, not to users.

---

[6] https://tools.ietf.org/html/rfc4430

There was some confusion among early users of IPv6 that IPsec was mandatory on all connections. It *is* mandatory for it to be supported in an IPv6 implementation, but it is *not* mandatory to use it on every connection. Therefore, with regard to IPsec, IPv6 is no more or less secure than IPv4. Either version of IP is more secure if IPsec is used and less secure if IPsec is not used.

We also covered an alternative to IKE called KINK.

CHAPTER 8

Transition Mechanisms

This chapter covers a variety of protocols and mechanisms that were created to simplify the introduction of IPv6 into the Internet. The goal is not to make an abrupt transition from all-IPv4 to all-IPv6 on some kind of "flag day" (as happened in the transition from the First Internet to the Second Internet*). That would be unbelievably disruptive and unlikely to succeed. The goal is to gradually add new capabilities that take advantage of IPv6, or work far better over it (e.g., IPsec VPN, SIP, IPTV,[1] and most other multicast), while continuing to use IPv4 for those things that work tolerably well over IPv4 with NAT (e.g., web, email, FTP, SSH,[2] and most client-server with intermediary servers). This allows immediate alleviation of the most grievous problems caused by widespread deployment of NAT and other shortcomings of IPv4 while allowing a longer, more controlled migration of those protocols that do not benefit as much from IPv6. Eventually, all protocols and applications will be migrated (with a few exceptions – likely Skype can never be ported to IPv6, being heavily based on NAT traversal), and IPv4 can quietly be dropped from operating systems and hardware. However, this will probably be 5–10 years from now. As more and more applications are transitioned to IPv6, that will take the pressure off the remaining stock of IPv4 addresses.

Most of these transition mechanisms are defined in RFCs as part of the IPv6 standards. There are many mechanisms, some with confusingly similar names, such as "6in4," "6to4," and "6over4," which are all quite different. Most deployments of IPv6 will use one or more of these transition mechanisms; none will use all of them. Some of the transition mechanisms are designed for use in the early phases of the transition, where there is an "ocean" of IPv4 with small (but growing) islands of IPv6 (e.g., 6in4 tunneling). Some are for use in the later stages of the transition, where the Internet has flipped into an "ocean" of IPv6, with small (and shrinking) islands of IPv4 (e.g., 4in6 tunneling, Dual-Stack Lite). Some are for use in the end stages of the transition where some networks are

---

[1] https://en.wikipedia.org/wiki/IPTV

[2] https://en.wikipedia.org/wiki/Secure_Shell

L. E. Hughes, *Third Generation Internet Revealed*, https://doi.org/10.1007/978-1-4842-8603-6_8

"IPv6-only" with no IPv4 present (e.g., NAT64/DNS64 to allow reaching legacy external IPv4-only servers from an IPv6-only node).

Since 2010, Teredo, ISATAP, and 6over4 have fallen out of favor, while 6in4, 6rd, and NAT64/DNS64 have become more widely used. 6in4 has the disadvantage that the user must have at least one public IPv4 address in their network to serve as one endpoint of the tunnel. These are becoming extremely difficult to obtain. No phones have them, few residential accounts have any, and even business accounts are getting fewer and fewer of them over time. Again, the transition was *supposed* to be complete by 2010, *before* IPv4 public addresses were totally depleted. 6rd works relatively well even without a public IPv4 address at the customer site.

A new standard, 464XLAT, has emerged for mobile devices, which allows telcos to deploy IPv6-only service to customer phones while allowing legacy (IPv4-only) apps to still work. All recent Android phones include support for 464XLAT. This approach is being widely deployed in the United States today.

Relevant Standards for Transition Mechanisms

RFCs related to transition mechanisms (except for Softwires) can be found in the following.

RFCs from the Softwires working group (Dual-Stack Lite, MAP-E, MAP-T, 4in6) can be found under Softwires.[3]

> RFC 2473, "Generic Packet Tunneling in IPv6 Specification," December 1998 (Standards Track) [4in6]

> RFC 2529, "Transmission of IPv6 over IPv4 Domains Without Explicit Tunnels," March 1999 (Standards Track) [6over4]

> **RFC 3053, "IPv6 Tunnel Broker," January 2001 (Informational)**

> **RFC 3056, "Connection of IPv6 Domains via IPv4 Clouds," February 2001 (Standards Track) [6to4]**

> RFC 3089, "A SOCKS-Based IPv6/IPv4 Gateway Mechanism," April 2001 (Informational)

---

[3] #_heading=h.4i7ojhp

RFC 3142, "An IPv6-to-IPv4 Transport Relay Translator," June 2001 (Informational)

RFC 3964, "Security Considerations for 6to4," December 2004 (Informational) [6to4]

RFC 4038, "Application Aspects of IPv6 Transition," March 2005 (Informational)

RFC 4213, "Basic Transition Mechanisms for IPv6 Hosts and Routers," October 2005 (Standards Track) [Dual Stack, 6in4]

RFC 4241, "A Model of IPv6/IPv4 Dual Stack Internet Access Service," December 2005 (Informational)

RFC 4380, "Teredo: Tunneling IPv6 over UDP Through Network Address Translations (NATs)," February 2006 (Standards Track) [Teredo]

RFC 4798, "Connecting IPv6 Islands over IPv4 MPLS Using IPv6 Provider Edge Routers (6PE)," February 2007 (Standards Track)

RFC 4942, "IPv6 Transition/Co-existence Security Considerations," September 2007 (Informational)

RFC 5158, "6to4 Reverse DNS Delegation Specification," March 2008 (Informational) [6to4]

RFC 5214, "Intra-Site Automatic Tunnel Addressing Protocol (ISATAP)," March 2008 (Informational) [ISATAP]

RFC 5569, "IPv6 Rapid Deployment on IPv4 Infrastructures (6rd)," January 2010 (Informational) [6rd]

RFC 5572, "IPv6 Tunnel Broker with the Tunnel Setup Protocol (TSP)," February 2010 (Experimental) [TSP]

RFC 5579, "Transmission of IPv4 Packets over Intra-Site Automatic Tunnel Addressing Protocol (ISATAP) Interfaces," February 2010 (Informational)

RFC 5902, "IAB Thoughts on IPv6 Network Address Translation," July 2010 (Informational)

RFC 6052, "IPv6 Addressing of IPv4/IPv6 Translators," October 2010 (Proposed Standard)

RFC 6127, "IPv4 Run-Out and IPv4-IPv6 Co-Existence Scenarios," May 2011 (Informational)

RFC 6146, "Stateful NAT64: Network Address and Protocol Translation from IPv6 Clients to IPv4 Servers," Aprille 2011 (Proposed Standard)

RFC 6147, "DNS64: DNS Extensions for Network Address Translation from IPv6 Clients to IPv4 Servers," April 2011 (Proposed Standard)

RFC 6180, "Guidelines for Using IPv6 Transition Mechanisms During IPv6 Deployment," May 2011 (Informational)

RFC 6219, "The China Education and Research Network (CERNET) IVI Translation Design and Deployment for the IPv4/IPv6 Coexistence and Transition," May 2011 (Informational)

RFC 6324, "Routing Loop Attack Using IPv6 Automatic Tunnels: Problem Statement and Proposed Mitigations," August 2011 (Informational)

RFC 6343, "Advisory Guidelines for 6to4 Deployment," August 2011 (Informational)

RFC 6384, "An FTP Application Layer Gateway (ALG) for IPv6-to-IPv4 Translation," October 2011 (Proposed Standard)

RFC 6535, "Dual Stack Hosts Using the Bump-In-the-Stack Technique (BIS)," February 2012 (Informational)

RFC 6586, "Experiences from an IPv6-Only Network," April 2012 (Informational)

RFC 6654, "Gateway-Initiated IPv6 Rapid Deployment on IPv4 Infrastructures (GI 6rd)," July 2012 (Informational)

RFC 6889, "Analysis of Stateful 64 Translation," April 2013 (Informational)

RFC 7021, "Assessing the Impact of Carrier-Grade NAT on Network Applications," September 2013 (Informational)

RFC 7050, "Discovery of the IPv6 Prefix User for IPv6 Address Synthesis," November 2013 (Standards Track)

RFC 7051, "Analysis of Solution Proposals for Hosts to Learn NAT64 Prefix," November 2013 (Informational)

RFC 7084, "Basic Requirements for IPv6 Customer Edge Routers," November 2013 (Informational)

RFC 7225, "Discovering NAT64 IPv6 Prefixes Using the Port Control Protocol (PCP)," May 2014, (Proposed Standard)

RFC 7269, "NAT64 Deployment Options and Experience," June 2014 (Informational)

RFC 7648, "Port Control Protocol (PCP) Proxy Function," September 2015 (Proposed Standard)

RFC 7857, "Updates to Network Address Translation (NAT) Behavioral Requirements," April 2016 (Best Current Practice)

RFC 7915, "IP/ICMP Translation Algorithm," June 2016 (Standards Track)

RFC 8215, "Local-Use IPv4/IPv6 Translation Prefix," August 2017 (Informational)

RFC 8219, "Benchmarking Methodology for IPv6 Transition Technologies," August 2017 (Informational)

Transition Mechanisms

There are four general classes of transition mechanisms to help us get from all-IPv4 through a mixture of IPv4 and IPv6 ("dual stack") to eventually all-IPv6.

Co-existence (Dual Stack and Dual-Stack Lite)

Co-existence involves all client and server nodes supporting both IPv4 and IPv6 in their network stacks. The only mechanisms in this group are dual stack and Dual-Stack Lite. This is the most general solution but also involves running essentially two complete networks that share the same infrastructure. It does not double network traffic, as some administrators fear. Any new connection over IPv6 is typically one less connection over IPv4. Over time, an increasing percentage of the traffic on any network will be IPv6, but the only increase in overall traffic will be from the usual suspects (increasing number of applications, users, and/or customers), not from supporting dual stack. In fact, at some point you will see the total amount of IPv4 traffic begin to decrease. You may see an increase in incoming customer connections (on devices that support IPv6) due to the ability of every IPv6 to now also *accept* connections. When YouTube started accepting connections over IPv6, there was an enormous and almost instant jump in IPv6 traffic on the backbone. Many nodes are ready to begin using IPv6 as soon as content is available, because of automated tunneling. In many cases, the end users might not even have been aware that they were now connecting over IPv6.

As an example, Facebook reports that over 90% of connections from US mobile phones are now over IPv6. Few of these uses are even aware that they have IPv6 service.

There is a recent variant of the dual-stack concept called *Dual-Stack Lite* that uses the basic dual-stack design but adds in IP-in-IP tunneling and ISP-based Network Address Translation to allow an ISP to share precious IPv4 addresses among multiple customers. It is defined in RFC 6333,[4] "Dual-Stack Lite Broadband Deployments Following IPv4 Exhaustion," August 2011. There is additional information in RFC 6908,[5] "Deployment Considerations for Dual-Stack Lite," March 2013, and in RFC 7870,[6] "Dual-Stack Lite (DS-Lite) Management Information Base (MIB) for Address Family Transition Routers (AFTRs)," June 2016. In my previous book (*The Second Internet*), there was only an Internet Draft on Dual-Stack Lite.

---

[4] https://tools.ietf.org/html/rfc6333
[5] https://tools.ietf.org/html/rfc6908
[6] https://tools.ietf.org/html/rfc7870

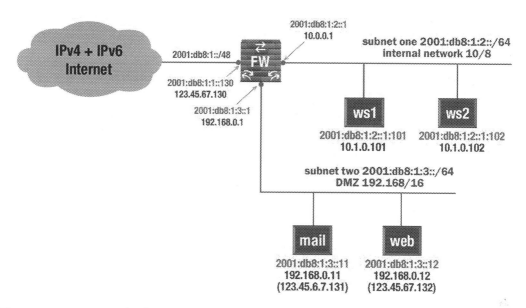

Figure 8-1. *Example dual-stack network*

Tunneling

Tunneling involves creating IP-in-IP tunnels with a variety of mechanisms to allow sending IPv6 traffic over existing IPv4 infrastructures by adding an IPv4 packet header to the front of an entire IPv6 packet. This treats the entire IPv6 packet, including IPv6 packet header(s), TCP/UDP header, and payload fields, as a "black box" payload of an IPv4 packet. In the later phases of the transition, it reverses this: it treats an entire IPv4 packet, including IPv4 packet header and options, TCP/UDP header, and payload fields, as a "black box" payload of an IPv6 packet. Some of these tunnel mechanisms are "automatic" (no setup required). Others require manual setup. Some require authentication, while others do not. The benefit is to leverage the existing IPv4 infrastructure as a transport for IPv6 traffic, without having to wait for ISPs and equipment vendors to support IPv6 everywhere before anyone can start using it. This allows early adopters to deploy nodes and entire networks today, regardless of whether or not their ISP supports IPv6 today. In some cases (e.g., tunnels to a gateway router or firewall), when the ISP does provide dual-stack service, it is a simple process to change from tunneled service to direct service, and the process is largely transparent to inside users. There are several organizations providing *free* tunneled IPv6 service (using various tunnel mechanisms) during the transition, to help with the adoption of IPv6. Tunneling mechanisms include 6in4, 4in6, 6to4, 6over4, and Teredo. TSP has fallen by the way. There are many operating system features and installable client software available to make use of these tunneling mechanisms.

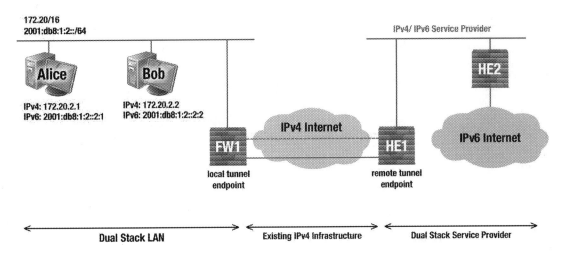

Figure 8-2. *Typical 6in4 tunnel*

Translation

This is basically Network Address Translation (with all its attendant problems), this time between IPv4 and IPv6 (as opposed to the more traditional NAT, which is IPv4 to IPv4). An IPv6-to-IPv4 translation gateway allows an IPv6-only internal node to access external IPv4-only nodes and allow replies from those legacy IPv4 nodes to be returned to the originating internal IPv6 node. Connections from an internal IPv6-only node to external IPv6-only or dual-stack nodes would be done as usual over IPv6 (without going through the translation gateway). This would be useful for deploying IPv6-only nodes in a predominantly IPv4 world. An IPv4-to-IPv6 gateway would allow an IPv4-only internal node to access external IPv6-only nodes and allow replies from those external IPv6 nodes to be returned to the internal IPv4-only node. Connections from an internal IPv4-only node to external IPv4-only nodes, or to dual-stack nodes, would be done as usual over IPv4 (without going through the translation gateway). This would be useful for deploying IPv4-only nodes in a predominantly IPv6 world. Some of these mechanisms require considerable modification to (and interaction with) DNS, such as NAT-PT and NAT64 + DNS64.

There are two broad classes of Network Address Translation between IPv4 and IPv6 – those that work at the IP Layer and are transparent to upper layers and protocols and those that work at the Application Layer (i.e., Application Layer gateways, also called proxies). The IP Layer mechanisms need only be implemented once, for all possible Application Layer protocols. Unfortunately, they also have the most technical issues.

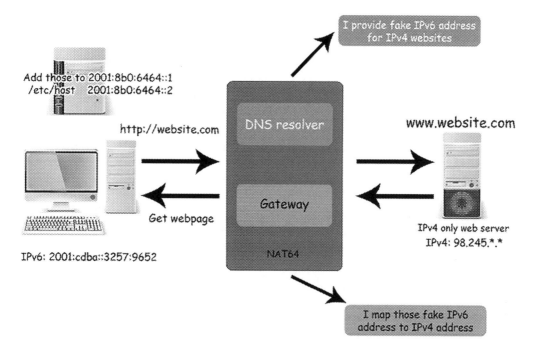

Figure 8-3. *Typical NAT64/DNS64 translation*

There has been a lot of work since 2010 on NAT64/DNS64, to provide access to legacy IPv4 nodes from otherwise IPv6-only networks via a NAT64 gateway on the network border. This was experimental and not very useful at the time of my previous book. NAT64 is specified in RFC 6146,[7] "Stateful NAT64: Network Address and Protocol Translation from IPv6 Clients to IPv4 Servers," April 2011. There is more information available in RFC 7269,[8] "NAT64 Deployment Options and Experience," June 2014. There are several commercial and open source implementations of NAT64 gateways. NAT64 requires use of DNS64 by all clients using the gateway. DNS64 is a variant of DNS, specified in RFC 6147,[9] "DNS64: DNS Extensions for Network Address Translation from IPv6 Clients to IPv4 Servers," April 2011.

464XLAT is specified in RFC 6877,[10] "464XLAT: Combination of Stateful and Stateless Translation," April 2013.

---

[7] https://tools.ietf.org/html/rfc6146

[8] https://tools.ietf.org/html/rfc7269

[9] https://tools.ietf.org/html/rfc6147

[10] https://tools.ietf.org/html/rfc6877

As pointed out in the 2014 OECD report, the big benefits from IPv6 deployment will come when you can phase out IPv4 (at least in the main network). There will be legacy (IPv4-only) nodes for some time to come that you might want to connect to, but that can be handled by a NAT64/DNS64 gateway at the border of an IPv6-only network. Even though there are problems with NAT64[11] (as with any NAT), where there are problems (e.g., VoIP, IPsec), people can switch to IPv6 for those protocols, while the easy stuff will work via NAT64. Over time, as more and more external sites support IPv6, there will be less and less need for NAT64. Meanwhile, we can get IPv4 out of our product networks, which will make network management and security much better and cheaper.

The home and corporate networks of the near future will be IPv6-only with access to legacy nodes via NAT64/DNS64.

Proxies (Application Layer Gateways)

The other kind of translation mechanism takes place at the Application Layer. They are called proxies, because they do things "on behalf of" other servers, much like a stock proxy voter will vote your stock on your behalf. They are also called Application Layer gateways (ALGs) because they are gateways (they do forwarding of traffic from one interface to another), and they work at the Application Layer of the TCP/IP four-layer model. They don't have the serious problems found in IP Layer translation mechanisms, such as dealing with IP addresses embedded in protocols (like SIP or FTP). However, there are some problems unique to proxies.

A proxy must be written for every protocol to be translated, and often even different proxies for incoming and outgoing traffic, even for a given protocol (e.g., "SMTP in" and "SMTP out"). Typically, each proxy is a considerable amount of work. Often only a handful of the most important protocols will be handled by proxies, while all other protocols are handled by packet filtering.

Writing a proxy involves implementing most or all of the network protocol, although sometimes in a simplified manner (e.g., there is no need to store incoming email messages in a way suitable for retrieval by POP3 or IMAP; they just need to be queued by destination domain for retransmission by SMTP).

Proxies can support SSL/TLS, but the secure connection extends only from client to proxy and/or from proxy to server (not directly from client to server). This includes

---

[11] www.cisco.com/web/learning/le21/le39/docs/TDW_130_Prezo.pdf

both encryption (the traffic will be in plain text on the proxy) and authentication (authentication is only from server to proxy and/or proxy to client, not from server to client). Typically, another digital certificate is required for the proxy server if it supports SSL/TLS (in addition to the one for the server).

Proxies can't work with traffic secured in the IP Layer (IPsec ESP), without access to the keys necessary to decrypt the packets.

Throughput is typically lower than with a packet filtering firewall, due to the need to process the protocol. Of course, the security is much better – it won't let through traffic that is not a valid implementation of the specific protocol, while packet filtering might let through almost anything so long as it uses the right port. There is typically no problem dealing with IP addresses embedded in a protocol.

In many cases, the proxies are not transparent, which means the client must know that it is talking not directly to a server, but via an intermediate proxy. Many protocols support this kind of operation, for example, HTTP provides good support for an HTTP proxy. Basically, there must be a way for a client to specify not only the nodename of the final server but also the address or nodename of the proxy server. In a browser (HTTP client), the nodename of the final server is specified as usual, and the address of the proxy server is specified during the browser configuration ("use a proxy, which is at address w.x.y.z"). When configured for proxy operation, the browser actually connects to the proxy address and relays the address of the final server to the proxy. The proxy then makes an ongoing connection to the final web server. Some protocols have no support for proxy-type operation (e.g., FTP). It is possible for a firewall to recognize outgoing traffic over a given port and automatically redirect it to a local proxy.

Application Layer gateways (e.g., for SIP, HTTP, and SMTP) work quite well. Basically, they accept a connection on one interface of a gateway and make a second "ongoing" connection (on behalf of the original node) via another interface of the same gateway. It is easy for the two connections to use different IP versions (e.g., translate IPv4 traffic to IPv6 traffic or vice versa). In some ALGs an entire message might be spooled onto temporary storage (e.g., email messages) and then retransmitted later. In other cases, the ongoing connection would be simultaneous with the incoming connection and bidirectional (e.g., with HTTP). This would correspond to a human "simultaneous translator" who hears one language (e.g., Chinese), translates, and simultaneously speaks another language (e.g., English).

Another example of this is an outgoing web proxy, which could accept connections from either IPv4-only or IPv6-only browsers and then make an ongoing connection to external servers using whatever version of IP those servers support (based on DNS queries). Again, this is a traditional (forward) web proxy, with the addition of IP version translation. This would allow IPv4-only or IPv6-only clients to access any external web server, regardless of IP version they support. Such a proxy could of course also provide any services normally done by an outgoing web proxy, such as caching and URL filtering.

Another example of this is a dual-stack façade that would accept incoming connections from outside over either IPv4 or IPv6 and make an ongoing connection over IPv4 to an internal IPv4-only (or over IPv6 to an IPv6-only) web server. It would relay the web server's responses using whatever version of IP was used in the original incoming connection to the client. This is a typical "reverse" web proxy, with the addition of IP version translation. This kind of translation can help you provide dual-stack versions of your web services quickly and easily, without having to dual-stack the actual servers themselves. The same technique could allow you to make your email services dual stack without having to modify your existing mail server.

Dual Stack

Dual stack is defined in RFC 4213,[12] "Basic Transition Mechanisms for IPv6 Hosts and Routers," October 2005. A dual-stack node should include code in the Internet Layer of its network stack to process both IPv4 and IPv6 packets. Typically, there is a single Link Layer that can send and receive either IPv4 or IPv6 packets. The Link Layer also contains both the IPv4 Address Resolution Protocol (ARP) and the IPv6 Neighbor Discovery (ND) protocol. The Transport Layer has only minor differences in the way IPv4 and IPv6 packets are handled, primarily concerning the way the TCP or UDP checksum is calculated (the checksum also covers the source and destination IP addresses from the IP header, which of course is different in the two IP versions). The Application Layer code can make calls to routines in the IPv4 socket API, the IPv6 basic socket API, and the IPv6 advanced socket API. IPv4 socket functions will access the IPv4 side of the IP Layer, and IPv6 socket functions will access the IPv6 side of the IP Layer.

---

[12] https://tools.ietf.org/html/rfc4213

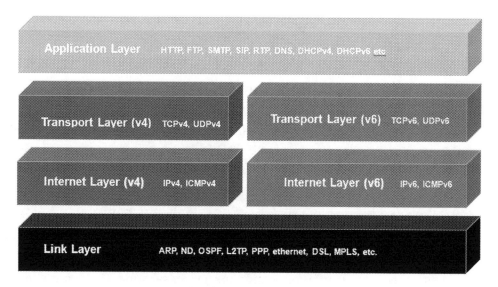

Figure 8-4. *Four-layer network model for dual stack*

The node should include the ability to do conventional IPv4 network configuration (including a node address, default gateway, subnet mask, and addresses of DNS servers, all as 32-bit IPv4 addresses). This configuration information can be done manually, via DHCPv4, or some combination thereof. The node should *also* include the ability to do conventional IPv6 network configuration (including a link-local IP address, one or more global unicast addresses, a default gateway, the subnet length, and the addresses of DNS servers, all 128-bit IPv6 addresses). This configuration information can be done manually, automatically via Stateless Address Autoconfiguration, automatically by DHCPv6, or by some combination thereof. There is usually a way to disable either the IPv6 functionality (in which case the node behaves as an IPv4-only node) or the IPv4 functionality (in which case the node behaves as an IPv6-only node). There may or may not also be some tunneling mechanism involved. If the node is in a native dual-stack network, no tunnel mechanism is seen by the user (any tunnel involved will be between the user's Customer Premises Equipment and the IPv6 service provider, not inside the network). If the node is in an IPv4-only or an IPv6-only network, there will need to be a tunnel mechanism to bring in traffic of the other IP version (typically 6in4, 4in6, or 6rd).

IPv4-only and *IPv6-only* applications (client, server, and peer-to-peer) will work just fine on a dual-stack node. They will make calls to system functions on only one side of the network stack. They will not gain any new ability to accept or make connections over the other IP version just because they are running on a dual-stack node.

A *dual-stack* client can connect to IPv4-only servers, IPv6-only servers, or dual-stack servers. A dual-stack server can accept connections from IPv4-only clients, IPv6-only clients, or dual-stack clients. Dual stack is the most complete and flexible solution. The only issues are the additional complexity of implementation and deployment and the additional memory requirements. For very small devices (typically clients), dual stack may not be an option. Some critics of IPv6 claim that dual stack is not viable because we are running out of IPv4 addresses. What they are missing is that there are *plenty* of private IPv4 addresses for use behind NAT, and the IPv4 side of dual-stack systems can be used only for protocols where this is not a problem while using their IPv6 side for those protocols that are incompatible with NAT (IPsec VPN, SIP, P2P, etc.) or can benefit from other IPv6 features, which are superior to their IPv4 equivalents, such as multicast and QoS (for SIP, IPTV, conferencing, P2P Direct, etc.). Also, any application running on that node that needs to accept a connection from external nodes (e.g., your own web server) can use a global unicast IPv6 address (for IPv6-capable clients). If you want to accept connections from IPv4 clients, you would have needed a globally routable IPv4 address for that anyway or would need to deploy NAT traversal (with or without dual stack). Dual stack cannot create more globally routable IPv4 addresses. It can, however, allow you to easily make use of an almost unlimited number of globally routable IPv6 addresses (both unicast and multicast). It is common for only a few nodes in a dual-stack network to have IPv4 public addresses (or forwarding via NAT from a border node with a public IPv4 address), but *every* node can have a public (global) IPv6 address. If incoming connections are not blocked at a firewall, those nodes are accessible over IPv6 from anywhere on the global IPv6 Internet.

A key part of a dual-stack network is a correctly configured dual-stack DNS service. It should not only be able to handle both A and AAAA records (as well as reverse PTR records for IPv4 and IPv6); it should also be able to accept queries and do zone transfers over both IPv4 and IPv6. A dual-stack network typically uses DHCPv4 to assign IPv4 addresses to each node and either Stateless Address Autoconfiguration and/or DHCPv6 to assign IPv6 addresses to each node. A dual-stack firewall can bring in either direct dual-stack service (both IPv4 and IPv6 traffic) from an ISP (if available), routing both to the inside network; or it can bring in direct IPv4 traffic from an ISP and terminate tunneled IPv6 traffic (from a "virtual" ISP usually different from the IPv4 ISP) and route both IPv4 and IPv6 into the inside network. In either case (direct dual-stack service or tunneled IPv6 with endpoint in the gateway), inside nodes appear to have native dual-stack service and require no support for tunneling.

The DNS support does not require any modifications to a standard DNS server (e.g., BIND). Virtually all current DNS servers and appliances have (at least some) support for IPv6. DNS just needs to be able to perform its normal forward and reverse lookups with either IPv4 (A/PTR) or IPv6 (AAAA/PTR) resource records. There is no need for the DNS server to do nonstandard mappings between IPv4 and IPv6 addresses as is required with most IP Layer translation schemes (e.g., NAT64 + DNS64).

Migrating IPv4-only client or server applications to IPv6-only is quite simple. There is essentially a one-to-one mapping of function calls from the IPv4 socket API to similar ones in the IPv6 basic socket API. Of course, more storage is required for each IP address in data structures (4 bytes for IPv4 addresses, 16 bytes for IPv6 addresses).

Modifying either IPv4-only clients or IPv4-only servers to *dual-stack* operation is somewhat more complicated. A dual-stack client must be modified to retrieve multiple addresses from a forward lookup (IPv4 and/or IPv6) and try connections sequentially to the returned address list until a connection is accepted. The default (assuming IPv6 connectivity is available) is to attempt connections over IPv6 first. If DNS advertises an IPv6 address and the node supports IPv6, but for some reason the client is unable to connect over IPv6 (e.g., the tunnel is down), there will be a 30-second timeout and then a fallback to IPv4. A dual-stack server must listen for connections on *both* IPv4 *and* IPv6 and process connections from either. It is also possible to deploy two copies of each server, one being IPv4-only and the other IPv6-only. This might involve cross-process file locking on any shared resource, such as a message store. Either approach to providing dual-stack servers will work fine, and the user experience will be the same. Conditional compilation could be used to have a single source code tree create both an IPv4-only and an IPv6-only executable (depending on settings of system variables at compilation time). For most server designs (process per connection or thread per connection), the split model (an IPv4-only server and an IPv6-only server) would roughly double the memory footprint compared with a single dual-stack server.

There has been an improvement on this scheme since 2010 called "**Happy Eyeballs.**" The first version of this was specified in RFC 6555,[13] "Happy Eyeballs: Success with Dual-Stack Hosts," April 2012. The second version of this was specified in RFC 8305,[14] "Happy Eyeballs Version 2: Better Connectivity Using Concurrency," December 2017. This mechanism is actually implemented in clients, especially web browsers. It usually connects over both IPv4 and IPv6 and uses whichever one responds first (with some

---

[13] https://tools.ietf.org/html/rfc6555
[14] https://tools.ietf.org/html/rfc8305

allowance for a slightly slow IPv6 response). The results of this measurement are stored, and that IP version is used for future connections to that server for some time. I have found that sometimes even if I have IPv6 and the server is IPv6, Happy Eyeballs will choose to connect over IPv4 (which violates the prior standard that IPv6 is preferred). This also impacts the statistics on server access over IPv6 (as seen on Google IPv6 stats). It would be nice if there were some way to disable Happy Eyeballs on a browser for people who know what they are doing, but no browser offers that option. It is permanently ON, on every browser I've tested. There is an implicit assumption that there is no difference in IPv6 and IPv4 (at least for web), which may not always be true. I may want to provide a better or more complete experience to people who connect over IPv6, but with Happy Eyeballs, the user has not control over this, unless they disable IPv4 on their node, which may cause other issues.

Most open source servers today have good support for dual-stack operation. These include the Apache web server, Postfix SMTP server, Dovecot IMAP/POP3 mail access servers, etc. If you are a developer and want to see examples of how to deploy dual-stack servers, there are numerous examples available in open source. Most open source client software also has good support for IPv6 and dual stack. These include the Firefox web browser, Thunderbird email client, etc. The open source community has done an excellent job of supporting the migration to IPv6. Both the original IPv4-only socket API and the newer IPv6 socket APIs are readily available on all UNIX and UNIX-like platforms. The documentation for the newer IPv6 socket APIs is in RFC 3493,[15] "Basic Socket Interface Extensions for IPv6," and RFC 3542,[16] "Advanced Sockets Application Program Interface (API) for IPv6." There is also RFC 5014,[17] "IPv6 Socket API for Source Address Selection," and RFC 4584,[18] "Extension to Sockets API for Mobile IPv6."

Virtually all Microsoft server products (since 2007) have had good support for dual-stack operation. The Azure Cloud service has been promising IPv6 support from the beginning, with very little progress. You can use load balancers to map an IPv6 address to an IPv4 address on the Azure VM, but you cannot configure an IPv6 address or make connections to IPv6 nodes, from an Azure node. AWS has provided at least some support for IPv6[19] for some time. Microsoft products that support IPv6 well include Windows

---

[15] https://tools.ietf.org/html/rfc3493

[16] https://tools.ietf.org/html/rfc3542

[17] https://tools.ietf.org/html/rfc5014

[18] https://tools.ietf.org/html/rfc4584

[19] http://thirdinternet.com/recipes-for-amazon-web-services-aws/

Server 2008 R2 and later (and all its components, such as DNS, file and printer sharing, etc.), Exchange Server 2007 or later, and many others. Their client operating systems have had good support for IPv6 since Vista. For Microsoft developers, both the original IPv4-only socket API (Winsock) and the new IPv6 socket APIs (basic and advanced) are available as part of the standard Microsoft developer libraries.

Tunneling

Tunneling is very different from translation – the packets from the foreign IP are sent, complete with packet headers, as the Data field of packets of the other IP. For example, 6in4 packets have an IPv4 header, followed by an IPv6 header and IPv6 body. 4in6 packets have an IPv6 header, followed by an IPv4 header and IPv4 body. Once they reach the end of the tunnel, the extra header is stripped off, and the inside packet is routed on its way.

6in4 Tunneling – Network Diagram

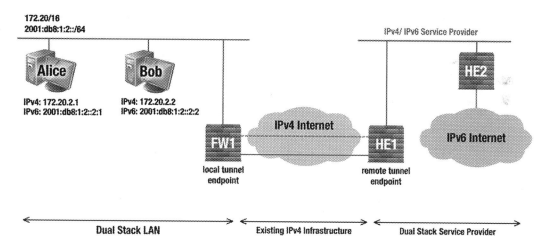

Figure 8-5. *How 6in4 tunnels work*

```
⊞ Frame 1: 114 bytes on wire (912 bits), 114 bytes captured (912 bits)
⊟ Ethernet II, Src: IntelCor_1d:c1:57 (00:1b:21:1d:c1:57), Dst: ZyxelCom_ab:4b:cf (00:23:f8:ab:4b:cf)
  ⊞ Destination: ZyxelCom_ab:4b:cf (00:23:f8:ab:4b:cf)
  ⊞ Source: IntelCor_1d:c1:57 (00:1b:21:1d:c1:57)
    Type: IP (0x0800)
⊟ Internet Protocol, Src: 122.52.125.243 (122.52.125.243), Dst: 184.105.238.1 (184.105.238.1)
    Version: 4
    Header length: 20 bytes
  ⊞ Differentiated Services Field: 0x00 (DSCP 0x00: Default; ECN: 0x00)
    Total Length: 100
    Identification: 0x7607 (30215)
  ⊞ Flags: 0x00
    Fragment offset: 0
    Time to live: 30
    Protocol: IPv6 (41)
  ⊞ Header checksum: 0x87d7 [correct]
    Source: 122.52.125.243 (122.52.125.243)
    Destination: 184.105.238.1 (184.105.238.1)
⊟ Internet Protocol Version 6, Src: 2001:470:3d:3000::2:1 (2001:470:3d:3000::2:1), Dst: 2001:470:20::2 (2001:470:20::2)
  ⊞ 0110 .... = Version: 6
  ⊞ .... 0000 0000 .... .... .... .... .... = Traffic class: 0x00000000
    .... .... .... 0000 0000 0000 0000 0000 = Flowlabel: 0x00000000
    Payload length: 40
    Next header: ICMPv6 (0x3a)
    Hop limit: 127
    Source: 2001:470:3d:3000::2:1 (2001:470:3d:3000::2:1)
    Destination: 2001:470:20::2 (2001:470:20::2)
⊟ Internet Control Message Protocol v6
    Type: 128 (Echo request)
    Code: 0 (Should always be zero)
    Checksum: 0x5b62 [correct]
    ID: 0x0001
    Sequence: 0x0052
  ⊟ Data (32 bytes)
      Data: 6162636465666768696a6b6c6d6e6f7071727374757677761...
      [Length: 32]
```

Figure 8-6. *6in4 tunneling – capture of IPv6 Echo Request showing nested structure*

If tunneled service is brought into the network by a gateway device (typically the gateway router or firewall), which contains the tunnel endpoint, the internal network is a native dual-stack network from the viewpoint of all internal nodes. No internal node needs to have support for any tunneling mechanism. If at some point the tunneled service is replaced with direct service (both IPv4 and IPv6 service direct from your ISP), a minor reconfiguration at the gateway is all that is required. Internal nodes will probably not require any reconfiguration at all. They will typically have a new IPv6 prefix (unless you were getting tunneled service from your ISP), so you will likely have to update all forward and reverse address references in your DNS server (only for IPv6 addresses), to reflect the new IPv6 prefix. If your DNS server supports *instant prefix renumbering* like Sixscape DNS, this is a quick, painless process. If you are using DHCPv6 in stateful mode (where it assigns IP addresses) in conjunction with dynamic DNS registration, even DNS changes due to change of IPv6 prefix may happen automatically.

A tunnel mechanism has both a server side and a client side. The server side typically can accept one or more connections from tunnel clients. It is also commonly called a *Tunnel Broker*. A tunnel client typically makes connections to a single tunnel server. Some such connections (e.g., with 6in4) are not authenticated (although the server can typically be restricted to accepting connections only from specific IP addresses or address ranges). Some such connections include authentication of the client to the server before the tunnel will begin operation. Some connections (e.g., 6in4) require a globally routable IPv4 address on the client (although this can be the same address as the hide-mode NAT address). Other tunnel clients (e.g., 6rd) will work behind NAT, even with a private address. These include a NAT traversal mechanism in the client, and typically all tunneled packets are carried over UDP. Once a tunnel is created, it is bidirectional. Packets can be sent either upstream or downstream. From a hop count perspective, the tunnel counts as one hop, no matter how many hops the tunneled packets traverse.

Typical Product Support for Tunneling: pfSense Open Source Dual-Stack Firewall

As an example of a typical product that includes support for tunneling, pfSense[20] is an open source dual-stack firewall. On the IPv4 side, it includes typical firewall capabilities including routing, filtering by port and address, stateful packet inspection, and various forms of NAPT (hide mode, BINAT or 1:1, and port forwarding). On the IPv6 side, it includes all that (except for NAPT), plus a Router Advertisement Daemon (to enable Stateless Address Autoconfiguration) and 6in4 server and client modes. You could use the 6in4 client mode to bring in IPv6 tunneled service from any 6in4 virtual ISP (e.g., Hurricane Electric[21]). You could create your own IPv6 virtual ISP using pfSense's 6in4 tunnel server mode. For example, you could provide tunneled IPv6 service from your HQ or collocation facility to various branches, using the 6in4 tunnel server at HQ and the 6in4 tunnel clients at each branch. You can carve off any number of "/64" subnets into each branch office. For example, you could split a "/48" block into 16 "/52" blocks and route one "/52" block into each branch office.

Because the client-mode tunnel endpoint is located inside a firewall node, incoming IPv6 packets from the tunnel can be filtered and routed into any inside network(s). Outgoing IPv6 packets from any internal network can be filtered and routed out the tunnel to the outside world (via the same 6in4 tunnel).

The server-mode tunnel endpoint is also located inside a firewall node, so the firewall's routing capabilities allow you to easily route any block of addresses from the outside world into any tunnel (and hence to branch offices) and outgoing packets (from tunnels coming from branch offices) to the outside world. Currently there is no support

---

[20] www.pfsense.org/

[21] https://tunnelbroker.net/

for OSPFv3 or BGP4+, so you would need to relay outgoing IPv6 traffic onward via an ISP (or virtual ISP) that can do further routing.

Because the tunnel mechanism used (6in4) is an IETF standard,[22] pfSense's tunnels will interoperate with server- or client-mode 6in4 tunnel endpoints on any other vendor's products or even on other open source routers or firewalls.

6in4 Tunneling

RFC 4213[23] (in addition to specification for dual stack) specifies 6in4 tunneling (unfortunately they use the term "6over4" when you might recognize "6in4," which is very confusing). Technically, *6in4* is a *tunneling* mechanism. *6over4* is a *transition* mechanism that uses 6in4 tunneling to create a virtual IPv6 link over an IPv4 multicast infrastructure (see RFC 2529,[24] "Transmission of IPv6 over IPv4 Domains Without Explicit Tunnels," March 1999). This book will use the term *6in4* unless we are specifically talking about 6in4 tunnels over IPv4 multicast. 6in4 is also sometimes referred to as "Protocol 41" tunneling. 6in4 tunneling requires both ends of the tunnel to have globally routable IPv4 addresses (neither tunnel endpoint can be behind NAT). It is possible for a firewall that is using a globally routable IPv4 address for HIDE-mode NAT (with multiple internal nodes hidden behind it) to use that same address as one endpoint of a 6in4 tunnel.

6in4 Encapsulation

This process is done to "push packets into the tunnel" for packets going from either end of the tunnel to the other. The basic idea is to prepend a new IPv4 packet header to a complete IPv6 packet (which itself consists of the basic IPv6 header, zero or more extension headers, a TCP or UDP header, and a payload) and treat the entire IPv6 packet as a "black box" payload for the IPv4 packet.

The encapsulation of an IPv6 datagram in IPv4 for 6in4 tunneling is shown in the following.

---

[22] https://tools.ietf.org/html/rfc4213
[23] https://tools.ietf.org/html/rfc4213
[24] https://tools.ietf.org/html/rfc2529

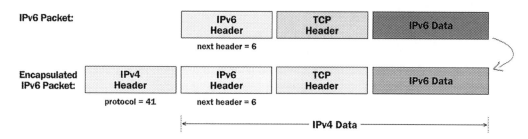

Figure 8-7. *Example of 6in4 encapsulation*

The new IPv4 packet header is constructed as follows (from the RFC):

IP Version

> 4 (the encapsulating packet is IPv4)

IP Header Length

> 5 (in 32-bit words, so 20 bytes, and no IPv4 options are used in the encapsulating header)

Type of Service

> 0 unless otherwise specified (see RFC 2983 and RFC 3168 for details)

Total Length

> IPv6 payload length plus IPv6 header length (40) plus IPv4 header length (20), so IPv6 payload length + 60

Identification

> Generated uniquely as for any IPv4 packet

Flags

> DF (Don't Fragment) flag set as specified in section 3.2 of RFC 4213

> MF (More Fragments) flag set as necessary if fragmenting

Fragment Offset

> Set as necessary if fragmenting

Time To Live (TTL)

Set as described in section 3.3 of RFC 4213

Protocol

41: This is the defined payload type for IPv6 tunneled over IPv4 and is used regardless of whether the IPv6 transport is UDP or TCP.

Header Checksum

Calculated as usual for an IPv4 packet header

Source Address

An IPv4 address of the encapsulator: either configured by the administrator or an address of the outgoing interface

Destination Address

IPv4 address of the tunnel endpoint (i.e., the client side of the tunnel)

6in4 Decapsulation

This is done for all packets received over the tunnel from the other end. The basic idea is to strip the outer (IPv4) packet header off (and discard it) and then handle what is left (the original IPv6 packet) as native IPv6 traffic.

From the RFC: When a dual-stack node receives an IPv4 datagram that is addressed to one of its own IPv4 addresses (or a joined multicast group address), which has a Protocol field of 41 (tunneled IPv6), the packet must be verified to belong to a configured tunnel interface (according to source/destination addresses), be reassembled (if it was fragmented), and have the IPv4 header removed, and then the resulting IPv6 datagram is submitted to the IPv6 layer on the node for further processing.

The decapsulation process for 6over4 tunneling is shown in the following.

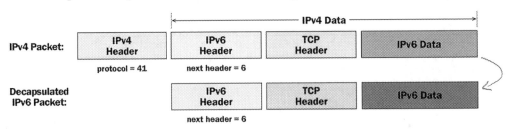

Figure 8-8. *Sample 6in4 decapsulation*

According to RFC 4213, section 3.2, the MTU of the tunnel *must* be between 1280 and 1480 bytes (inclusive) but *should* be 1280 bytes. Section 3.3 specifies that the tunnel counts as a *single* hop to IPv6, regardless of how many hops the underlying IPv4 packet traverses. The actual TTL value in the outer IP header should be set as for any IPv4 packet (see RFC 3232 and RFC 4087).

RFC 4213 section 3.4 specifies how to handle errors that happen while the encapsulated packet is *inside* the tunnel. Unfortunately, older routers may not return enough of the packet to include both source and destination IPv6 addresses of the encapsulated packet, so it may not be possible to construct a correct ICMPv6 error message. Newer routers typically include enough of the failed packet for correct ICMPv6 error message creation.

6over4 Tunneling

6over4 tunneling is defined in RFC 2529,[25] "Transmission of IPv6 over IPv4 Domains Without Explicit Tunnels," March 1999. It is a transition mechanism that uses 6in4 tunneling over an IPv4 multicast–capable network. The term 6over4 is sometimes confusingly used for 6in4 tunneling. Due to the requirement for IPv4 multicast, which is very difficult to deploy, 6over4 is not commonly used. You can deploy a basic 6in4 tunnel without IPv4 multicast.

6to4 Tunneling

6to4 tunneling is described in the following RFCs:

- **RFC 3056, "Connection of IPv6 Domains via IPv4 Clouds," February 2001**

- RFC 3068, "An Anycast Prefix for 6to4 Relay Routers," June 2001 (deprecated by RFC 7526, in May 2015)

- **RFC 3964, "Security Considerations for 6to4," December 2004 (Informational)**

- RFC 5158, "6to4 Reverse DNS Delegation Specification," March 2008

---

[25] https://tools.ietf.org/html/rfc2529

6to4 is a transition mechanism that provides tunneled IPv6 over IPv4 without explicitly configured tunnels. With the original 6to4 mechanism, the IPv4 addresses involved must be valid globally routable IPv4 addresses (not behind NAT). Teredo is a variant of 6to4 tunneling that will work even behind NAT.

6to4 does not provide general translation to IPv4 addresses for interoperation between IPv6 hosts and IPv4 hosts (it is not a *translator* – it is a *tunneling* scheme). It uses automatically created tunnels over IPv4 to facilitate communication between any number of IPv6 hosts.

A "6to4 host" is a regular IPv6 host that also has at least one 6to4 address assigned to it.

A "6to4 router" is a regular IPv6 router that includes a 6to4 pseudo interface. It is normally a border router between an IPv6 site and a wide-area IPv4 network.

A "6to4 relay router" is a 6to4-capable router, which is also configured to support transit routing between 6to4 addresses and native IPv6 addresses.

Without 6to4 relay routers, you can communicate with other nodes that use 6to4 tunneling over IPv6 (even though your ISP does not yet support IPv6). To communicate with IPv6 users who are *not* using 6to4, you need to relay your traffic through a 6to4 relay router. You can create your own relay router. It must have both a 6to4 pseudo interface and native (not 6to4) IPv6 connectivity to the IPv6 Internet.

A 6to4 router will send an encapsulated packet directly over IPv4 if the first 16 bits of an IPv6 destination address are 2002, using the next 32 bits as the IPv4 destination (which must be another 6to4 node that will unpack the IPv6 packet being sent and use it or relay it to other IPv6 hosts). For all other IPv6 destination addresses, a 6to4 router will forward the packet to the IPv6 address of a well-known relay router that has access to native IPv6 (or simply send it to the IPv6 anycast address 2002:c058:6301::/128, which will send it to the nearest available 6to4 relay router).

For details on how to configure a FreeBSD node with 6to4 tunneling, see `www.kfu.com/~nsayer/6to4`.

An IPv6 address for use with 6to4 tunneling looks like the following:

```
| 3 |  13  |    32     |   16   |             64 bits              |
+---+------+-----------+--------+----------------------------------+
|FP | TLA  | V4ADDR    | SLA ID |           Interface ID           |
|001|0x0002|           |        |                                  |
+---+------+-----------+--------+----------------------------------+
```

Essentially the IPv6 prefix for all 6to4 addresses is 2002:(ipv4addr)::/48. RFC 2374 defines SLA ID as follows:

> The SLA ID field is for a Site Level Aggregator Identifier. This can be used by individual organizations to create its own local addressing hierarchy and to identify subnets. It is analogous to subnets in IPv4, except that each organization has a much greater number of subnets.

RFC 3056 defines a 6to4 pseudo interface as follows:

> 6to4 encapsulation of IPv6 packets inside IPv4 packets occurs at a point that is locally equivalent to an IPv6 interface, with the link layer being the IPv4 unicast network. This point is referred to as the pseudo-interface. Some implementers may treat it exactly like any other interface, and others may treat it like a tunnel endpoint.

Teredo

Teredo is one extension of basic 6to4 tunneling. It adds encapsulation over UDP datagrams and uses a simplified version of STUN NAT traversal, allowing a Teredo client to be behind NAT. It is defined in RFC 4380,[26] "Teredo: Tunneling IPv6 over UDP Through Network Address Translations (NATs)," February 2006. The name "Teredo" is part of the Latin name for a little worm that bores holes through wooden ship hulls. This gives you a pretty good idea of what the Teredo protocol does to your firewall. Teredo is installed and enabled by default in Windows Vista and Windows 7. It is possible to disable it, which everyone should do!

---

[26] https://tools.ietf.org/html/rfc4380

There is an open source Teredo client for Linux, BSD, and Mac OS X called *Miredo.*[27] It can act as a client, relay, and server.

There are publicly available Teredo "relay routers" that allow any node with Teredo to access the IPv6 Internet. Microsoft makes several very large ones available for use from Windows nodes. Windows nodes are preconfigured to use these relay servers. Unlike 6to4 and some other tunnel mechanisms, Teredo can only provide a single "/128" IPv6 address per tunnel endpoint. Teredo allows you to let *one* node connect to the IPv6 Internet, not an entire network.

Teredo uses a different IPv6 address block than basic 6to4 tunneling. The rest of the Teredo address is defined differently as well:

- Bits 0–31 contain the Teredo prefix, which is 2001:0000::/32. You might want to block this range for both incoming and outgoing connections on your border firewall.

- Bits 32–63 contain the IPv4 address of the Teredo server used.

- Bits 64–79 contain some flags. Currently only bit 64 is used. If set to 1, the client is behind a cone NAT; otherwise, it is 0. More of these flag bits are used in Vista, Windows 7, and Windows Server 2008.

- Bits 80–95 contain the obfuscated UDP port number (port number that is mapped by NAT, with all bits inverted).

- Bits 96–127 contain the obfuscated IPv4 address of the node (public IPv4 address of the NAT with all bits inverted).

As an example, a Teredo address might be 2001::4136:e378:8000:63bf:3fff:fdd2, which broken into fields is as follows:

- *Bits 0–31*: 2001:0000 – the Teredo prefix

- *Bits 32–63*: 4136:e378 – IPv4 address 65.54.227.120 in hexadecimal

- *Bits 64–79*: 8000 – cone-mode NAT

- *Bits 80–95*: 63bf – obfuscated port number 40000

- *Bits 96–127*: 3fff:fdd2 – obfuscated public IPv4 address of the node (192.0.2.45)

---

[27] https://en.wikipedia.org/wiki/Miredo

Hurricane Electric, as of Q1 2009, had deployed 14 public Teredo relays (via anycast), in Seattle, Washington; Fremont, California; Los Angeles, California; Chicago, Illinois; Dallas, Texas; Toronto, Ontario; New York, New York; Ashburn, Virginia; Miami, Florida; London, England; Paris, France; Amsterdam, Netherlands; Frankfurt, Germany; and Hong Kong SAR.

Usage of Teredo has dropped off to virtually zero as native IPv6 and 6in4 tunnels have become more common.

6rd: IPv6 Rapid Deployment

6rd is another extension of 6to4 tunneling that adds reliable routing. Normal 6to4 tunnels use the standard 2002://16 prefix and in theory scale to the entire world. Unfortunately, there is no way to control who can connect to 6to4 public servers, and there is no incentive to provide quality service. Also there is no guarantee that any 6to4 node will be reachable. The same is true of Teredo.

6rd instead works only within the confines of a single ISP, and instead of the 2000://16 prefix, each ISP uses a prefix that they own and control and runs the relay router. They can ensure quality service and reachability of all nodes within their network.

6rd was deployed by a French ISP called "Free" (in spite of the name, this is a commercial ISP). This was done in 5 weeks starting in December 2007. This gave France the second highest IPv6 penetration in the world, 95% of which was via Free's 6rd. RFC 5569 discusses Free's 6rd deployment. The current Internet Draft that defines 6rd (draft-ietf-softwire-IPv6-6rd-08, "IPv6 via IPv4 Service Provider Networks '6rd,'" March 23, 2010) should be approved soon. Meanwhile, you can read the draft.

In January 2010, Comcast (a large US ISP) announced plans to do a trial deployment of IPv6 using 6rd. SoftBank (a large Japanese ISP) also has announced that they will roll out IPv6 using 6rd.

Intra-site Automatic Tunnel Addressing Protocol (ISATAP)

ISATAP is a transition mechanism that allows transmission of IPv6 packets between dual-stack nodes on top of an IPv4 network. It is similar to 6over4, but it uses IPv4 as a virtual non-broadcast multiple-access (NBMA) network Link Layer and does not require IPv4 multicast (which 6over4 does require). It is discussed in RFC 5214, "Intra-Site Automatic Tunnel Addressing Protocol (ISATAP)."

ISATAP specifies a way to generate a link-local IPv6 address from an IPv4 address, plus a mechanism for performing Neighbor Discovery on top of IPv4.

The generated link-local address is created by appending the 32-bit IPv4 address onto the 96-bit prefix *fe80:0:0:0:0:5efe::*. For example, the IPv4 address *192.0.2.143* in hexadecimal is *c000028f*. Therefore, the corresponding ISATAP link-local address is *fe80::5efe:c000:28f*.

The Link Layer address for ISATAP is not a MAC address, but an IPv4 address (remember IPv4 is used as a virtual Link Layer). Since the IPv4 address is just the low 32 bits of the ISATAP address, mapping onto the "Link Layer" address simply involves extracting the low 32 bits (ND is not required). However, router discovery is more difficult without multicast. ISATAP hosts are configured with a *potential routers list* (PRL). Each of the routers on this list is probed by an ICMPv6 Router Discovery message, to determine which of them are functioning and to then obtain the list of on-link IPv6 prefixes that can be used to create global unicast IPv6 addresses.

Current implementations create their PRL by querying the DNS. DHCPv4 is used to determine the local domain. Then a DNS query is done for isatap.<localdomainname>. For example, if the local domain is *demo.com*, it would do a DNS query for *isatap. demo.com*.

ISATAP avoids circular references by only querying DNS over IPv4, but it is still a lower-layer protocol that is using a higher-layer function (DNS). This is a violation of network design principles.

ISATAP is implemented in Windows XP, Windows Vista, Windows 7, Windows Mobile, and Linux (since Kernel 2.6.25). It is not currently implemented in *BSD[28] due to a potential patent issue.

Softwires (Includes Dual-Stack Lite, MAP-E, MAP-T, and 4in6)

The IETF has a very active *Softwires* working group. Essentially, they are trying to create standards for tunneling IPv6 over IPv4 networks and for tunneling IPv4 over IPv6 networks. There are two basic models for this; one is called *hub and spoke*. This is similar to the way that airlines have a few large *hub* airports and many *spokes* or local flights radiating from those hubs to smaller communities nearby. For example, Atlanta

---

[28] *BSD refers to the family of BSD variants: FreeBSD, NetBSD, OpenBSD.

International Airport is a hub for the entire Southeastern United States. If you fly in or out of that region, you will likely interchange in Atlanta. There are several schemes that vary in exactly what part of the network path the softwire is deployed:

- From ISP to customer modem/router

- From ISP via customer modem/router to an inside softwire router

- From ISP via customer modem/router to an end-user node

All the components necessary to deploy the various schemes are widely available, including

- *LNS*: Large ISP-based *L2TP Network Server*

- *Dual AF CPE*: Customer Premises Equipment modem/router with support for L2TPv2 softwires

- *Dual AF router*: Customer premise dual-stack router with support for L2TPv2 softwires

- *Dual AF host*: Client software for end-user nodes with support for L2TPv2 softwires

In the preceding, "Dual AF" means Dual *Address Family*, in other words, IPv4 + IPv6, or dual stack.

The other softwire architecture is called *mesh*. This involves several peer nodes, with multiple connections between them. If all nodes are connected to all other nodes, that would be a fully meshed network.

The term *softwire* refers to a tunneled link between two or more nodes. In early RFCs related to this technology, sometimes the term *pseudowire* is used instead. Softwires are assumed to be long-lived, and the setup time is expected to be a very small fraction of the total time required for the startup of the Customer Premises Equipment/Address Family border router. The goal is to make cost-effective use of existing facilities and equipment where possible.

Current softwire solutions are mostly based on L2TPv2, which is defined in RFC 2661,[29] "Layer Two Tunneling Protocol 'L2TP,'" August 1999. L2TPv1 was defined in RFC 2341,[30] "Cisco Layer Two Forwarding (Protocol) 'L2F,'" May 1998. L2TPv2 is layered on

---

[29] https://tools.ietf.org/html/rfc2661
[30] https://tools.ietf.org/html/rfc2341

PPP, which is defined in RFC 1661,[31] "The Point-to-Point Protocol (PPP)," July 1994. All L2TPv2 connections use UDP encapsulation. There are already some very large deployments of softwires on L2TPv2 in ISPs today. L2TPv2 meets all *IPv6-over-IPv4* softwire requirements today. It is 99% ready for *IPv4-over-IPv6* softwire today.

Future softwire solutions will be based on L2TPv3, which is defined in RFC 3931,[32] "Layer Two Tunneling Protocol – Version 3 (L2TPv3)," March 2005. L2TPv3 *can* be layered on PPP, but in v3 it is optional (it can layer directly on IP). UDP encapsulation is also optional in v3. UDP encapsulation is useful for NAT traversal, but it increases overhead and lowers throughput and reliability. If no NAT needs to be traversed, turning off the UDP encapsulation can lower overhead. Session ID and Control Connection IDs are 32 bits (vs. 16 in L2TPv2). L2TPv3 also provides better user authentication and data channel security through use of optional *cookies*. An L2TPv3 cookie is an up to 64-bit cryptographically generated random value, included in every packet. L2TPv3 is close to meeting all softwire requirements.

Relevant Standards for Softwires

RFC 4925, "Softwire Problem Statement," July 2007 (Informational)

RFC 5512, "The BGP Encapsulation Subsequent Address Family Indicator (SAFI) and the BGP Tunnel Encapsulation Attribute," April 2009 (Standards Track)

RFC 5543, "BGP Traffic Engineering Attribute," May 2009 (Standards Track)

RFC 5549, "Advertising IPv4 Network Layer Reachability Information with an IPv6 Next Hop," May 2009 (Standards Track)

RFC 5565, "Softwire Mesh Framework," June 2009 (Standards Track)

RFC 5566, "BGP IPsec Tunnel Encapsulation Attribute," June 2009 (Standards Track)

---

[31] https://tools.ietf.org/html/rfc1661
[32] https://tools.ietf.org/html/rfc3931

RFC 5571, "Softwire Hub and Spoke Deployment Framework with Layer Two Tunneling Protocol Version 2 (L2TPv2)," June 2009 (Standards Track)

RFC 5619, "Softwire Security Analysis and Requirements," August 2009 (Standards Track)

RFC 5640, "Load-Balancing for Mesh Softwires," August 2009 (Standards Track)

RFC 5969, "IPv6 Rapid Deployment on IPv4 Infrastructures (6rd) – Protocol Specification," August 2010 (Standards Track)

RFC 6333, "Dual-Stack Lite Broadband Deployments Following IPv4 Exhaustion," August 2011 (Standards Track)

RFC 6334, "Dynamic Host Configuration Protocol for IPv6 (DHCPv6) Option for Dual-Stack Lite," August 2011 (Standards Track)

RFC 6519, "RADIUS Extensions for Dual-Stack Lite," February 2012 (Standards Track)

RFC 6674, "Gateway-Initiated Dual-Stack Lite Deployment," July 2012 (Standards Track)

RFC 6908, "Deployment Considerations for Dual-Stack Lite," March 2013 (Informational)

RFC 7040, "Public IPv4-over-IPv6 Access Network," November 2013 (Informational)

RFC 7596, "Lightweight 4over6: An Extension to the Dual-Stack Lite Architecture," July 2015 (Standards Track)

RFC 7597, "Mapping of Address and Port with Encapsulation (MAP-E)," July 2015 (Standards Track)

RFC 7598, "DHCPv6 Options for Configuration of Softwire Address and Port-Mapped Clients," July 2015 (Standards Track)

RFC 7599, "Mapping of Address and Port Using Translation (MAP-T)," July 2015 (Standards Track)

RFC 7600, "IPv4 Residual Deployment via IPv6 – A Stateless Solution (4rd)," July 2015 (Experimental)

RFC 7785, "Recommendations for Prefix Binding in the Context of Softwire Dual-Stack Lite," February 2016 (Informational)

RFC 7856, "Softwire Mesh Management Information Base (MIB)," May 2016 (Standards Track)

RFC 7870, "Dual-Stack Lite (DS-Lite) Management Information Base (MIB) for Address Family Transition Routers (AFTRs)," June 2016 (Standards Track)

RFC 8026, "Unified IPv4-in-IPv6 Softwire Customer Premises Equipment (CPE): A DHCPv6-Based Prioritization Mechanism," November 2016 (Standards Track)

RFC 8114, "Delivery of IPv4 Multicast Services to IPv4 Clients over an IPv6 Multicast Network," March 2017 (Standards Track)

RFC 8115, "DHCPv6 Option for IPv4-Embedded Multicast and Unicast IPv6 Prefixes," March 2017 (Standards Track)

RFC 8389, "Definitions of Managed Objects for Mapping of Address and Port with Encapsulation (MAP-E)," December 2018 (Standards Track)

RFC 8513, "A YANG Data Model for Dual-Stack Lite (DS-Lite)," January 2019 (Standards Track)

Dual-Stack Lite

The IETF Softwires working group has come up with a variant on the basic dual-stack network design, which is described in RFC 6333,[33] "Dual-Stack Lite Broadband Deployments Following IPv4 Exhaustion," August 2011. There is additional information on Dual-Stack Lite in RFC 6908,[34] "Deployment Considerations for Dual-Stack Lite," March 2013.

---

[33] https://tools.ietf.org/html/rfc6333
[34] https://tools.ietf.org/html/rfc6908

Clients using Dual-Stack Lite will still need to support both IPv4 and IPv6, but the service from the ISP to the customer will be IPv6-only, with IPv4 service tunneled over IPv6 in both directions. If you examine the traffic between the CPE and the ISP, there will only be IPv6 packets, but some of them will contain IPv4 packets as the Data field. The IPv4 addresses provided to the customer will be RFC 1918 *private addresses,* provided by a giant Carrier-Grade NAT (CGN) at the ISP. The NAT involved actually uses the customer's IPv6 address to tag the private IPv4 addresses used by the client, which would allow multiple ISP clients to use the same private address range (e.g., all of them could use 10.0.0.0/8, and the LSN would keep each organization's addresses separate based on their unique IPv6 address). There is a special new private address range (100.64/10) that is used in the carrier-based mapping. So, if the address assigned to the WAN node on your CPE is in 100.64/10, you are behind CGN. This is becoming more and more common as unallocated IPv4 public addresses vanish. According to the CGN RFC (6598[35]), no one should deploy CGN without also deploying IPv6, but many telcos and ISPs ignore this and deploy CGN without *any* IPv6, in order to keep providing their customers with IPv4 service.

IPv6-only or dual-stack nodes at the client would be able to connect to any IPv6 node in the world directly, via the ISP's IPv6 service. IPv4-only or dual-stack nodes at the client would be able to connect to any IPv4 node in the outside world via IPv4 tunneled over IPv6, with addresses from the ISP's Carrier-Grade NAT. There is no 6to4 translation that would allow an IPv6-only node to connect to external IPv4 nodes or 4to6 translation that would allow an IPv4-only node to connect to external IPv6 nodes. Any internal node that needs to connect to external IPv4 nodes should be configured to support dual stack. The tunneling of IPv4 packets inside the outgoing IPv6 packets takes place inside the CPE, as does the de-tunneling of IPv4 packets from the incoming IPv6 packets. It's basically 6in4 upside down. This scheme can be deployed for a very long time compared with basic dual stack.

The way this differs from basic dual-stack operation is that there is no direct IPv4 service provided, and the IPv4 addresses used at the client are private and managed by infrastructure at the ISP. This allows the ISP to share a relatively small number of precious real IPv4 addresses among a large number of customers and also allows the ISP to run IPv6 only to the customer. A major advantage of DS Lite is that no 6to4 or 4to6 translation is required. The downside is that all nodes on the internal network are still

---

[35] https://tools.ietf.org/html/rfc6598

dual stack – you must still manage two sets of IP addresses (IPv6 and IPv4). It is much cleaner and less expensive to eliminate IPv4 altogether in the internal network, other than via NAT64 border gateways.

This will require a firmware upgrade (or replacement) of the Customer Premises Equipment (CPE), which is typically a DSL or cable modem, with embedded router and NAT.

The Internet Systems Corporation (who also supplies the BIND DNS server and dhcpd DHCPv4 server) has created a freeware implementation of the ISP-side facilities to support DS Lite, called AFTR[36] (Address Family Transition Router). This includes IPv4-over-IPv6 tunneling, DHCPv4, DHCPv6, and some other pieces.

The CPE device for DS-Lite[37] is called B4 (Basic Bridging BroadBand Element). There is an open source implementation of this for the Linksys WRT-54GL. Some network vendors are beginning to produce DS Lite–compatible CPE now.

PET (Prefixing, Encapsulation, and Translation)

PET is one of the emerging softwire standards, which is trying to work out the optimal combination of tunneling and translation mechanisms to provide a workable framework for IPv4/IPv6 co-existence. The types of tunnels discussed are

- IP-in-IP tunnels (RFC 2893, RFC 4213)

- GRE tunnel (RFC 1702)

- 6to4 tunnel (RFC 3056)

- 6over4 tunnel (RFC 2529)

- Softwire transition technique (RFC 5565)

The translation mechanisms discussed include

- SIIT (RFC 2765)

- NAT-PT (RFC 2766 – deprecated)

- BIS (RFC 2767)

- SOCKS64 (RFC 3089)

---

[36] www.isc.org/downloads/aftr/

[37] www.isc.org/blogs/ds-lite-architecture-overview-and-automatic-configuration/

- BIA (RFC 3338)

- IVI (RFC 6219)

These standards discuss various combinations of the preceding tunneling and translation mechanisms to accomplish different kinds of co-existence. The recommended tunneling scheme is the softwire transition technique (RFC 5565). It also notes that DNS may have to interact with the co-existence solution using a DNS Application Layer gateway, such as DNS64.

Translation

Translation between IPv4 and IPv6 is by far the most complex transition mechanism. It has all the issues of IPv4-to-IPv4 Network Address Translation, plus new issues that complicate it even further. There is a great deal of activity in the IETF trying to create standards that will be implementable and deployable.

Since IPv4 addresses are running out, many ISPs would like to deploy IPv6-only service to their customers (as opposed to dual stack with both IPv4 and IPv6 services). Without translation, an IPv6-only node cannot access legacy IPv4-only nodes on the Second Internet (which currently includes most online sites). Over time, more and more sites and services will be dual stack, which will make IPv6-only nodes more useful. Until that time, translation gateways will be needed for IPv6-only nodes. It will be far simpler and cheaper, resulting in a superior user experience if both IPv4 and IPv6 are deployed, even if the IPv4 service is heavily NATted. However, ISPs seemed to be obsessed with deploying translation. There are a variety of ways that this can be accomplished, but most are quite complex and likely to be major sources of problems.

Tunneling cannot achieve IPv4-to-IPv6 interworking, but it's highly transparent and lightweight, can be implemented by hardware, and can keep IPv4 routing and IPv6 routing separated. It allows existing infrastructure (whether IPv4 or IPv6) to be used as a transport to link two nodes (or networks) using the other version of IP.

Translation achieves direct intercommunication between IPv4 and IPv6 nodes or networks by means of converting the semantics between IPv4 and IPv6. However, it has limitations in operational complexity and scalability. Like any NAT, it may have serious issues with transparency (some protocols may not work through it). Correct translation requires

- Address or (address, port) tuple substitution

- MTU discovery

- Fragmentation when necessary

- Translation of both IP and ICMP fields

- ICMP address substitution in payloads (e.g., with SIP)

- IP/TCP/UDP checksum recomputation

- Application Layer translation when necessary

Stateless translation consumes IPv4 addresses to satisfy IPv6 hosts, which does not scale (for one thing we are running out of IPv4 addresses; for another, there are lots more IPv6 addresses than IPv4 addresses). It can be implemented in hardware, but any ALG translation is too complex for hardware.

Stateful translation requires maintaining complex state for dynamic mapping of (address, port) tuples and cannot be implemented in hardware.

NAT64/DNS64

This transition mechanism requires both a NAT64 gateway and either a DNS server that supports DNS64 mapping or a DNS ALG that supports DNS64. What follows is a highly simplified description of operation. The full details are covered in the RFCs (there is quite a bit of complexity involved in the real operation).

The NAT64 gateway should have two interfaces, one connected to the IPv4 network (with a valid IPv4 address on that network) and the other connected to the IPv6 network (with a valid IPv6 address on that network). IPv6 traffic from a node on the IPv6 network going to an IPv4 node is sent in IPv6 and routed to the NAT64 gateway. The gateway does address translation and forwards the translated packets to the IPv4 interface, from which they are routed to the destination node. Reply packets from the IPv4 node are sent in IPv4 to the gateway and are translated into IPv6 and forwarded to the IPv6 interface, from which they are routed back to the original sender. This process requires state, binding an IPv6 address and TCP/UDP port (referred to as an IPv6 transport address) to an IPv4 address and TCP/UDP port (referred to as an IPv4 transport address).

Packets that originate on the IPv4 side cannot be correctly translated, because there would be no state from the packets coming through the gateway in the v6->v4 direction. NAT64 is not symmetric. For traffic initiated by an IPv6 node, everything works right.

Once the binding is created, that traffic flow can continue (from the IPv6 node to the IPv4 and back).

For the traffic originating on the IPv4 side to be translated to IPv6, it requires some additional mechanism, such as ICE or a static binding configuration.

This mechanism depends on constructing *IPv4-converted IPv6 addresses*. Each IPv4 address is mapped into a different IPv6 address by concatenating a special IPv6 prefix assigned to the NAT64 device (Pref64::/n).

It also uses a small pool of IPv4 addresses, from which mappings will be created and released dynamically, as needed (as opposed to permanently binding specific IPv4 addresses to specific IPv6 addresses). This implies that NAT64 does both address and port translation.

When an IPv6 initiator does a DNS lookup to learn the address of the responder, DNS64 is used to synthesize AAAA resource records from A resource records. The synthesized AAAA resource records are passed back to the IPv6 initiator, which then initiates an IPv6 connection with the IPv6 address that is associated with the IPv4 receiver. The packet will be routed to the NAT64 device, which will create the IPv6-to-IPv4 address mapping as described before.

In general, dual-stack nodes should not use DNS64. If they get a synthesized IPv6 address *and* a native IPv4 address, the rule to prefer IPv6 will cause the dual-stack host to do the access via the NAT64 gateway instead of direct using IPv4. If you deploy DNS64, it should be used *only* by IPv6-only nodes, and there should be a regular DNS for use by any dual-stack nodes.

IVI

This address translation scheme is being used on a large scale between CERNET (IPv4-only) and CERNET2 (IPv6-only) for nodes on either side to connect to nodes on the other side, as well as allowing IPv6-only nodes to connect to IPv4 nodes out on the public Internet.

The pros of using IVI are as follows:

- It is stateless, so it scales to a large number of nodes better than NAT64/DNS64.

- The translation is decoupled from DNS.

- It is symmetric, so can be used for connections initiated on either side of the gateway (IPv4 to IPv6 side or IPv6 to IPv4 side).

- There is an open source implementation of the IVI gateway and DNS64 ALG available on Linux.

The cons of using IVI are as follows:

- An ALG is still required for any protocol that embeds IP addresses in the protocol, such as SIP.

- It restricts the IPv6 hosts to use a subset of the addresses inside the ISP's IPv6 block. Therefore, IPv6 Stateless Address Autoconfiguration cannot be used to assign IPv6 addresses to nodes. You must either manually assign addresses or use stateful DHCPv6.

- There are still some issues with end-to-end transparency, address referrals, and incompatible semantics between protocol versions.

- You still need a DNS64 ALG for DNS.

Preferred Network Implementation Going Forward: IPv6-Only

As the 2014 OECD report points out, the real benefits of IPv6 only come once you remove IPv4, except for gateway access to legacy IPv4-only nodes outside your network.

One very interesting discussion of this approach can be found in "Microsoft Works Toward IPv6-Only Single Stack Network,"[38] by Veronika McKillop (Microsoft CSEO), April 3, 2019.

This is a large-scale, real-world deployment of IPv6-only and will be fully realized globally over time. It is already far enough along to provide some very good insights into doing this from actual experience.

Here are key points from this writeup:

- IPv4 address depletion is already a serious problem, and not just *public addresses*. Microsoft is now having problems allocating even

---

[38] https://teamarin.net/2019/04/03/microsoft-works-toward-ipv6-only-single-stack-net work/?fbclid=IwAROiqDbK8uehUouCC-NA1Da55RfjiHRPHFSky4jRyKZxB3TeF4Uh6IR54no

private addresses company-wide. They have predicted that they will no longer be able to use the 10/8 private address block in around 2–3 years. They have explored reclaiming unused IPv4 addresses, with little success.

- There are big benefits to a single stack network in troubleshooting, security, and QoS policies. Dual stack still involves having to work with NAT44, which has many problems.

- Since all companies today use private IPv4, there are always problems of address conflict in acquisitions, requiring even more NAT44 and address renumbering. These problems are not present in IPv6, even with ULA.

- Industry pressure is growing, such as Apple's decision to require IPv6 in all apps submitted to the App Store. It is critical that app developers have an IPv6-only environment to test apps. MA currently has 12 locations for this kind of work.

- A good IPv6 address plan is critical. The one they created in 2006 has required very minor changes (one in 2015 and another in 2018). They started with one /32 from ARIN and then in 2013 added /32s from RIPE and APNIC.

- It is important that both DHCPv6 and RDNSS (IPv6 addresses for DNS via RA messages) must be implemented everywhere, since some nodes only support one way or the other.

- Extensive training in IPv6 for engineering staff is critical.

- Working with outside vendors often requires forcing them to support IPv6 well.

- Clouds are still mostly IPv4-only, which causes major problems for cloud-based security.

- Global routing works better with IPv6.

- Currently, 20–30% of their internal traffic is IPv6.

- NAT64/DNS64 is essential but still problematic.

- They have a "scream test" that involves removing all IPv4 from a network temporarily and seeing who screams and what about.

- "My advice is to take your deployment bit by bit. Focus on things that give you the biggest benefit, the biggest learning, the biggest impact on the largest group of users."

- "Dual stack is only a temporary solution. The ultimate solution is IPv6-only."

Supporting IPv6 for Developers at Sixscape

We develop products for Windows, MacOS, Android, and iOS. All must fully support IPv6 and even work in IPv6-only environments (where possible). This means our developers must have access to three different network architectures, IPv4-only, dual stack, and IPv6-only.

Most of our developers use notebooks, and even a desktop can be provided with a Wi-Fi network adapter, so we chose to implement multiple Wi-Fi networks (and for both 2.4 GHz and 5.0 GHz). So we have six SSIDs in our office:

- V4-2.4: IPv4-only, 2.4 GHz, 172.18/16

- V4-5.0: IPv4-only, 5.0 GHz, 172.18/16

- DS-2.4: IPv4 + IPv6, 2.4 GHz, 172.17/16 and 2001:470:xxxx:1000::/64

- DS-5.0: IPv4 + IPv6, 5.0 GHz, 172.17/16 and 2001:470:xxxx:1000::/64

- V6-2.4: IPv6-only, 2.4 GHz, 2001:470:xxxx:2000::/64

- V6-5.0: IPv6-only, 5.0 GHz, 2001:470:xxxx:2000::/64

Depending on what Wi-Fi adapter your computer has, you may see only the 2.4 GHz or both 2.4 GHz and 5.0 GHz SSIDs. From the visible SSIDs, choose the subnet you want to test with. 5.0 GHz has higher speeds (up to 867 Mbps internally, although our ISP connection is only 500 Mbps).

DHCPv4, DHCPv6, and RDNSS are all implemented. Static routes and firewall rules allow the 172.18/16 and 2001:470:xxxx:2000::/64 subnets access anything in the 172.17/16 and 2001:470:xxxx:1000::/64 subnets (and vice versa). The public IPv4 Internet is accessible from the V4-only and DS subnets, while the public IPv6 Internet is accessible from the V6-only and DS subnets. All internal nodes can configure appropriate internal IP addresses and addresses of DNS and find the default gateways. We can open incoming ports to any nodes on the V6-only or DS subnets and provide limited access via incoming connections via our 6 public IPv4 addresses (either BINAT or port mapped). I have IPv6 at home as well and often access the node at my desk via RDP from home – it is just like being in the office. If needed, we can make a wired Ethernet connection from any subnet to any internal node, but typically the wired connections are only to the DS subnet.

Our firewall (pfSense based) supports four NICs – one for WAN, one for V4-only LAN, one for DS LAN, and one for V6-only LAN. Those NICs are connected to three Wi-Fi routers (via the LAN taps, not WAN taps). This bridges the Wi-Fi networks to the wired networks. I have not yet implemented NAT64 on the V6-only subnet but will do that soon. It is interesting currently (without NAT64) to see how much outside stuff works on the V6-only subnet. Most things do, amazingly (Google, FB, DynDNS, etc.).

As an example, our DNSSEC appliance is fully functional in a V6-only network – most have some IPv4 dependency (e.g., NTP, SNMP, etc.). I believe we need to extend IPv6-ready certification to include working in an IPv6-only subnet (without NAT64), as a higher-level certification.

Our ISP does not currently offer native IPv6, so we use 6in4 to bring a /48 block in from Hurricane Electric (there is a tap here in Singapore, so performance is quite good). Even if we only had a single IPv4 public address, we would be able to use that for both cone NAT and our endpoint of the 6in4 tunnel. We route one /64 block to the DS subnet and another to the V6-only subnet. The fact that we obtain IPv6 via a tunnel does not impact this setup at all. However, if we had an ISP that only provided one /64, we would not be able to do this.

Summary

In this chapter, we covered the many transition mechanisms intended to help during the transition from all-IPv4 to all-IPv6. Some of these (dual stack, 6in4 tunneling, etc.) have been successful and are still in use. Some of these (6over4, Teredo, ISATAP) were used in the early days but due to various problems have fallen out of use. We covered those here in case you tun into an old implementation of them.

Most of the translation mechanisms have not worked very well. The only one still in use had to be severely restricted in terms of how it was deployed for it to actually work (NAT64/DNS64). It now only supports connections from IPv6-only nodes in an IPv6-only subnet to external IPv4 servers. This can help during the deployment of IPv6-only subnets.

One of the hot topics today is finally doing away with IPv4 in entire subnets (IPv6-only). The US DoD has now mandated that new equipment must work in dual-stack and IPv6-only subnets. This means there can be no IPv4 dependencies (e.g., using IPv4 versions of ancillary protocols such as NDP or SNMP).

CHAPTER 9

IPv6 on Mobile Devices

My telco in Singapore (M1) was providing IPv6 service on their cellular dataplan if you knew how to configure your phone. The trick on Android was to change your service type to "LTE/3G/2G" and set the APN protocol to "IPv4/IPv6." On iPhone no special settings were required – it just worked. In the United States, my service from AT&T includes IPv4 and IPv6 with no configuration required – it just works out of the box. They allocate a /64 block for every phone. My phone currently has block 2600:380:b0d0:f919::/64 allocated. Note that with AT&T I can see both Wi-Fi and dataplan IPv6 addresses at the same time.

It is a bit tricky to find your mobile IP addresses on iOS – get the HE.NET app from the App Store (shown in the following). In Android the Network Info II app allows you to see IPv6 addresses. I will show examples of this later.

Android

The IPv6 address *2401:7400:6000:93d9:1:1:9366:9705* is configured on my phone, along with the private IPv4 address *10.194.78.202*. The external public IPv4 address is *246.106.56.119*. Note that on Android, if the Wi-Fi configures an IPv6 address, no additional IPv6 addresses are configured on the dataplan. This screenshot required disabling the Wi-Fi. Note that on my ISP, if the phone goes to sleep, it will keep the same /64 prefix when it wakes back up. If I reboot, I get a new /64 prefix. Your mileage may differ.

Note that the phone actually gets a /64 block, not a single /128 address. If you set up a hot spot, both IPv4 and IPv6 are shared, and devices that connect to the hot spot get an IPv6 address in the same /64 block as the address on the phone. This may be significant as we start running servers on phones! Sixscape has created a way to securely register your IPv6 address from a mobile device (using IRP).

© Lawrence E. Hughes 2022
L. E. Hughes, *Third Generation Internet Revealed,* https://doi.org/10.1007/978-1-4842-8603-6_9

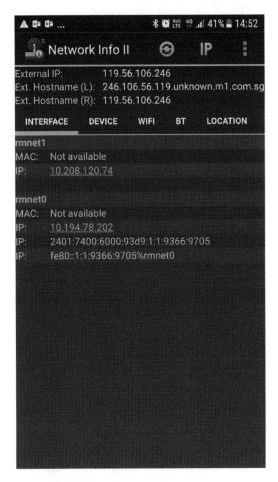

Figure 9-1. *Network IP address allocation on Android*

Figure 9-2. *IPv6test.com test of IPv6 on Android*

Note that I can even ping this address from my Windows node:

```
C:\Windows\system32>ping 2401:7400:6000:93d9:1:1:9366:9705

Pinging 2401:7400:6000:93d9:1:1:9366:9705 with 32 bytes of data:
Reply from 2401:7400:6000:93d9:1:1:9366:9705: time<1ms
Reply from 2401:7400:6000:93d9:1:1:9366:9705: time=51ms
Reply from 2401:7400:6000:93d9:1:1:9366:9705: time=107ms
Reply from 2401:7400:6000:93d9:1:1:9366:9705: time=104ms
Ping statistics for 2401:7400:6000:93d9:1:1:9366:9705:
    Packets: Sent = 4, Received = 4, Lost = 0 (0% loss),
Approximate round trip times in milli-seconds:
    Minimum = 0ms, Maximum = 107ms, Average = 65ms
```

iPhone

Figure 9-3. *IP address allocation on iPhone 7 using HE network application*

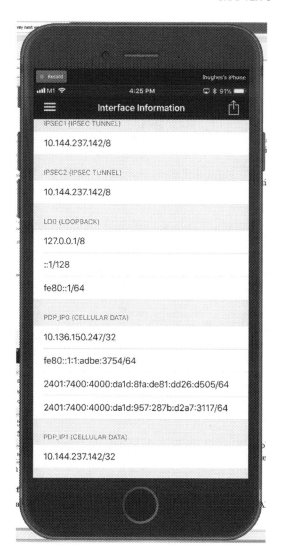

Figure 9-4. *Rest of IP allocation info on iPhone 7*

On the Wi-Fi interface, a link-local IPv6 address and two global IPv6 addresses have automatically configured.

On the dataplan (cellular data), another link-local IPv6 address and two more global IPv6 addresses have automatically configured.

For outgoing IPv6, here is the result from surfing to ipv6-test.com on the iPhone (with Wi-Fi disabled, so the dataplan IPv6 address was used).

Figure 9-5. *Ipv6-test.com site on iPhone 7*

As on Android, I can ping both of these addresses from my Windows node:

```
C:\Windows\system32>ping 2401:7400:4000:da1d:957:287b:d2a7:3117

Pinging 2401:7400:4000:da1d:957:287b:d2a7:3117 with 32 bytes of data:
Reply from 2401:7400:4000:da1d:957:287b:d2a7:3117: time=258ms
Reply from 2401:7400:4000:da1d:957:287b:d2a7:3117: time=74ms
Reply from 2401:7400:4000:da1d:957:287b:d2a7:3117: time=92ms
Reply from 2401:7400:4000:da1d:957:287b:d2a7:3117: time=106ms

Ping statistics for 2401:7400:4000:da1d:957:287b:d2a7:3117:
Packets: Sent = 4, Received = 4, Lost = 0 (0% loss),
Approximate round trip times in milli-seconds:
```

```
Minimum = 74ms, Maximum = 258ms, Average = 132ms

C:\Windows\system32>ping 2401:7400:4000:da1d:8fa:de81:dd26:d505

Pinging 2401:7400:4000:da1d:8fa:de81:dd26:d505 with 32 bytes of data:
Reply from 2401:7400:4000:da1d:8fa:de81:dd26:d505: time=703ms
Reply from 2401:7400:4000:da1d:8fa:de81:dd26:d505: time=90ms
Reply from 2401:7400:4000:da1d:8fa:de81:dd26:d505: time=111ms
Reply from 2401:7400:4000:da1d:8fa:de81:dd26:d505: time=100ms

Ping statistics for 2401:7400:4000:da1d:8fa:de81:dd26:d505:
Packets: Sent = 4, Received = 4, Lost = 0 (0% loss),
Approximate round trip times in milli-seconds:
Minimum = 90ms, Maximum = 703ms, Average = 251ms
```

On Android, if there is an IPv6 address on Wi-Fi, any IPv6 address on the dataplan is ignored. On iOS, it will configure IPv6 from *both* Wi-Fi and dataplan, but if there is an IPv6 Wi-Fi address available, connections will use it rather than the dataplan address.

What Are the Implications of This?

For the first time ever since Internet access has been available on mobile devices, there are finally *public IP addresses* for mobile devices. That is a major game changer. Up until now, all IP addresses on mobile devices have been private (RFC 1918) addresses, behind NAT (in some cases, behind multiple layers of NAT, with CGN). That means you can only make *outgoing* connections from your mobile device to nodes with public addresses, typically servers at telcos, ISPs, large content providers, etc. There is no way for external nodes to make connections to your phone. So the only apps you can get or run on your phone are *clients*.

Now, with IPv6 on mobile devices, you have *public* IP addresses (accessible only by IPv6 users, but definitely public addresses). These addresses can not only be used for *outgoing* connections; they will also accept *incoming* connections, from any IPv6 node in the world (unless the ports involved are blocked somewhere along the way). You can now run *servers* (FTP, web, email, LDAP, etc.) on your mobile device. If you publish this address in DNS, anyone you allow can connect to those servers.

This also opens up the possibility of *end-to-end direct connections*: straight from my node to yours, with no intermediary server between us. The Second Internet (using IPv4 with NAT) forces us to use intermediary servers running on nodes with public addresses. If I want to chat with you, with IPv4 I cannot make a connection directly to your node; we both have to make *outgoing* connections to some intermediary server (XMPP and so on). That server then relays messages back and forth between our two outgoing connections. If the connections are secured by TLS, the messages will be in plain text on that intermediary server. Most snooping takes place on intermediary servers, not in transit between end users and the server. You can achieve end-to-end privacy and sender-to-recipient authentication with S/MIME, but that requires all parties to have mutually trusted client certificates and a PKI to manage them. S/MIME is primarily for email, although I have made it work over FTP as well. I don't know of any chat protocol that supports S/MIME currently, although in theory it could be done (not with XMPP).

With end-to-end direct messaging, there is only *one link* involved and no intermediary server(s). In this case, TLS is actually end to end. If both ends use a client cert, this approach achieves strong *mutual authentication* (we both know for certain who the other party is) and *end-to-end privacy*. I call this *PeerTLS*. Messages are encrypted in my node and decrypted in yours. Most people are not aware that TLS can work with a client cert at both ends, but I have verified that it works great. When I message you, I am not really interested in verifying your nodename; I want to know *your* identity (in other words the information in your client cert, not in a server cert). Note that this *will* work over IPv4 but only within a private Internet (e.g., your company LAN) – it cannot cross a NAT gateway. With IPv6, there are no NAT gateways, so assuming port 4605 is open from my node to yours, my node can connect directly to yours, regardless of where we are in the world.

Note that decentralized messaging uses existing protocols, just implemented in an end-to-end model, using PeerTLS. This has been tested with chat, email (SMTP), file transfer (FTP), and VoIP (SIP/RTP). With this model, every user has both a client and a (personal) server running on their node (even on a phone). There is no need for a centralized intermediary server (or an account from an ISP or telco). Anyone can exchange messages directly with any other user (assuming both have IPv6). This provides true end-to-end encryption (only one link is involved) and mutual strong authentication (both parties know for certain who they are communicating with). This may be one of the biggest wins of the migration to IPv6.

This is 5G-style decentralized messaging, not possible on the *Second Internet* because of NAT.

Decentralized Messaging

There are many advantages to decentralized messaging over current client/server via intermediary servers:

- As long as there is network connectivity between our nodes, it doesn't matter if our connection to the Internet is up or not. For comparison, even when I'm chatting with the person next to me on Skype, if the ISP connection goes down, the chat stops.

- The speed of the connection is limited only by the bandwidth on the shortest network path between us.

- Traffic never leaves the shortest network path between us, although that could involve going to our ISP if the other party is not local.

- PeerTLS with both ends using a client cert provides strong mutual authentication in addition to end-to-end encryption.

- You are not dependent on any intermediary servers being up and running or having an account on them (you will need an IRP account if you want to publish and obtain IP addresses and/or client certs for other parties, but the messaging traffic does not go through this path).

- No one but the parties communicating have any control over the message content.

- It is very difficult for anyone to snoop on decentralized connections, especially if they are encrypted end to end – there can be no man-in-the-middle attack if there is no "middle."

Summary

IPv6 deployment is very far along with mobile service providers, as there are major and obvious benefits over trying to splice together many IPv4 10.x.x.x subnets. The largest IPv4 subnet is only a /8, which provides some 16 million private IP addresses. Most mobile service providers have many more than 16M subscribers. With IPv6, a single subnet can provide /64 blocks for every subscriber with no problem.

If you check on your phone today (and you may need to download an application, like *Network Info II* on Android or HE.NET on iPhones), you can see the IPv6 addresses on your phone. Even without those applications, if you surf to `https://ipv6-test.com`, you can see whether you have IPv6 today. Most of the people reading this probably already do have it. For mobile device users, the future has already arrived.

If your Wi-Fi supports IPv6, both Android and iOS phones can get IPv6 over Wi-Fi, in addition to over your dataplan. If your service provider is not currently providing IPv6 for some reason, but you have IPv6 running in your local subnet, you can still connect to nodes in the Third Internet via Wi-Fi.

This is one of the most exciting areas of IPv6 deployment. For the first time since phones were given access to the Internet, they now can get public IP addresses. I'm already pioneering several aspects of this, including direct end-to-end connections and PeerTLS.

CHAPTER 10

DNS

DNS (the Domain Name System) is a critical part of today's Internet. Without it, we would have to keep massive (and always out-of-date) directories (like telephone books), where you could look up the name of some site (such as Dell's pages about their PCs) and then find the "telephone number" (IP address) of that page, which you would then "dial" (type into your browser). This is clearly not very practical. DNS is such a complex and critical topic for both IPv4 and IPv6 that I have included a chapter just for it.

How DNS Evolved

Various schemes have been used to keep track of nodenames and their corresponding IP addresses. The end result is a remarkably powerful and flexible system called DNS.

Host Files

In the early days of TCP/IP, a list of known "hosts" (computers, routers, firewalls, etc.) was kept in a special file called *hosts* in the */etc* directory of all UNIX computers (the complete filename was */etc/hosts*). This file included one or more lines, each of which contained an IP address, followed by one or more nodenames (e.g., www) or even fully qualified nodenames (e.g., `www.ibm.com`). If you told your copy of UNIX to use the hosts file for "name resolution," when you used one of the nodenames listed in your *hosts* file, it would use the IP address associated with that name. This still works today – you can override DNS with your *hosts* file if you specify it first in the search order. Even Windows systems have a *hosts* file, typically located at

- C:\Windows\System32\drivers\etc\hosts

© Lawrence E. Hughes 2022
L. E. Hughes, *Third Generation Internet Revealed*, https://doi.org/10.1007/978-1-4842-8603-6_10

A typical *hosts* file might look like

```
172.20.0.11   ws1 ws1.hughesnet.local
172.20.0.12   ws2 ws2.hughesnet.local
172.20.0.13   us1 us1.hughesnet.local
```

Network Information Service (NIS)

In organizations with many UNIX computers, and especially once people started linking networks together with TCP/IP, it became necessary to keep everyone's *hosts* file up to date and synchronized. This was done manually for a while. Then NIS (Network Information Service) was created by Sun to automatically distribute copies of the official *hosts* file (in addition to other important configuration files for UNIX) to every node, on a periodic basis.

DNS Was Invented

Soon even this became unwieldy, so in 1983, at the request of Jon Postel, Paul Mockapetris designed DNS as a distributed database engine with distributed data. We are still using this system today. You will find that with minor extensions, it even supports IPv6 and dual-stack networks. There is a gigantic, worldwide hierarchical system of DNS servers that allows each network administrator to manage the names and IP addresses of nodes in their network that users anywhere in the world might need to know about (e.g., that organization's web servers, email servers, etc.). DNS is also used to keep track of the nodenames and IP addresses of all *internal* nodes in an organization's private network, which only users in that organization need to know about (the organization's file servers, network printers, intranet web servers, other users' workstations, etc.).

Domain Names

Domain names refer to the hierarchical namespace as defined by RFC 1035 (above); RFC 1123, "Requirements for Internet Hosts – Application and Support," October 1989; and RFC 2181, "Clarifications to the DNS Specification," July 1997. Briefly, domain names consist of a list of names (e.g., *atlanta, usa, exampleco,* and *com*), in most specific to least specific order, separated by periods (e.g., *atlanta.usa.exampleco.com*). Here, *com* is the

TLD (or top-level domain) name for commercial organizations. The name *exampleco* is the name of a hypothetical company, which is a commercial organization, and all parts of it use the domain *exampleco.com*. Within ExampleCo, there is a branch in the United States, which uses the subdomain *usa.exampleco.com*. Finally, there is an office in Atlanta, Georgia, which uses the subdomain *atlanta.usa.exampleco.com*. If there is a web server named *www* in that office, it would have a fully qualified domain name of `www.atlanta.usa.exampleco.com`. There is no way to tell (without more information) if the first name in such a string is a node's name or the first component of a domain name.

Top-Level Domain Names

There are a number of TLDs including *generic* ones that have been in use for a long time:

com: For commercial organization (company)

org: Noncommercial organization

edu: Educational organization

net: Internet related, for example, ISP

gov: Government related

mil: Military related

Recently, this was opened up – anyone with sufficient funds can now have their own top-level Domain and manage subdomains under it. There are now over 1500 top-level domains.

There are also many ccTLDs (country code top-level domains), each of which uses the ITU two-letter code for the country, such as *us, uk, jp,* and *ph*). There are a few exceptions to the ITU code usage, for example, the ITU code for Great Britain is *gb*, while their ccTLD is *uk*. Each country manages subdomains under their ccTLD as they see fit. Certain ccTLDs appear to have other meanings, like *tv* for the country of Tuvalu, which sells domains in their space to people who want to use it to mean *television*. Under ccTLDs, there are usually (but not always) second-level domains, such as *co* for commercial, *or* for organization, etc. Actual organization names would then be *third*-level domain names. Hence, a UK-based commercial entity called Warmbeer, Ltd. might have the domain name *warmbeer.co.uk*. Their web server might be `www.warmbeer.co.uk`.

A few ccTLDs, like *ph* for the Philippines, use the full three-letter code for organization type, instead of the more common two-letter codes, as a second-level domain name, for example, *infoweapons.com.ph*.

Internationalized Domain Names

There are also *Internationalized Domain Names* (IDNs) that use 16-bit Unicode characters to allow domain names in languages that have non-Latin alphabets. Your browser will translate these Unicode domain names into strings in UTF-8, using the *punycode* algorithm (shown in the last column in the following). This is defined in RFC 3492, "Punycode: A Bootstring Encoding of Unicode for Internationalized Domain Names in Applications (IDNA)," March 2003. For example, the following (believe it or not) are syntactically valid URLs (although they do not currently point to real sites):

http://пример.испытание/ e1afmkfd.xn--80akhbyknj4f/	Russian	http://xn--
http://例子.测试/ fsqu00a.xn--0zwm56d/	Chinese	http://xn--
http://실례.테스트/ -9n2bp8q.xn--9t4b11yi5a/	Korean	http://xn-
http://例え.テスト/ r8jz45g.xn--zckzah/	Japanese	http://xn--
http://उदाहरण.परीक्षा/ p1b6ci4b4b3a.xn--11b5bs3a9aj6g/	Indian	http://xn--

NS Resolver

All operating systems today include a DNS client, called a *resolver*. All network applications use the resolver to look up nodenames and obtain their corresponding IP addresses, whether those nodes are local (on the organization LAN) or external (out on the Internet). The resolver contacts one of the DNS servers specified in their TCP/IP configuration (either local or at your ISP). If that server is *authoritative* for the requested domain names, it returns the addresses immediately. Otherwise, that server can either return a hint of where to look ("I don't have that information. Try here") or do a recursive lookup ("I didn't have that information, but I went and found it for you"). Of course, the lookup could fail ("I couldn't find that domain name anywhere").

DNS Server Configuration

The full process of setting up DNS servers (usually two or more) for an organization and populating them with node information is too complicated to cover in this book. If you are using the Microsoft DNS server (included free with Windows Server), see their documentation for details. If you are using BIND (the freeware DNS server from the Internet Systems Corporation), see O'Reilly's *DNS and BIND,* fifth edition, for details. If you have a DNS appliance, consult their documentation or online help for details.

In general, though, you define both "forward zones" that map nodenames to IP addresses and "reverse zones" that map IP addresses onto nodenames. You also have to inform all client computers of the IP addresses of at least two DNS servers that they can use for resolving nodenames to IP addresses (or vice versa). Client computers can be informed of these DNS server addresses either via manual configuration or automatically via DHCPv4 or DHCPv6. If your client computer doesn't know where to find DNS servers, you may have full Internet connectivity but no name resolution. You can ping (or even surf to) nodes anywhere in the world by specifying their numeric IP addresses (e.g., *http://64.170.98.32* – try it!). However, most people would consider such a computer to not be very useful. This gives you a very good idea of how important DNS is to the Internet (even the Third Internet).

DNS Protocol

DNS is an *Application Layer* protocol. It uses UDP port 53 (for most queries and responses) or TCP port 53 (for zone transfers between DNS servers). It was originally defined in RFC 882, "Domain Names – Concepts and Facilities," November 1983, and RFC 883, "Domain Names – Implementation and Specification," November 1983. Those were replaced by RFC 1034, "Domain Names – Concepts and Facilities," November 1987, and RFC 1035, "Domain Names – Implementation and Specification," November 1987. There have been numerous updates to these, including RFCs 1101, 1183, 1348, 1876, 1982, 1995, 1996, 2065, 2136, 2137, 2181, 2308, 2535, 2845, 3425, 3658, 4033, 4034, 4035, 4343, and 4592.

DNS Resource Records

The data in DNS servers is kept in *resource records*. In forward zones, it is possible to have any of the following resource records (the following list is not comprehensive):

Name	Contents
A	"A" – IPv4 address associated with a domain name
AAAA	"Quad-A" – IPv6 address associated with a domain name
MX	"Mail eXchange" – domain name of a mail server for the domain
SRV	"Service" – domain name of servers for other protocols, such as SIP and LDAP
CNAME	"Alias" – provide an alternative domain name for another domain name
HINFO	"Host Info" – any arbitrary info you want to provide about a host
NAPTR	"Naming Authority Pointer" – used mostly in ENUM
NS	"Name Server" – name of a valid DNS server for this domain
SOA	"Start of Authority" – start of a zone in configuration files, includes default TTL
SPF	"Sender Policy Framework" – used in anti-spam technology
TSIG	"Transaction Signature" – symmetric cryptographic key used in zone transfers
TXT	"Text" – any arbitrary text information (not interpreted by DNS)

In reverse zones, typically only the following resource records are found:

NS   "Name Server" – name of a valid DNS server for this domain

SOA "Start of Authority" – same as in forward zones

PTR "Pointer Record" – IP address for a specific node, in reverse order

The following examples show how typical resource records look:

ws1	IN A	172.20.0.11
ws1	IN AAAA	2001:418:5403:3000::c
	IN MX	10 ws1.hughesnet.org
11.0.20.172	IN PTR	ws1.hughesnet.org

c.0.0.0.0.0.0.0.0.0.0.0.0.0.0.0.0.0.0.0.3.3.0.4.5.8.1.4.0.1.0.0.2   IN
PTR   ws1.hughesnet.org

In general, it is a pain to manually create reverse PTR records, and any change to IP addresses (e.g., from changing ISPs) requires changes to all forward and reverse resource records in DNS. Here again, an appliance with a GUI can help by automatically generating reverse PTR resource records. This is especially useful for IPv6 reverse PTR records.

In the Sixscape DNS appliance, you can define *named networks* for IPv6. When you define a network, which will create the associated reverse zone, you can assign that network a name, which has the value of the network's prefix. You can then define node addresses in terms of the network name. First, this fills in the first 64 bits of each address, which reduces errors and saves time. However, if you ever change ISPs, you can simply redefine the prefix for the network, and all forward and reverse resource records created from the nodes specified using that network name will be updated with the new prefix. This is called *instant prefix renumbering*. There was an *a6* resource record created at one point for IPv6 forward resource records that was supposed to accomplish this, but there were so many problems with it. It has now been deprecated (you are not supposed to use it anymore). It is much better to do this in an appliance that has a GUI and database and generates only the standard AAAA resource records.

DNS Servers and Zones

A given DNS server can have any number of *zones* defined on it. A given zone can be a *forward zone* (for mapping domain names to IP addresses) or a *reverse zone* (for mapping IP addresses to domain names). There is usually one forward zone for each *domain* for which the DNS server contains information (e.g., *hughesnet.org*) and one reverse zone for each *network* that the DNS server contains information for (e.g., *172.20.0.0/16*). So the forward zone for hughesnet.org might contain mappings for *ws1. hughesnet.org* to *172.20.0.11, for us1.hughesnet.org* to *172.20.0.13*, and so on. The reverse zone for *172.20.0.0/16* might contain mappings from *172.20.0.11* to *ws1.hughesnet.org*, from *172.20.0.13* to *us1.hughesnet.org*, *172.20.0.91* to *us1.v6home.org*, and so on.

Any zone (forward or reverse) can be a *primary zone* or a *secondary zone*. A primary zone is one of which the DNS administrator manages the contents (either via a GUI interface or via editing BIND configuration files). A secondary zone is one whose contents are automatically transferred from a corresponding primary zone of the same name on a different DNS server (no management is required for a secondary zone, once that zone is created). When you create a secondary zone, you specify the IP address of the DNS server that contains the corresponding primary zone. Usually there is one primary zone (on one DNS server) and one or more secondary zones (each on other DNS servers) for a given set of records. A given DNS server can have any mix of primary zones and secondary zones. Sometimes the terms *primary* and *secondary* are used for entire DNS servers, especially if all zones on a server are all primary zones or all secondary zones, but technically the terms refer to zones, not servers. The transfer of all records from a primary zone on one DNS server to a secondary zone on another DNS server is called a *zone transfer*. Typically, a primary zone is configured to allow zone transfers only to secondary zones on authorized DNS servers (by IP address). There is also a cryptographic authentication scheme called TSIG that can restrict zone transfers to only authorized secondary zones. Otherwise, a hacker could perform a zone transfer from one or more of your primary zones and obtain information useful in attacking your network (effectively, a "map" of at least part of your network). Typically, zone transfers from primary zones to secondary zones are done automatically on a periodic basis. If a hacker changes data in a *secondary* zone, the correct data would be automatically restored as of the next zone transfer. If a hacker changes data in a *primary* zone, the hacker's changes will be automatically and securely transferred to all secondary zones via the regular zone transfers. It is *very* important to secure your primary zones.

It is possible for all the zones on a given DNS server to be accessible by one or more clients for performing DNS resolutions (lookups), in which case it is a *resolving* server. A primary server that is not accessible for resolutions by any client (or other DNS servers) is called a *stealth server*. It is only ever used to do zone transfers to secondary servers (hence need not be very powerful). Access via UDP port 53 can be completely disabled (zone transfers take place over TCP port 53), and even those can be restricted by IP address. Use of a stealth server lowers the possibility of hackers being able to attack your primary DNS server. There would be no real use for a "stealth secondary server."

Different Types of DNS Servers

There are different types of DNS servers based on how they are populated with data.

Authoritative DNS Servers

A DNS server that contains a primary zone or a secondary zone is said to be *authoritative* for the domain (or network) defined in that zone. All resolving servers cache (temporarily store) the results of any query they perform on behalf of clients.

If a client makes a query of a resolving server that currently has the required information (either because it is authoritative or because it has cached it from a previous query), it responds with that information to the client immediately. If a resolving server is asked for information it does not currently have, it can either return a *reference* ("I don't know; go ask *this* server"), or it can do a *recursive* query on the client's behalf ("I didn't know, but I went and found out for you by making client queries myself on your behalf, and here is what I found"). A recursive query can go through several servers before the requested information is finally obtained and returned to the client that asked for it in the first place. Any server involved in the process typically caches the retrieved information. Every record published by a DNS server has a *Time To Live* (TTL) defined for it. When a record is cached, it is kept on the caching server only for the defined *Time To Live* for that record, after which it is considered *stale* and is discarded. Once a DNS server discards stale information, if it is asked for again, it must do another recursive query, at which point it again caches the record. This caching and expiration scheme keeps the data current, but means that a change to authoritative information may take a while to *propagate* to all other servers (often 24–48 hours, depending on *Time To Live* values chosen).

When a client obtains information from an authoritative server, it is reported as an authoritative answer. When it obtains information that has been cached, it is reported as a non-authoritative answer. This doesn't mean it is any less trustworthy, just that it obtained the information at "second hand" (out of some DNS server's cache) instead of directly from the authority on the subject (an authoritative server).

Caching-Only Servers

A resolving server that has no defined primary or secondary zones is called a *caching-only* server and typically, once set up and configured, requires little or no management.

Client Access to DNS

In a typical network, every client should have the addresses of at least two valid resolving DNS servers configured. If a connection to one of them fails, the client will automatically try the other configured address. This increases the robustness of the network. In a small network (e.g., home connection), the specified servers may be located at and managed by the ISP. In some cases, the DSL or cable modem might provide a *DNS proxy* function, which allows DNS queries to be submitted to the default gateway address. The modem relays such requests to the DNS servers configured in the modem and returns the replies to the internal client that made the request.

Any network can have one or more *local* DNS servers (assuming they can make outgoing queries via UDP port 53). To run an authoritative server on a network, that server must be accessible by relevant clients and other servers. If any of those clients or servers is *external*, then the authoritative server must have a globally accessible "external" IP address (not a *private* IP address). For example, I run an authoritative DNS server for my domain *hughesnet.org* in my home, on a DNS server that has a valid external IP address. I also run other servers (email and web) that have globally accessible *external* IP addresses (in my case, *both* IPv4 and IPv6 addresses). I can access these services from anywhere on the Internet, just like using servers at ISPs. This sort of thing is far simpler and less expensive with IPv6 than with IPv4.

Recursive DNS Queries

A single DNS query (e.g., "lookup the IP addresses for node *ws1.hughesnet.org*") can actually require several resolutions. If the server already has information for *ws1.hughesnet.org*, either because it is authoritative for that information or because it has still valid cached information, it returns the requested information immediately ("*ws1.hughesnet.org* has an IPv4 address, which is *172.20.0.11*, and an IPv6 address, which is *2001:df8:5403:3000::c*"). It is up to the client which of these is used. If it is a dual-stack client (supports both IPv4 and IPv6), it should use the IPv6 address by preference.

If information for *ws1.hughesnet.org* is not present on the resolving DNS server, that server must find the authoritative server for the domain *hughesnet.org*. The server that is authoritative for the domain *org* can tell it this information. To locate *that* server, the resolving server can ask any *root* DNS server. To locate a root DNS server, the resolving server can look in its *root hints* file. Any of these things could already be in cache (and typically are, if any other nodename ending in .org or hughesnet.org has been looked up by any client recently). If none of them are in cache, then first, the resolving DNS server will ask a root DNS server "who is authoritative for domain *org*." It will cache the response it gets and ask the returned server for *org* "who is authoritative for domain *hughesnet.org*." It will cache that response also and ask the returned server for *hughesnet. org* "what is the IP address of *ws1.hughesnet.org*." It will cache that response as well and return the answer to the client, who has been patiently waiting. Most DNS servers have a way to empty (or "dump") the cache if you would like to watch all this happen with a network sniffer (this would require root-level access on the computer running your DNS server).

The Root DNS Servers

All DNS queries eventually chain up to one of the 13 *root DNS servers* (or the cached data from them). In reality, "DNS anycast" is employed so that there are actually quite a few copies of most of the 13 root servers distributed around the world (see the following table). The current information on the root servers (from which I made the following table) is always available at

```
www.root-servers.org
```

Every DNS server includes a file with the current anycast addresses of the 13 root servers (a.root-servers.net to m.root-servers.net), as summarized in the following table. A copy of the official current file (in BIND format) can always be found at

```
www.internic.net/zones/named.root
```

All DNS server operators from time to time obtain the current copy of this file and update their server(s) *root hints* file with it. The information in this file allows a DNS server to locate a root server when it needs one.

The only thing the DNS root servers publish is the information in a short file that is maintained by IANA that helps other DNS servers locate the DNS hierarchy layer just below that of the DNS root servers (i.e., the servers that are authoritative for the top-level domains such as *com, net, org, uk, jp,* etc.). All root servers publish the same information, so only one ever needs to be asked (typically chosen at random from the 13 available). A copy of the current version of this information (in BIND format) can always be found at

```
www.internic.net/domain/named.root
```

Only operators of DNS root servers ever actually need to obtain this file and update their DNS root servers with it. In reality, due to DNS caching, the actual root servers are only rarely involved in a typical DNS query. A typical non-root DNS server only needs to access a root server about once every 48 hours. It would normally have the information published by the root servers cached in memory from previous inquiries. Only once the *Time To Live* expires for a given resource record obtained from an actual root server would the DNS server have to go back and obtain more up-to-date information from a DNS root server (which it would again cache to use in future lookups). Most of the time, this new information will just be the same information that just expired.

Current root servers (all in the domain "root-servers.net")

Name	Organization	Count	IPv4 Address	IPv6 Address
A	VeriSign, Inc.	6/3	198.41.0.4	2001:503:ba3e::2:30
B	Information Sciences Inst.	1/1	192.288.79.201	2001:478:65::53
C	Cogent Communications	6/0	192.33.4.12	-
D	University of Maryland	1/0	128.8.10.90	-
E	NASA Ames Research Center	1/0	192.203.230.10	-
F	Internet Systems Consort.	49/22	192.5.5.241	2001:500:2f::f
G	US DoD Network Info Ctr	6/0	192.112.36.4	-
H	US Army Research Lab	1/1	128.63.2.53	2001:500:1::803f:235
I	Autonomica	34/0	192.36.148.17	-
J	VeriSign, Inc.	70/6	192.58.128.30	2001:503:c27::2:30

K	RIPE NCC	18/10	193.0.14.129	2001:7fd::1
L	ICANN	3/3	199.7.83.42	2001:500:3::42
M	WIDE Project	6/5	202.12.27.33	2001:dc3::35

In the preceding table, the first number in the Count field (before the slash) is the total number of anycast servers for that name, regardless of IP version. The second number (after the slash) is the number of anycast servers for that name that can accept queries over IPv6. Currently, all root servers will accept queries over IPv4 (this may not always be the case). All root servers can return A and/or AAAA records for the servers authoritative for top-level domains. One of the watershed events for IPv6 happened in February 2008, when VeriSign enabled IPv6 access on enough of the root servers that a client doing queries over IPv6 would always be able to complete a query without having to fall back to IPv4. Since then, clients that access DNS over IPv6 (IPv6-only nodes or dual-stack nodes) can resolve names to addresses as effectively as IPv4-only nodes have been able to since the introduction of DNS. Eventually all root servers will support queries over IPv6.

Total root server names = 13

Total root server names that accept connections over IPv6 = 8 (61.5% of names)

Total deployed root servers = 202

Total deployed root servers that accept connections over IPv6 = 51 (25.2% of total)

MX and SRV Resource Records

In addition to providing nodename to IP address lookup (forward resolution) and IP address to nodename lookup (reverse resolution), DNS servers can also advertise the preferred servers for various functions, such as email (SMTP), VoIP (SIP), etc.

The MX (Mail eXchange) record can advertise one or more email server names, with priorities. Other mail servers when they want to deliver mail to your domain will do a DNS query asking for the MX record(s) for your domain. The sending server will try to make connections over port 25 (SMTP) to the advertised nodenames, in decreasing priority, until it either has a connection accepted (in which case it will deliver all the mail it has for your domain) or it runs out of advertised nodenames (in which case it will try again on some schedule, until it succeeds or decides your domain is not currently online). Thus, a client can send messages to any name at your domain (fred@hughesnet.org), and it will be delivered to one of your preferred mail servers, which will then deliver it to that person's mailbox or return it as undeliverable.

The SRV record (defined in RFC 2782,[1] "A DNS RR for specifying the location of services (DNS SRV)," February 2000) can be used to specify preferred servers for your domain for services other than email, such as VoIP (SIP), Jabber Instant Messaging (XMPP), and Directory Services (LDAP). In the same style as email addresses, it is possible to specify a new style "phone number" as a SIP URI, for example, *sip:fred@ hughesnet.org*. A good SIP client could take that URI, do a DNS SRV query to determine the preferred SIP server for the domain *hughesnet.org*, and then attempt to connect to user *fred* on that server. The SRV mechanism of DNS will make VoIP as scalable and decentralized as the current Internet email system is. The same mechanism would allow clients anywhere in the world to contact your Jabber IM client (*im:fred@hughesnet.org*) or retrieve your directory entry from your organization's external LDAP server.

ENUM

During the transition from legacy telephony service (using ITU E.164 numeric phone numbers) to more general SIP URIs, there is a need for a transition mechanism. One has been developed called ENUM (which stands for *E.164 Number Mapping*, where E.164 is the ITU standard for conventional numeric telephone numbers), including the international country codes (e.g., +1 for the United States, +63 for the Philippines, +852 for Hong Kong). ENUM is implemented as a sub-function of DNS.

With ENUM, you do the equivalent of reverse DNS lookups, but from ITU telephone numbers to one or more URIs, typically including a SIP URI. This allows people with legacy telephone customer equipment that only has a ten-key pad to map E.164 telephone numbers onto the complex alphanumeric SIP URIs (e.g., *sip:fred@hughesnet.org*). In addition to SIP URIs, you can also map a single E.164 number onto other URIs, including instant messaging (*im:*), email (*mailto:*), web (*http:*), etc. This would allow a smart client that supports unified messaging via various protocols to contact you via any of your addresses using a single numeric "phone number." These URIs are associated with an E.164 number using the NAPTR DNS resource record. In theory, the long list of various kinds of "addresses" on your business card whereby people can contact you

---

[1] https://tools.ietf.org/html/rfc2782

(telephone, fax, email, IM, web, etc.) could be replaced with a single ENUMber. There are standalone ENUM clients that can allow someone to simply view the mappings for any ENUMber, for use with devices such as legacy telephone handsets.

Essentially, you dial a person's ENUMber. Then your client (or your VoIP server) does a DNS query to map that onto one or more URIs, which it then uses to contact that person. ENUM can be run at the national level, typically by a large telco or a government agency, using ENUMbers that start with the actual country code. In this case, the general public DNS is used to do the mapping. It can also be done on a private basis, using any "country code" you want (even made-up ones), but requires clients to specify a custom DNS server to do the custom mapping (all ENUM-compliant clients allow specification of which DNS server to use for ENUM lookups).

If traditional DNS management is used to manage the URI mappings for a large number of people, this could be an enormous workload and lead to long delays in making additions or changes to URI mappings. It is better to allow the end users to view and modify their own mappings. This requires user authentication, which most often would be linked to an existing authentication server, such as Microsoft Active Directory (within an organization) or RADIUS (for a telco or ISP).

Because most phones from 3.5G onward will be based on IPv6, it is important that an ENUM-capable DNS server supports queries over IPv6. There will be some legacy phones that still use IPv4 for some time, so it must also support queries over IPv4. In other words, the DNS server used for ENUM must be dual stack.

ENUM is defined in RFC 3761,[2] "The E.164 to Uniform Resource Identifiers (URI) Dynamic Delegation Discovery System (DDDS) Application (ENUM)," April 2004.

There is more information on ENUM on the following website:

```
www.ripe.net/enum
```

DNSSEC (Secure DNS)

As mentioned before, it is possible for a hacker to tamper with the information in the DNS. You may think you are surfing to your bank's website, but it could be a clever mockup in some hacker's basement. It will show the correct URI in your browser, but anything you input (like your login) can be easily captured by the hacker and then used

---

[2]`www.ietf.org/rfc/rfc3761.txt`

to gain access to your account on the real site. A hacker could also trick people into reading fake news stories that appear to be from legitimate websites.

There are some patches to BIND that make it more difficult for a hacker to alter DNS data, but the only surefire way to detect it is to add digital signatures into the DNS. The following RFCs define such an extension to DNS, which is called DNSSEC:

RFC 3833, "Threat Analysis of the Domain Name System (DNS)," August 2004

RFC 4033, "DNS Security Introduction and Requirements," March 2005

RFC 4034, "Resource Records for the DNS Security Extensions," March 2005

RFC 4035, "Protocol Modifications for the DNS Security Extensions," March 2005

RFC 4431, "The DNSSEC Lookaside Validation (DLV) DNS Resource Record," February 2006

RFC 4641, "DNSSEC Operational Practices," September 2006

RFC 5074, "DNSSEC Lookaside Validation (DLV)," November 2007

Currently, various parts of the DNS namespace (the *org* domain, the *gov* domain, various ccTLDs) are being secured with DNSSEC. Once a top-level domain is signed, this simplifies signing of domains under it. Eventually the root of the entire DNS tree (".") will be signed, and this will unify DNSSEC for all domains. Until that time, there will be a need for DNS servers to obtain the root key material for any part of the DNS space for which you want to verify signatures.

DNSSEC depends heavily on digital signatures and public key digital certificates. You will need to understand these concepts to follow this discussion. Basically though, a digital signature is created by a cryptographic algorithm that produces a numeric result that is derived from a specific plain text (in this case, a single DNS resource record) and an asymmetric cryptography *private key*. These signatures are generated, encoded into ASCII characters, and inserted into the BIND configuration files after each resource record. When DNS data are retrieved, the digital signatures flow along with the resource records. The user's resolving DNS server can verify the signature of any retrieved records by using the *public key* associated with the *private key* used to sign the records. Only if the signature is verified does the resolving server return the retrieved record(s) to the client. If records are compromised, they appear to not be available to the user. Eventually all client software will contain code to verify the DNSSEC signatures, in which case the resolving DNS server will just return the records and signatures to the client, which can notify the user that the records were found but had been compromised (if the signature fails).

Administrators using BIND directly (without a GUI front end) will find that DNSSEC is extraordinarily difficult to deploy and requires massive time and effort and extensive knowledge of PKI to produce signed resource records, even using the public domain PERL scripts. A good DNS appliance (like Sixscape DNS) can totally automate both the signing and validation processes, which vastly speeds up signing an entire domain, eliminates many possible errors, and requires little or no security or cryptographic expertise on the part of the DNS administrator.

Ideally the private key used to sign records should be kept in an HSM (Hardware Storage Module), which the private key never leaves. The object to be signed is sent into the HSM, the private key is used to do the signing inside the HSM, and the result is retrieved and used. For highest security, the HSM should be certified by FIPS 140-2 or equivalent. FIPS is the US Federal Information Processing Standards. FIPS 140-2 is the "Security Requirements for Cryptographic Modules," May 2001. Most commercially available HSMs are quite expensive, and their performance is very low compared with doing the same algorithm in software on modern processors (e.g., 1200 signatures per second for a very good HSM, as opposed to 12,000 signatures per second in software on entry-level hardware). Typically, an HSM is needed only on a *signing DNSSEC server*.

Digital signature verification (used in DNSSEC validation) uses only the *public key* corresponding to the private key used to create the signatures, so it can be done without an HSM, hence at much higher performance. The entry-level model of Sixscape DNS can do about 30,000 queries per second without DNS verification or about 27,000 per second with DNSSEC verification. Any DNS server can do DNSSEC validation, with no need for an HSM.

DNSSEC is not strictly an IPv6 technology and is equally applicable to IPv4. It is however being deployed at the same time as IPv6, and of course, it is important that any system deployed to support DNSSEC is able to sign and validate both IPv4- and IPv6-related resource records and can support queries over both IPv4 and IPv6. DNSSEC is a very important part of the Third Internet.

Essentially, DNSSEC introduces *trust* into the DNS.

Summary

Almost everyone *uses* DNS, but not many people understand how it works. This chapter explained how DNS in general works. It has been in use with IPv4 for quite a while.

The good news for IPv6 users is that DNS was extended (with AAAA records) to support IPv6 in parallel to IPv4 (which uses A records). The reverse lookups support IPv6 as well.

You can also configure a DNS server to accept queries over only IPv4, only IPv6, or both (dual stack). Of course, the DNS server would need to be in a subnet that supports those versions of IP, and the DNS server itself would need to support IPv6 and have an IPv6 address configured to allow IPv6 nodes to connect to it. This is one of the first things you would do when you migrate from IPv4-only to dual stack (supports both IPv4 and IPv6 on your DNS servers).

On client nodes, you can configure both IPv4 and IPv6 addresses for the DNS servers in your subnet.

You can publish only the IPv4 address of a server, only the IPv6 address, or both IPv4 and IPv6. In the latter case, if the client supports IPv6, it will try to connect over that first and then fall back to IPv4 if the IPv6 connection fails. There are not currently many servers that only publish IPv6 addresses, as there would be no way for users with only IPv4 to find them.

For more advanced usage of DNS, DNSSEC (secure DNS) fully supports IPv6 as well.

Just as there are public DNS servers for IPv4 (e.g., 4.2.2.2), there are public DNS servers for IPv6 (e.g., 2001:470:20::2).

CHAPTER 11

The Future of Messaging with No NAT

In *the* Second Internet (the one being used by most people today, based on IPv4), most nodes do not have *public* (globally routable) IP addresses. There are simply not enough of these to go around. Those addresses have mostly all been allocated. Today most Internet users are second-class netizens, with only *private* addresses. These are addresses that work only in their subnet and cannot accept incoming connections. This has a major impact on messaging.

First off, let's define *public* and *private* IP addresses.

Private IPv4 Addresses

A *private* IPv4 address is one that is valid only in your own local subnet. It should fall within one of the following address ranges (as per RFC 1918 – "Address Allocation for Private Internets"):

```
10/8          10.0.0.1 - 10.255.255.254
172.16/12     172.16.0.1 - 172.31.255.254
192.168/16    192.168.0.1 - 192.168.255.254
```

Any *private Internet* can use addresses from any of these three address ranges. Any assigned address must be unique within its private Internet, but they need not be *globally unique*. There might be thousands of nodes around the world using the address 10.1.2.3, in different private Internets. On the other hand, public addresses must be *globally unique*. Only one node in the entire public IPv4 Internet can be using the public address 123.45.67.89.

© Lawrence E. Hughes 2022
L. E. Hughes, *Third Generation Internet Revealed*, https://doi.org/10.1007/978-1-4842-8603-6_11

If your node has an IPv4 address in one of the RFC 1918 ranges (e.g., 172.20.2.1), you are a *second-class netizen*. You can only make outgoing connections from your private Internet. You cannot run a server on your node that can be accessed from anywhere in the entire IPv4 Internet. You can't allow people to connect directly to your node (they have to do so via intermediary servers). If you had a public IPv4 address, you would be a *first-class netizen* and could do those things. Too bad that public IPv4 addresses are pretty much all allocated to other people. Ask your ISP nicely if you can have one please. Be prepared for all kinds of excuses. Or they might say, "For only four times the price, you can have a commercial account and get ONE public address." They really don't have many more to allocate. Their source for more IPv4 public addresses (their RIR) has dried up. Their RIR's source for new public IPv4 addresses (IANA) ran out back in early 2011.

A private IPv4 address is similar to an extension in a company phone system (PBX) like "x101," but since private addresses *look* just like public addresses, they confuse some people. They are 32 bits long and are shown in dotted decimal notation. Telephone extensions (e.g., x101, x102) *look* different from "real" telephone numbers, so nobody confuses them with real telephone numbers. Before RFC 1918, the preceding address ranges WERE public addresses. Those address blocks were repurposed and combined with Network Address Translation (NAT) to keep IPv4 alive for a few more years, while the real solution (IPv6) was being developed and deployed.

Any company can use the same set of internal extensions (e.g., x100–x199) for their extension phones. The company will have few "real" (global) telephone numbers that anyone can call, but maybe hundreds or thousands of internal extensions. If you try calling a public phone number from your extension phone, you have to dial "9" first, which connects you to one of the real company phone numbers (you hear an "outside" dial tone). If all your company's real numbers are already in use, you get a busy signal. Otherwise, you then dial the external number. If someone can see the number you are calling from (with "caller ID"), it will be one of the real company phone numbers, not your extension number. This is similar to outgoing network connections from behind NAT. With NAT, all internal users will appear to be connecting from the single public IPv4 address on the NAT gateway, not from your private address. This messes up website logging and law enforcement tracing of bad guys.

If someone with a real telephone number tried dialing your extension on their real phone, how would the phone company figure out which of the possibly thousands of x101s in various companies to connect you to? You have to first dial the company's real phone number and then somehow convince a receptionist (live or automated) there to connect you to the extension 101 in that company. Unfortunately, there is no equivalent to a receptionist in a NAT gateway. You can't make incoming connections from the public Internet to a node behind NAT. So how does Skype appear to do just that? Something called NAT traversal, which is a very bad design.

There is no way packets from a private Internet can be routed globally. If a packet with the private address 10.1.2.3 was released onto the backbone, which of possibly thousands of private Internets that have a node with that address would it be routed to? It's a moot point anyway since most border routers (between your subnet and the backbone) would never let that packet pass onto the backbone. Private addresses are *local* to your private Internet. Occasionally packets with internal addresses do "leak" onto the Internet backbone (due to misconfigured border routers). They are like lost souls that haunt the Internet for a short while, unable to find any useful routing information, until they finally "cross over" (die) when they reach their hop limit, or some other backbone router kills them ("what's THIS doing here?"). This is one reason to use only RFC 1918 addresses in private Internets instead of public addresses. If public addresses from a private Internet leak onto the Internet backbone, they can cause conflicts.

I say *private Internet*, instead of *private subnet*, since you can actually have multiple subnets behind a NAT gateway, separated from each other by routers (without NAT). Those subnets would constitute a private routing domain. Packets from any node in any of those domains could be sent to the global Internet via the one NAT gateway, and the reply makes its way back to the sender. Within the routing domain, all traffic is bidirectional (from any node to any node, regardless of subnet). The one-way limitation comes only at the NAT gateway. All these internal subnets use private IPv4 addresses. For example, you could have subnets 172.20.0.0/16, 172.21.0.0/16, and 172.22.0.0/16 connected together with routers, all behind a single NAT gateway.

Public IPv4 Addresses

There are a few blocks of the IPv4 address space reserved for special purposes:

Loopback	127/8	127.0.0.1 - 127.255.255.255
RFC 1918	10/8	10.0.0.1 - 10.255.255.254
	172.16/12	172.16.0.1 - 172.31.255.254
	192.168/16	192.168.0.1 - 192.168.255.254

RFC 6598 (CGN)	100.64/10	100.64.0.1 – 100.127.255.254
Multicast	224/4	224.0.0.0 – 239.255.255.255
Experimental	240/4	240.0.0.0 – 255.255.255.255

All other IPv4 addresses are *public* and can be allocated to users. Actually, most of these *have been* allocated.

A packet with a *public* address can be globally routed. That means it can be delivered to any node in the world that is connected to the public IPv4 Internet via a public IPv4 address (not through NAT). A node with a public address can make *outgoing* connections to any other node on the public IPv4 Internet that has a public address and accept *incoming* connections from any other node on the IPv4 Internet, even ones behind NAT. That is a first-class netizen. Before 1995 or so, everyone was a first-class netizen. We had a monolithic global Internet – any node could connect to any other node (subject to port blocking by routers or firewalls).

The *public IPv4 Internet* used to be the entire Internet (before NAT). Today it is only a small part of the total IPv4 Internet. It consists of those nodes connected directly to the IPv4 backbones, as opposed to via NAT gateways. Every such node has a public IPv4 address. The remaining public IPv4 Internet exists mostly in telcos, ISPs, hosting facilities, and cloud providers (AWS, Azure) today and in smaller pieces in corporate networks. As any example, at home I have ONE public IPv4 address, or about 0.000000023% of the total theoretical address space. My company has SIX public IPv4 addresses (for which we have to pay a lot). In comparison, several companies that got "class A" blocks in the early days each have about 16.7M addresses, or about 0.39% of the theoretical address space (AT&T and HP each have two class A blocks).

Network Address Translation

If an IPv4 private Internet is connected to the public IPv4 Internet, it must be via a "cone-mode" NAT gateway, which hides the internal private addresses behind a single globally unique public IPv4 address. Basically, a NAT gateway maps private IPv4 addresses to a single public IPv4 address for outgoing packets, in a reversible manner, so that the gateway can later route the reply back to the internal node that sent the outgoing packet. This is why incoming connections don't work for nodes behind NAT – the NAT gateway can't *undo* a mapping it never did.

There is another kind of NAT called "1:1 NAT," in which a single private IPv4 address is mapped (reversibly) to a single public IPv4 address. The reply packets go directly to the corresponding private node. Each such mapping requires one of your previous public IPv4 addresses. Any external node can make incoming connections to such internal nodes, using any port (subnet to port blocking in routers or firewalls). In my company, we have a few servers behind 1:1 NAT. They have internal addresses (e.g., 172.17.0.11), and connections to the associated public address (e.g., 66.96.216.18) get relayed to the internal address (e.g., 172.17.0.11). The reply comes from the private address but gets mapped to the public address. Outgoing connections from this node appear to be coming from the 1:1 public address for this node (66.96.216.18), not the general NAT public address.

Yet another kind of NAT called "port mapping" allows me to redirect incoming connections to a public address used for NAT, to a specific port, to one internal node. For example, I could map port 80 onto internal node 172.17.2.1. A given port (e.g., 80) can only be mapped onto a single internal node. But I could map incoming connections on port 25 to the same or a different internal node. So even with only a single public IPv4 address, I could run a server on internal nodes via port mapping. If the WAN interface of your ISP router has a private address on it, you are out of luck.

NAT Gateways Can Run Out of Port Numbers

NAT gateways do their mapping with "port shifting." For example, let's assume my node is at 172.20.2.1. If the source port on an outgoing packet is 20123, the IP address gets changed to the public IPv4 address, and the source port gets shifted to another port not currently in use (e.g., 30345). An entry in a NAT table keeps track of that mapping. When the reply comes back, the destination IP address will be the NAT public address, and the destination port will be 30345. The NAT gateway looks that port up and determines that the original packet came from 172.20.2.1 and the source port was 20123. It changes the IP address to 172.20.2.1 and the destination port to 20123 and delivers the reply to me.

A given node (in this case, the one running the NAT gateway) has about 30,000 ports available. Sounds like a lot until you realize there might be 1000 people using the NAT gateway (so there are about 30 ports per person). Some recent apps, like Google Maps, can use as much as 300 ports for one user (to improve performance). If a lot of your users are using such apps, you can use up all available port numbers on the NAT gateway. When a NAT gateway can't find an available port number, it just quietly drops packets. There is no indication to the user or even the NAT gateway admin that this has happened. But some of the traffic, for example, some part of the map, just disappears. This can be very difficult to debug.

The Need for Centralized Servers in the IPv4+NAT Internet

Since with NAT there are no incoming connections, nodes behind NAT gateways can only make outgoing connections, and only to nodes with public IP addresses. This means that to allow people in one private Internet to communicate with people in another private Internet, we must have *intermediary* servers that have public IPv4 addresses.

Alice's node cannot connect directly to Bob's if there is a NAT gateway between them. More typically BOTH of them are behind NAT; neither one of them can connect to the other. But Alice *can* make an outgoing connection to an intermediary server with a public address, and so can Bob. So they both connect to an intermediary server (which has a public address) and send messages back and forth through it:

Alice's Node -> Intermediary Server<- Bob's Node

Someone has to set up and run that intermediary server, and they must both have "accounts" on it. Accounts on intermediary servers are usually not free. Whoever runs that server may have access to their traffic (if they don't encrypt it end to end) and may have policies that limit what they can send (amount of traffic, types of traffic, etc). They might even charge them based on how much traffic they run through that intermediary node.

It's likely that Alice and Bob do not both have accounts on the same intermediary server, so there may be *two* intermediary servers and *three* links involved – Alice's node to her server, her server to and from Bob's server, and Bob's node to his server:

Alice's Node -> Alice's Server <-> Bob's Server <- Bob's Node

This is complicated to set up, and it forces centralization on the otherwise decentralized Internet. So we wind up with a small number of centralized intermediary servers (that may service thousands or millions of users). This affects reliability (if one of these servers goes down, it affects a lot of people, and even if they stay up, they can easily get overloaded) and privacy (someone could put snooping software on the intermediary servers and see all traffic going back and forth). Even if you use TLS to secure the links, on the intermediary servers the data is in plain text (unencrypted). You also lose authentication from the original sender to the final recipient. It is possible to do end-to-end privacy and authentication via multiple links, using S/MIME, but this requires issuing client digital certificates to every user and having software that supports S/MIME.

This situation seems "normal" to most people (even network engineers) since we've been using this approach since the mid-1990s and most people assume that's the way it's always been and always WILL be.

Carrier-Grade NAT (NAT444)

Very few Internet users today have global IP addresses on their nodes. At best they may have one on their router with all internal nodes hidden behind it with NAT44. Many home users now don't even have *one* public address in their subnet. They are behind *two* layers of NAT – the first NAT, at their carrier, maps between the public IPv4 Internet and a special private Internet using address range 100.64/10 (100.64.0.1–100.127.255.254). You might get a private address 100.100.35.72 on your WAN interface. The second NAT, at your site, maps from that private address to an RFC 1918 block (e.g., 172.16/16). Addresses in 100.64/10 are not public addresses; they are a new type of private address that only carriers (ISPs) can use. See RFC 6598,[1] "IANA-Reserved IPv4 Prefix for Shared Address Space." This two-layer scheme is called CGN (Carrier-Grade NAT[2]), or NAT444. So outgoing packets go through one NAT mapping to the ISP and another from the ISP to the public IPv4 Internet. Replies have to go back through two reverse mappings to get to your node. If you are behind CGN, you are not even a *second-class* netizen; you are a *third-class* netizen.

How do you tell if you are behind CGN? Try to do a traceroute to 4.2.2.2:

```
C:\Windows\system32>tracert 4.2.2.2

Tracing route to b.resolvers.Level3.net [4.2.2.2]
over a maximum of 30 hops:

  1    <1 ms    <1 ms    <1 ms  fw.sg.sixscape.net [172.17.0.1]
  2     1 ms     2 ms     1 ms  3-193-96-66.myrepublic.com.sg [66.96.193.3]
  3     2 ms     1 ms     2 ms  103-6-148-41.myrepublic.com.sg [103.6.148.41]
  4     2 ms     2 ms     2 ms  116.51.31.45
```

---

[1] https://tools.ietf.org/html/rfc6598
[2] https://en.wikipedia.org/wiki/Carrier-grade_NAT

```
5      2 ms      2 ms      2 ms   ae-8.r20.sngpsi07.sg.bb.gin.ntt.net
                                  [129.250.4.174]
6      2 ms      2 ms      2 ms   ae-1.r01.sngpsi07.sg.bb.gin.ntt.net
                                  [129.250.3.100]
7      2 ms      4 ms     14 ms   ae-1.a01.sngpsi07.sg.bb.gin.ntt.net
                                  [129.250.2.240]
8      *         9 ms     10 ms   ae-0.level3.sngpsi07.sg.bb.gin.ntt.net
                                  [129.250.8.46]
9     10 ms      9 ms      9 ms   b.resolvers.level3.net [4.2.2.2]
```

Trace complete.

If there is another private IP address after the first hop (from 100.64/10 or the RFC 1918 address ranges), you are behind CGN. In my case the second hop was to a public IPv4 address (66.96.193.3), which is at my ISP. No CGN for me!

Centralization on the IPv4 Internet

The Internet was designed to be as decentralized as possible. Complexity at the edge, simplicity at the core. But with NAT, I can't connect directly to you. We can both make only outgoing connections, so if we want to communicate, we both have to make outgoing connections to some intermediary server(s) with a public IPv4 address (es). Those servers will relay messages back and forth between us. In some cases, those centralized servers may handle thousands or even millions of users. It is expensive to run such servers - you need lots of bandwidth and computing power. Today such servers are run by ISPs, telcos, and "hosting" companies (e.g., Rackspace).

Centralization creates "single points of failure" that should not be there. If one of these centralized servers goes down, a LOT of people might lose service. In a decentralized model, if my node goes down, only I (and anyone trying to communicate with me) am out of luck. The rest of the world can go on happily communicating with each other. Decentralization is good for reliability and availability.

Centralization also makes it easier for certain kinds of people to snoop on everyone. Snoops only need probes in the small number of centralized nodes. With decentralization, they need to have probes on every node in the Internet (or every network segment). This is not practical. They can still monitor traffic on the giant backbones, but this requires exotic equipment with massive bandwidth and computing power. And even so, if the traffic in question never goes on the backbones, even they can't see it. With decentralization, traffic between Alice and Bob only needs to go over the shortest network path between them. If they both happen to be in the same LAN (e.g., working for the same company), that traffic might never leave the local LAN. If they are on the same LAN, why would they want to clutter up their company's link to the ISP with traffic going out from Alice and right back into Bob? What if that ISP link goes down? With centralized servers, our communication would stop, even if we were in the same room. Do you think that Skype is real peer-to-peer? What happens when you are chatting with a neighbor in the same LAN and the ISP link goes down? Your chat *stops*. Skype has the illusion of peer-to-Peer messaging (due to NAT traversal) but not the reality. With decentralized messaging over IPv6, there would be no need for traffic between us to go anywhere but over the shortest network path between us. If the ISP link goes down, we can keep communicating!

Centralization can also impact performance. Our ISP connection might be 20 Mbps, but our LAN runs at 1 Gbps. With centralized messaging, traffic between Alice and Bob has to go from Alice out our ISP connection, to some centralized server, back into through our ISP connection, and finally into Bob (or vice versa), even if they are sitting next to each other. Throughput between Alice and Bob is limited to the company's ISP speed, or 20 Mbps. If they communicate directly to each other, throughput can be 1 Gbps. If Alice is communicating with someone in another network, traffic *has* to go over our ISP connection and hence will be limited to 20 Mbps. Of course, bandwidth on the ISP link should be reserved for traffic to and from people in other networks, not for traffic between people sitting next to each other.

With IPv4 today, only telcos and ISPs have public addresses. Only they can run servers for mail, web, chat, etc. They don't want you to run servers at home, and with only a private address, you *can't* have a server that accepts incoming connections. You also can't make direct connections to other people in other networks. You can only make connections to centralized servers that have public IPv4 addresses. With web hosting today, your website might share the same public IPv4 address with hundreds or thousands of other websites. They have ways to map a particular nodename onto a shared IPv4 public address. But if you have a hosted server, try connecting to it with its IPv4 public address. You can't. It won't work. With IPv6, the hosting facility could have thousands of hosted nodes, each with a real, distinct global IPv6 address. No address sharing.

In particular, phones and other mobile devices with Internet connectivity have *never* had public addresses with IPv4. You can only run client software on them. A server requires a public IPv4 address. No phone has ever had a public IPv4 address. By the time we started providing Internet connectivity to phones, public IPv4 addresses were mostly gone. It would be handy to have an FTP or SFTP server on a phone, so people could send you files directly, from anywhere. Sorry, no can do (with IPv4). I can't have my phone connect directly to yours. We both have to make outgoing connections to a common server.

But with IPv6 on a phone, I can run *servers*. I can run *user agents* (these are apps that can make outgoing connections or accept incoming connections). If you have a user agent on your phone, I can connect directly to your phone from mine. No need for a telco in between.

NAT breaks a lot of protocols, like VoIP, IPsec, and IKE. If you use these protocols from behind NAT, you must use NAT traversal.

But Doesn't NAT "Protect" My Network?

Many people believe they are "safe" if no one can make incoming connections to their nodes through NAT. NAT does *not* add any security to a network. That is a myth. Incoming and outgoing connections are blocked by router or firewall mechanisms that block traffic on specific ports. NAT really just complicates this process. All you need to breach the NAT "protection" is an "inside man" (NAT traversal). Get someone to run an app that makes an outgoing connection to a STUN server, and you can do anything you want to their network using that app, from anywhere on the Internet.

Hackers can also *hide* behind NAT. If they attack someone from behind NAT, there might be hundreds or even thousands of other people who appear to be coming from the same address. When law enforcement tries to track down the culprit, they can only see the public address that the hacker was hiding behind. Which one of the thousands of users behind that public address was the culprit? Unless you can log *all* mappings done by the NAT gateway (very difficult to do), it is pretty much impossible to find the hacker in the haystack. With CGN it is even harder – there could be tens of thousands of people behind that one IP address (and two mappings to unravel). NAT is the hacker's friend. Europol (a European version of Interpol) has asked that ISPs stop using CGN.[3]

Today, NAT is so widely used in the Second Internet it should be called the InterNAT (a "top of the hat" to Latif Ladid, president of the IPv6 Forum, who I believe originated that term).

NAT Traversal: How Skype Fakes Incoming Connections

So, if IPv4 nodes behind NAT cannot accept incoming connections, how does Skype work? It sure *looks* like I connect from my node to yours, even if you are behind NAT. I don't log into an intermediary server with my copy of Skype.

Skype (and many other similar apps) uses something called *NAT traversal*[4] to fake incoming connections.

My copy of Skype makes a long-lasting outgoing connection to a special node called a STUN[5] server. This server is on the global IPv4 Internet and has a public address. It basically allows you to use *its* public address to accept an incoming connection and then relays it through your outgoing connection to the STUN server.

So who owns and runs STUN servers? I didn't sign up for one when I installed Skype. ANYONE can deploy a STUN server – hackers, the Mob, the NSA, pimply faced 14-year-old kids in their parents' basement, etc. Skype does a DNS SRV lookup and then chooses one *at random*. Your incoming information from other Skype users goes through that server. The traffic between your node and the STUN server may be encrypted with TLS, but it will be in plain text on the STUN server. Whoever runs and controls the STUN server can monitor your traffic all they want. They could even modify it, and you would never know. They might even be able to take over the Skype client at the other end of this outgoing connection and, from that, take over your entire node. Sounds scary?

---

[3] www.europol.europa.eu/newsroom/news/are-you-sharing-same-ip-address-criminal-law-enforcement-call-for-end-of-carrier-grade-nat-cgn-to-increase-accountability-online

[4] https://en.wikipedia.org/wiki/NAT_traversal

[5] https://en.wikipedia.org/wiki/STUN

IT IS. Many companies who are security conscious BAN Skype. It's not an application; it's a HACKING TOOL, because of NAT Traversal.

This is kind of like robbing a bank by having an inside man, who unlocks the bank door to let you in, after hours.

What if Everyone Had Public Addresses?

If Alice and Bob somehow magically managed to get some of the precious remaining public IPv4 addresses, they could connect directly to each other!

Actually, if both of them were in a single private Internet (with no NAT gateway between them), they could communicate directly with each other. But neither of them could communicate directly with anyone else in any *other* private Internet (and there are millions of private Internets all over the world – one behind every global IPv4 address). While interesting, this is probably not very useful. For instance, in a home network you would only be able to connect directly to other people in your home.

There is no way to reestablish the pre-1995 monolithic global Internet based on IPv4. Once you go to NAT and private Internets, there is no going back. Each year we create another billion or so new devices that need IP addresses. There are basically no IPv4 public ones left. With telephones, if they run out of phone numbers, they add another digit to the phone number (e.g., seven-digit to eight-digit numbers). That is basically what IPv6 is – only we went from 9-digit numbers to 38-digit numbers in one giant leap. There was no simple way to make the 32-bit addresses just a little bigger, without changing almost everything at the IP Layer.

And as long as they had the streets dug up making the addresses bigger, they fixed a LOT of stuff that could have been done better way back in 1981 when IPv4 was specified. IPv6 incorporates a great deal of valuable experience in creating and running a global Internet. You can think of IPv6 as IPv4++. There are enough global IPv6 addresses for every device on earth (or that we will make in the next 500 years) to have one. So anyone in the world can communicate directly with anyone else in the world, so long as both of them have IPv6 service and no router or firewall is blocking the relevant ports between them. The entire global IPv6 Internet is now one giant monolithic network with billions of nodes on it. This is revolutionary. Today most people don't "get" the importance of this. In a few years, we are not going to believe that we ever got along without IPv6 and decentralized connections. Your phone will have servers on it. It will be able to connect directly to any other phone on earth or accept incoming connections, without any intermediate server or NAT traversal.

IPv6: The NAT-Less Internet

With IPv6, there is no need for NAT (at least for NAT66, mapping IPv6 addresses to other IPv6 addresses). The *only* reason for deploying NAT44 was to extend the lifetime of the IPv4 address space for just a few more years despite running out of public IPv4 addresses, while the successor protocol (IPv6) was designed and deployed. It has done a remarkable job of that – we now have more than 20 billion devices connected to an Internet that has less than 3 billion public IP addresses. But in the process, it has splintered the global IPv4 Internet into millions of private Internets, broken many protocols, and made dangerous NAT traversal necessary. It has also greatly complicated the design and implementation of messaging apps.

While there is no need for NAT66, there is a potential need for something called NAT64,[6] which maps IPv6 addresses to IPv4 addresses, to allow IPv6 nodes to access legacy (IPv4-only) nodes. Unfortunately, NAT64 has all the problems of NAT44 and then some. For one thing it requires a nonstandard version of DNS called DNS64,[7] for all nodes that use NAT64. Other nodes cannot use DNS64. NAT64 breaks a number of protocols also. A lot of people are trying to make NAT64/DNS64 work since then you can get rid of IPv4 in the LAN and provide border translation to legacy (IPv4-only) nodes.

Note that the scheme used in NAT64 involves the DNS64 server embedding the IPv4 address in a fabricated IPv6 address, which the NAT64 server uses to connect to the real IPv4 node. This cannot be done in the other direction ("NAT46") since you can't fit a 128-bit address into a 32-bit address.

For the web there are better ways to translate IPv6 to IPv4 and vice versa that don't need DNS64 and don't have the preceding problems. These involve deploying a reverse web proxy in a dual-stack network. The existing A record in DNS is left intact, so IPv4 traffic goes directly to the legacy server (does not go through the proxy). For IPv6 a new AAAA record is created pointing to the reverse proxy. The reverse proxy makes an ongoing connection to the legacy server (over IPv4) and returns the reply over IPv6. Only IPv6 traffic goes through the reverse proxy. Unlike NAT64, this scheme works in both directions – you can make an IPv6-only server dual stack the same way – you leave the old AAAA record in DNS but add an A record pointing to the reverse proxy. Unfortunately, this approach does not help with non-web protocols.

---

[6] https://en.wikipedia.org/wiki/NAT64
[7] https://tools.ietf.org/html/rfc6147

In IPv6 there is no shortage of public addresses – hence no need to extend the life of the IPv6 address space. We can get rid of NAT once and for all for people who have IPv6 global addresses.

VoIP and IPv6

VoIP is a bit different from a protocol like SMTP. VoIP involves *two* protocols: SIP and RTP. SIP handles call setup (finding the other party), and RTP handles real-time encoding of digitized audio and/or video.

SIP[8] (Session Initiation Protocol, RFC 3261[9]) sets up the call. It normally works over TCP but can be used over UDP (well-known port 5060). When used over TCP, it can be secured with TLS (SIPS, well-known port 5061). In fact, SIP looks and acts a lot like SMTP. SIP does not handle any digitized audio.

RTP[10] (Real-Time Transport Protocol, RFC 3550[11]) carries the encoded, digitized audio or video. RTP works over UDP, so cannot be secured with TLS, but you can use SRTP[12] to provide encryption and/or authentication of the digitized audio or video. Encryption in SRTP uses an AES key exchanged with ZRTP[13] or MIKEY.[14] There is no well-known port for RTP (or SRTP) – the ports used are chosen dynamically and communicated via the signaling protocol (e.g., SIP). The IETF recommends ports 6970–6999.

This two-protocol scheme, and the dynamic allocation of ports in RTP, does not work well with NAT. SIP can go through NAT okay, but RTP (or SRTP) can't be mapped like other protocols. Therefore, to cross a NAT gateway, NAT traversal must be used. Of course, with IPv6, there is no issue with NAT. There is no need for the clients and the servers to be in the same subnet (or private Internet). Setting up the NAT traversal is by far the most difficult part of deploying VoIP. It also introduces significant security risks as described elsewhere.

---

[8] https://en.wikipedia.org/wiki/Session_Initiation_Protocol
[9] https://tools.ietf.org/html/rfc3261
[10] https://en.wikipedia.org/wiki/Real-time_Transport_Protocol
[11] https://tools.ietf.org/html/rfc3550
[12] https://en.wikipedia.org/wiki/Secure_Real-time_Transport_Protocol
[13] https://en.wikipedia.org/wiki/ZRTP
[14] https://en.wikipedia.org/wiki/MIKEY

There is also peer-to-peer SIP.[15] This does away with any need for "VoIP servers." P2P VoIP clients are not clients in the true sense of the word (only make outgoing connections) – they are user agents (can make outgoing connections *and* accept incoming connections). Of course, to accept incoming connections with IPv4 from external users, NAT traversal is needed on each user agent. With IPv6, every user agent has a public IPv6 address, which allows incoming connections from anywhere on the IPv6 global Internet.

If you use VoIP over IPv6 in a client/server model, there are a few clients and a few servers that support IPv6 (but many of the most popular ones don't support IPv6 currently). This is amazing considering how many of the worse problems with VoIP are solved by using IPv6.

One of the most popular VoIP clients (Bria[16] from CounterPath) supports IPv6 today. This is available for Windows, Android, and iOS.

With VoIP servers, recent Cisco VoIP products[17] support IPv6. Two popular commercial server products (3CX[18] and Ozeki[19]) do not currently support it. The open source Asterisk,[20] Kamailio,[21] and OpenSIPS[22] VoIP servers all support IPv6. It is not unusual for open source products to be ahead of commercial products in IPv6 support.

If you want your VoIP clients to be able to make calls to PSTN[23] (Public Switched Telephone Network) nodes (and accept calls from them), you need to add a SIP Trunk[24] on your VoIP server. There are SIP service providers that can help you connect your VoIP server to legacy telephones, for a price. Few if any of these currently support IPv6. The good news is that if your VoIP server supports IPv6 and is on a dual-stack server, you can support clients over IPv6, but relay the calls to and from the PSTN network over IPv4 via an IPv4 trunk. Of course, this means you can't run your VoIP server on IPv6-only.

---

[15] https://en.wikipedia.org/wiki/Peer-to-peer_SIP

[16] https://blog.counterpath.com/ipv6/

[17] www.cisco.com/c/en/us/td/docs/ios-xml/ios/ipv6/configuration/15-2mt/ipv6-15-2mt-book/ip6-voip.html

[18] www.3cx.com/

[19] www.ozekiphone.com/

[20] www.asterisk.org/

[21] www.voip-info.org/wiki/view/Kamailio

[22] www.opensips.org/

[23] https://en.wikipedia.org/wiki/Public_switched_telephone_network

[24] https://en.wikipedia.org/wiki/SIP_trunking

Skype

Skype is a chat, voice, and video messaging app that uses a proprietary protocol. Its protocol is heavily based on IPv4 NAT traversal. It was created by the same people who did Kazaa, which was a file sharing app that used NAT traversal to work. On Windows if you disable IPv4, Skype stops working and says there is no Internet connection (even though there may be a perfectly good IPv6 connection).

With iOS 9, Apple required all apps in the App Store to support IPv6[25] and, with iOS 10, required them to work on an IPv6-only device. Skype is available in the App Store, so they have found some way to make it work over IPv6. That way is NAT64/DNS64, which is available in iOS. This allows the phone to map an IPv6 address to IPv4 to make outgoing connections to other Skype nodes, via STUN servers. It is not real IPv6 support but appears to be close enough that Apple allows it in the App Store. Even though I have native IPv6 on my phone (from M1), I cannot verify that Skype for Mobile really works over IPv6 myself since I can't disable IPv4 on my phone or snoop on either end of the connection with a network sniffer (like Wireshark).

The NAT64/DNS64 dodge works for iOS 9 because there are still IPv4 addresses on a dual-stack phone. The requirement for iOS 10 is a different matter. There are no IPv4 addresses to map to IPv6 on an IPv6-only phone. Skype is still in the App Store in iOS 10 (and 11), so perhaps Apple gave them a pass.

---

[25] https://forums.appleinsider.com/discussion/193049/apple-says-all-apps-must-support-ipv6-only-networking-by-june

On Android[26] there is a mechanism similar to NAT64/DNS64 called 464XLAT,[27] specified in RFC 6877.[28] This can tunnel IPv4 traffic through IPv6 traffic on an IPv6-only phone. At the phone, 464XLAT maps IPv4 to IPv6, which goes over the IPv6-only data path to the telco, where they map it back to IPv4 and route it onto the legacy IPv4 Internet. So legacy apps that only support IPv4 should work on Android even with only IPv6 service. For 464XLAT there is no need for IPv4 addresses on the phone. It does not appear that Apple supports 464XLAT.

WhatsApp

WhatsApp is another widely used chat, voice, and video messaging app that uses a proprietary protocol. WhatsApp will work over IPv6 and in fact will work on an IPv6- only device. However, it is in the App Store, so it appears that Apple accepts it working via NAT64/DNS64. As with Skype, WhatsApp works over IPv6 on Android via 464XLAT. There is no native WhatsApp application for Windows (only something based on web that really uses your WhatsApp account on your phone), so there is no issue with the Windows version and IPv6.

Email over IPv6

Most email clients (e.g., Outlook, Thunderbird) work fine over IPv6 (even from an IPv6-only node). There are also plenty of email servers that support IPv6 (MS Exchange,[29] open source Sendmail,[30] Postfix,[31] Dovecot,[32] etc.).

Hosted email providers are another matter. Rackspace[33] does not currently support SMTP or IMAP over IPv6. I can test this by checking the MX records for our domain (which is hosted at Rackspace):

---

[26] www.internetsociety.org/blog/2013/11/skype-on-android-works-over-ipv6-on-mobile-networks-using-464xlat/

[27] https://sites.google.com/site/tmoipv6/464xlat

[28] https://tools.ietf.org/html/rfc6877

[29] https://en.wikipedia.org/wiki/Microsoft_Exchange_Server

[30] https://en.wikipedia.org/wiki/Sendmail

[31] https://en.wikipedia.org/wiki/Postfix_(software)

[32] https://en.wikipedia.org/wiki/Dovecot_(software)

[33] www.rackspace.com/en-sg

```
C:\Users\lhughes>nslookup
Default Server:  ws2008a.hughesnet-sg.org
Address:   2001:470:ed3d:1000::11

> set q=mx
> sixscape.com
Server:  ws2008a.hughesnet-sg.org
Address:   2001:470:ed3d:1000::11

Non-authoritative answer:
sixscape.com    MX preference = 20, mail exchanger = mx2.emailsrvr.com
sixscape.com    MX preference = 10, mail exchanger = mx1.emailsrvr.com

mx1.emailsrvr.com        internet address = 184.106.54.1
>
```

Note that mx2.emailsrvr.com (the secondary MX record for sixscape.com) also resolves to only an IPv4 address (173.203.187.2). So mail sent to sixscape.com over IPv6 has no way of reaching us. Soon that will not be acceptable to many people. It does support Webmail over IPv6, so all is not lost.

On the other hand, Gmail via SMTP and IMAP works fine over IPv6.

In general email works great over IPv6. So long as your email server is dual stack, you can exchange email with anyone, whether their server is IPv4-only, IPv6-only, or dual stack. It is possible for a dual-stack email server to publish MX records that make dual-stack email work very well. You can advertise two MX records with the IPv6 address of your email server as the highest priority (lowest-priority number) and the IPv4 address as the lowest priority (highest-priority number). Then other mail servers will try to deliver to you first over IPv6 but fall back to IPv4 if they can't. Currently if you create an IPv6-only email server, you will probably not get a lot of messages from others (only from those with servers that support IPv6).

The Future of Messaging on the Third Internet

The good news is that an IPv6-only messaging app is a *lot* easier to create than Skype or WhatsApp was (no need to find ways to get around NAT). It will work much better and will not require NAT traversal (so it will be much more secure). Of course, you will only be able to communicate with other IPv6 users (some might say those are the only people worth chatting with, but that is currently a minority viewpoint). You could always use the legacy apps (Skype or WhatsApp) to communicate with laggards stuck in the Second Internet. It's possible a real IPv6-compliant messaging app could support IPv4 users to at least some level, via a dual-stack intermediary gateway. Hopefully, we can finally use industry standard protocols that will interoperate with other services (like legacy telephones via SIP gateways). This means using SIP for call setup, RTP for encoded analog, and SIMPLE for chat. It also opens the door for decentralized end-to-end direct messaging (no intermediary servers).

Imagine an IPv6-only messaging user agent app on your home node or even your phone. Any other IPv6 user with a compatible user agent would be able to connect directly to your node and communicate with you (including file transfer). Even basic TLS would be end-to-end secure in that case (since there is only one link involved). With Strong Client Authentication with a client certificate during the TLS handshake, this is very strong security and mutual authentication. Traffic would only go over the shortest network path between the two communicating parties. It could be based on peer-to-Peer SIP[34] (and SIMPLE[35]), or I could just deploy a personal SIP server right on my phone for handling incoming connections (even more than one connection for group messaging). It could include ITU-compliant voice and even video. The same user agent app could be designed to also connect to legacy VoIP servers or gateways for when you want to talk with (or accept calls from) someone on a legacy phone or app. Of course, most legacy phones don't support video, chat, or file transfer.

And phone numbers[36]? Those are going away. With SIP, the preferred "address" of someone is a SIP URI,[37] for example, *sip:lhughes@leh.sixscape.com*. I can create a subdomain under sixscape.com (for free) just for me and publish an SRV record under it pointing to my user agent (with IP address and port number). I could even publish multiple SRV records, one for each node where I run chat user agents. If there are several with the same priority, other user agents could try connecting to all of them until the first one answers. When two nodes connect, they could exchange digital certificates (so you know for certain who is calling and who answers) and even exchange a symmetric

---

[34] https://en.wikipedia.org/wiki/Peer-to-peer_SIP
[35] https://en.wikipedia.org/wiki/SIMPLE_(instant_messaging_protocol)
[36] https://en.wikipedia.org/wiki/E.164
[37] https://en.wikipedia.org/wiki/SIP_URI_scheme

session key so the entire call can be encrypted. And where is the ideal place for your certificate and private key? On a computer, in a hardware security token, or on a phone, in the SIM!

How will someone reach you from a phone that only has a 12-key dial pad? There is a DNS-based scheme called ENUM[38] that can map numeric phone numbers to SIP URIs. So you can obtain a unique telephone number that can be mapped to your SIP URI by an ENUM server. There are VoIP gateways that know how to use ENUM. Again, my company has a carrier-grade DNSSEC appliance that supports ENUM.

I can have a *personal SMTP server* running on my home node or phone. There would be a local message store for incoming messages and a *personal IMAP server* that accesses the local message store, so a standard email client can work with it. Unlike traditional SMTP and IMAP servers, these need only support a single user. Spam would be a problem, so you might want to implement whitelisting (only accept incoming messages from a list of addresses – anyone else gets a message to request being whitelisted). Again, it would be best to create a personal subdomain for each user, for example, leh.sixscape. net. The email address could be *mailto:lhughes@leh.sixscape.net*. You could publish an MX record in your personal subdomain pointing to your home node or phone. You could optionally provide a secondary MX record to accept incoming messages when your node is offline. When you go back online, you could send an ETURN command to the backup mail server, which would deliver the held messages to you. There could be a new market for telcos or ISPs to provide backup SMTP servers. All SMTP servers will keep trying to deliver for several days, so you could just wait until the original server retries. This would be completely decentralized, and you would not need to worry about your messages being stored on intermediary servers (unless you provide a backup MX record).

I can even deploy a personal FTP or SFTP server running on my node or *even my phone*. Why not? I've got a public address and can publish it in DNS. Anyone could use standard FTP or SFTP clients (e.g., WinSCP) to send files to my file transfer server or retrieve files from my file transfer.

I can publish my name and SIP URI (as well as email address, organization, S/MIME certificate, etc.) in LDAP (with fine-grained control over who can access it), and every user agent can know how to search a list of LDAP servers for other people's contact info.

---

[38] https://en.wikipedia.org/wiki/Telephone_number_mapping

All we need now is a way to securely update your current IPv6 address in DNS. My company has created a way to do just that using our Identity Registration Protocol (IRP). With that, within seconds of entering a coffee shop that has IPv6 on Wi-Fi, my phone would configure a new IPv6 address in their subnet and then securely register that in DNS. If I use a short TTL on that registration, within minutes, anyone will be able to connect directly to my phone.

5G: The Grand Convergence of the Internet and Telephony

5G[39] is coming to a telco near you! It is not just the generation after 4G or just higher-bandwidth radio – it finally eliminates the legacy telco infrastructure (giant telco switches) and protocols (Signaling System 7). The underlying infrastructure for 5G telephony is now routers, switches, submarine optical cables, etc. You know, the INTERNET. Voice and chat are now done with VoIP (Voice over Internet Protocol). On 5G, there is no need for a "voice plan" (or minutes) or "SMS." You will only get a "dataplan." Everything will be done by apps (on desktop/notebook computers or mobile devices).

Historically mobile devices and desktop devices were fairly separate. That is coming to any end. With IPv6 in broadband networks, the same app architectures will work on both mobile and desktop, and interoperation will improve. You will no longer have apps that work only on mobile (like WhatsApp). The Grand Convergence is happening. A VoIP app on desktop can easily connect to a VoIP app on mobile. With Xamarin and WPF, it is getting easier to create a cross-platform app that covers Windows, MacOS, Android, and iOS. And with 5G, they can all have the same basic connectivity.

There is no strict mandate from 3GPP[40] to use *only* IPv6 with 5G, but NAT is an idea whose time has passed. VoIP does not work through NAT (without NAT traversal). It works *great* over IPv6. It is becoming more and more difficult and expensive for telcos to keep IPv4 alive *just one more year*. Many telcos (especially in the United States) are going to IPv6 (and even IPv6-only). This is saving them significant cost and is far easier to manage.

---

[39] https://en.wikipedia.org/wiki/5G
[40] www.3gpp.org/

One of the things holding the telcos back from deploying IPv6 was the problem with legacy IPv4-only apps (like Skype) from working on IPv6-only dataplans. 464XLAT and NAT64/DNS64 have solved that problem, on Android and iOS, respectively. There are *many* benefits to providing IPv6 to mobile devices. The migration is well underway. See the following:

www.apnic.net/community/ipv6-program/ipv6-for-mobile-networks/

www.telecomasia.net/content/china-mobile-outlines-ipv6-migration-plans

www.itnews.com.au/news/telstra-claims-success-with-ipv6-on-mobile-network-486322

www.apnic.net/wp-content/uploads/2017/01/vzw_apnic_13462152832-2.pdf

www.internetsociety.org/blog/2015/05/verizon-wireless-nears-70-ipv6-att-crosses-50-more/

www.internetsociety.org/blog/2016/08/facebook-akamai-pass-major-milestone-over-50-ipv6-from-us-mobile-networks/

www.telecompetitor.com/mobile-ipv6-milestone-more-than-half-of-mobile-requests-now-ipv6/

Because of NAT, current mobile device IPv4-based apps use a "hub and spoke[41]" design, with centralized servers. With IPv6, that will still work, but is no longer required. It is kind of like watching SD content on your new HD television set. As 5G is deployed, you will see more and more decentralized apps that go beyond the hub and spoke model.

The wireless service providers that adopt pure IPv6 will have significant competitive advantages over providers who try to continue supporting IPv4. With 464XLAT and NAT64/DNS64, there is no real downside to going with IPv6 only. As decentralized "5G-style" apps become popular, those will work only on IPv6 carriers. This will accelerate the migration to IPv6 for wireless providers.

Summary

In this chapter we discussed a number of aspects of messaging that will likely be undergoing major changes with the deployment of IPv6.

First, we covered exactly what *private* and *public* IP addresses are and why most Internet users have only private IP addresses today.

---

[41] https://en.wikipedia.org/wiki/Spoke%E2%80%93hub_distribution_paradigm

We then covered NAT (Network Address Translation) and NAT traversal (two nodes in different subnets connecting despite both subnets being behind different NAT gateways). Some messaging apps do support IPv6 today, while some never will (Skype).

We discussed what life is like in the Third Internet (based on IPv6) where there *is* no NAT to cause problems. Every node, including mobile devices, can now have public IP addresses and hence can run servers or even connect directly end to end with no intermediary servers. Imagine being able to send files securely over FTPS direct from my phone to yours.

We introduced 5G, which is much more than just "faster 4G" – it is the Grand Convergence of telephone communications and the Internet. Most 5G nodes will have full IPv6 support. However, even many 4G nodes already have IPv6 support (as seen in Chapter 9).

Face it – so long as you have only IPv4, you are going to be a second-class netizen. If you want to use the cool new messaging apps, you'll need to have IPv6.

CHAPTER 12

IPv6-Related Organizations

There are quite a few international- and national-level organizations involved in making this transition from the Second Internet to the Third Internet work. This chapter lists the most prominent ones but does not claim to be comprehensive.

Internet Governance Bodies

The first group of organizations helps govern the Internet. There is no Internet Corporation or any UN Internet Authority. The Internet is something quite different from the kinds of entities most people are familiar with. Its ownership and management is as decentralized and transnational as the physical implementation of the Internet itself. For example, what country is the Internet located in? All of them!

Anyone who really wants to can join one of these organizations, and the various groups address a variety of aspects of creating the standards that others use to build the hardware and software that make up the physical Internet. Other groups help manage resources, such as domain names or Internet addresses. Others help resolve disputes and set policies that help the millions of owners of the various pieces of the Internet to get along and be willing to continue voluntarily connecting their networks to each other. Some national governments try to control or regulate the Internet, or the content on it, but the highly decentralized nature of it, and the difficulty of even pinning down what jurisdiction something on the Internet happened in, makes such control difficult at best. Any country whose people are forbidden access to the Internet is missing out on things that allow those who do have access to run circles around them competitively. It would be like a blind person and a fully sighted one having a sword fight or a race between someone on foot and someone in a Ferrari. It would all be over in seconds. China has had a very difficult time trying to maintain strict authoritarian communist rule while

© Lawrence E. Hughes 2022
L. E. Hughes, *Third Generation Internet Revealed*, https://doi.org/10.1007/978-1-4842-8603-6_12

enjoying the economic benefits of access to the Internet. They have tried to deploy what some call "the great firewall of China," but there are many ways for people who understand the technology to gain access to those parts of it that China doesn't want their people to see. If they are really good, they can do it without their government even being able to detect it.

So who *is* in charge? The easy answers are *no one* or *everyone*. It is possible to identify some organizations that clearly are in charge of some aspects of the Internet. Most of them are heavily involved in trying to help the users of the Internet survive a looming disaster (the depletion of the IPv4 address space) and migrate smoothly and safely to the wonderful new promised land, the Third Internet.

Internet Corporation for Assigned Names and Numbers (ICANN)

ICANN was formed in 1998, as a not-for-profit public-benefit corporation. Participants all over the world help keep the Internet secure, stable, and interoperable. ICANN does not try to control content on the Internet, stop spam, or control access to the Internet. It does do the following things:

- Oversee the generic top-level domain (gTLD) names and the country code top-level domain (ccTLD) names. They also oversee authorization of Internationalized Domain Names (IDNs) in various languages and scripts.

- Oversee operation of the DNS root servers.

- Draw up contracts with each Domain Name Registry.

- Oversee IANA (Internet Assigned Numbers Authority).

- Publish all corporate documents, bylaws, financial information, major agreements, policies, operating plan, and strategic plan, at www.icann.org/en/documents/.

- Hold monthly meetings of the ICANN board to address issues and set policy. The minutes of each meeting are made publicly available at www.icann.org/en/minutes/.

Internet Assigned Numbers Authority (IANA)

One of the key organizations with regard to both the old IPv4 addresses and the new IPv6 addresses is the Internet Assigned Numbers Authority (IANA). You can find their website at `www.iana.org`. They do the following things:

- Oversee the DNS Root Zone (creation and management of the generic TLDs and ccTLDs), as well as the *int* domain registry (for international organizations) and the *arpa* zone. The *arpa* TLD has several very important parts of DNS under it, such as the reverse zones for both IPv4 (*in-addr.arpa*) and IPv6 (*ip6.arpa*), plus the ENUM E.164 zones (*e164.arpa*). Some parts of these reverse hierarchies can be delegated to ISPs or even to end-user organizations, but IANA is in charge of the overall structure.

- Maintain the Interim Trust Anchor Repository (ITAR) for those parts of the Internet's domain space that have already been signed with DNSSEC. This is a temporary role until such time as the root of the entire domain space is signed. This is where you can find the public keys needed to verify DNSSEC signatures for signed DNS zones.

- Perform the top-level management of IPv4 addresses. IANA allocates giant blocks of IPv4 addresses (called "/8s") with about 16.7 million addresses each to the five Regional Internet Registries for the world, AfriNIC, APNIC, ARIN, LACNIC, and RIPE NCC. As of this writing, only 20 "/8" blocks are left to allocate (about 7.8% of the original 256). As we reach the "end days" for IPv4, the IANA will be the first to run out. They already have a plan for this. When they get down to five remaining "/8" blocks, they will allocate one of those to each of the five RIRs and then close shop (as far as allocation of IPv4 addresses goes). This will probably happen on or before September 2011, by best current estimates. The RIRs will probably run out within 6 months after that. When they're gone, they're gone.

- Perform the top-level management of IPv6 addresses. They perform the same basic allocation function with IPv6 addresses that they have done for many years with IPv4 addresses. The main difference is that there are a *lot* more IPv6 addresses. They allocate giant chunks of IPv6 addresses to the RIRs as needed. It is unlikely that IANA will ever run out, so long as there is something recognizably TCP/IP. There are enough IPv6 addresses just in the 2000::/3 block marked for allocation for every human alive today to get over 5,000 of the standard allocation blocks, which are "/48s." Each "/48" is large enough for the biggest organization on earth.

- Manage AS numbers. AS stands for autonomous systems. It refers to complete networks at the top level of the routing hierarchy. Below the AS level, Interior Gateway Routing Protocols are used (e.g., RIP2, EIGRP, etc.). At the AS level, Exterior Gateway Routing Protocols are used (e.g., BGP4 and BGP4+). Each autonomous system network has a unique number. They have been using 16-bit numbers (which allowed 65,536 possible ASs). Just like with IPv4, we are running out of AS numbers, so they are in the process of changing to 32-bit AS numbers. That is causing some issues, but nothing like the issues related to changing from 32-bit IPv4 addresses to 128-bit IPv6 addresses. There is no worldwide "32-bit AS number" forum or any need for one. The people affected are fairly savvy technically and are simply making the changeover quietly.

- Allocate and assign IPv4 and IPv6 multicast addresses.

- Allocate and assign IPv6 anycast addresses.

- Allow people to reserve and register port numbers and other assigned numbers related to Internet protocols.

IANA is heavily involved in promoting the adoption of IPv6 throughout the world. They know how close they are to the bottom of the barrel with IPv4 addresses. They encourage the Regional Internet Registries to promote the adoption of IPv6, and each of them is doing this.

Regional Internet Registries (RIRs)

There are five top-level registries directly below IANA, who set address allocation policy for their region and allocate blocks of both IPv4 and IPv6 addresses to ISPs and other interested parties. One way to obtain addresses is to join one of the registries and apply for addresses. Some regional registries charge for these; others provide them free. You can only obtain addresses from the registry in the region where you reside or where the HQ of your organization is based.

Each Regional Internet Registry provides the following services for Internet users in their part of the world:

- IPv4 and IPv6 address space allocation, transfer and record maintenance.

- Autonomous system (AS) number allocation, transfer and record maintenance.

- Provide online directories of registration transaction information (WHOIS database).

- Provide online information about routing (Internet Routing Registry).

- Management of reverse DNS for addresses assigned by the RIR.

- Hold periodic meetings and elections.

- Perform education and training on relevant topics (such as IPv6).

- Maintain policy discussions on email lists, conduct public policy meetings, and publish policy documents on their website.

The three largest RIRs (ARIN, RIPE NCC, and APNIC) are all aggressively advocating for adoption of IPv6. Like IANA, they know how many addresses are left and how rapidly they are being allocated. They know that the "end times" for IPv4 allocation are near. All are strongly encouraging all ISPs and organizations that obtain addresses from them to begin adoption of IPv6 *now*. If the major oil companies told people that there was not going to be any gas for new cars made after a certain date (less than 2 years off), there would be a mad scramble to create and sell cars that ran on something else. This is just as big a deal and, according to OECD, will have very serious economic consequences for every country and organization that has not prepared for the end of IPv4 allocations.

The five Regional Internet Registries and their coverage areas are as follows.

American Registry for Internet Numbers (ARIN): www.arin.net

ARIN provides services to Internet users in North America (including the United States, Canada, plus many Caribbean and North Atlantic islands).

ARIN runs an IPv6 wiki at *www.getIPv6info.info*. This site includes book reviews, self-education, IPv6 presentations and documents, survey results, planning information, management tools, etc.

On May 7, 2007, the ARIN Board of Trustees passed the following resolution:

RESOLUTION OF THE BOARD OF TRUSTEES OF ARIN ON INTERNET PROTOCOL NUMBERING RESOURCE AVAILABILITY

WHEREAS, community access to Internet Protocol (IP) numbering Resources has proved essential to the successful growth of the Internet; and,

WHEREAS, ongoing community access to Internet Protocol version 4 (IPv4) numbering resources cannot be assured indefinitely; and,

WHEREAS, Internet Protocol version 6 (IPv6) numbering resources are available and suitable for many Internet applications,

BE IT RESOLVED, that this Board of Trustees hereby advises the Internet community that migration to IPv6 numbering resources is necessary for any applications which require ongoing availability from ARIN of contiguous IP numbering resources; and,

BE IT ORDERED, that this Board of Trustees hereby directs ARIN staff to take any and all measures necessary to assure veracity of applications to ARIN for IPv4 numbering resources; and,

BE IT RESOLVED, that this Board of Trustees hereby requests the ARIN Advisory Council to consider Internet Numbering Resource Policy changes advisable to encourage migration to IPv6 numbering resources where possible.

Implementation of this resolution will include both internal and external components. Internally, ARIN will review its resource request procedures and continue to provide policy experience reports to the Advisory Council. Externally, ARIN will send progress announcements to the ARIN community as well as the wider technical audience, government agencies, and media outlets. ARIN will produce new documentation, from basic introductory fact sheets to FAQs on how this resolution will affect users in the region. ARIN will focus on IPv6 in many of its general outreach activities, such as speaking engagements, trade shows, and technical community meetings.

Réseaux IP Européens Network Coordination Centre (RIPE NCC): www.ripe.net

RIPE NCC provides services to Internet users in Europe, the Middle East, and Central Asia. This includes

- *Southwest Asia*: Azerbaijan, Bahrain, Cyprus, Georgia, Iran, Iraq, Israel, Jordan, Lebanon, Saudi Arabia, Syria, Turkey, UAE, and Yemen

- *Central Asia*: Kazakhstan, Kyrgyzstan, Tajikistan, Turkmenistan, and Uzbekistan

- *Europe*: Albania, Armenia, Austria, Belarus, Belgium, Bosnia-Herzegovina, Bulgaria, Croatia, Czech Republic, Denmark, Estonia, Finland, France, Germany, Greece, Hungary, Iceland, Ireland, Italy, Latvia, Lithuania, Macedonia, Moldova, Montenegro, Norway, Netherlands, Poland, Romania, Russia, Serbia, Slovakia, Spain, Sweden, Switzerland, Turkey, Ukraine, United Kingdom, and Yugoslavia

- *North America*: Greenland

RIPE NCC runs the "IPv6 Act Now" site (www.ipv6actnow.org) with lots of information on IPv6 for small businesses, enterprises, ISPs, and governments.

On October 26, 2007, RIPE NCC issued the following warning, which is included here verbatim:

During the RIPE 55 meeting in Amsterdam, the RIPE community agreed to issue the following statement on IPv4 depletion and the deployment of IPv6.

"Growth and innovation on the Internet depends on the continued availability of IP address space. The remaining pool of unallocated IPv4 address space is likely to be fully allocated within two to four years. IPv6 provides the necessary address space for future growth. We therefore need to facilitate the wider deployment of IPv6 addresses.

While the existing IPv4 Internet will continue to function as it currently does, the deployment of IPv6 is necessary for the development of future IP networks.

The RIPE community has well-established, open and widely supported mechanisms for Internet resource management. The RIPE community is confident that its Policy Development Process meets and will continue to meet the needs of all Internet stakeholders through the period of IPv4 exhaustion and IPv6 deployment.

We recommend that service providers make their services available over IPv6. We urge those who will need significant new address resources to deploy IPv6. We encourage governments to play their part in the deployment of IPv6 and in particular to ensure that all citizens will be able to participate in the future information society. We urge that the widespread deployment of IPv6 be made a high priority by all stakeholders."

RIPE NCC issued another warning concerning IPv4 address space depletion, on April 10, 2008:

Currently, 180 of 256 blocks of "/8" have already been allocated. Of the remaining 76, 35 are already reserved for the Internet Engineering Taskforce (IETF) and the remaining 41 blocks are held in the Internet Assigned Numbers Authority (IANA) pool for future allocation to the RIRs.

As IPv6 provides the necessary address space for future growth, RIPE NCC is urging business and government leaders to ease the path for wider deployment of IPv6 addresses. Failure to adopt these new resources could mean a slowing in the pace of Internet innovation.

"Now is the time to recognize that sustainable growth of the IPv4-based Internet is coming to an end, and that it is time to move on, with IPv6 ready as the successor.

"In order to sustain the impressive speed of Internet innovation and ensure a healthy Internet economy for the future, we recommend that content providers make their services available over IPv6," comments Axel Pawlik, Managing Director at RIPE NCC.

"We view governments as key players in Internet growth and urge them to play their part in the deployment of IPv6 and in particular to lead by example in making content available in IPV6. Ultimately, we urge that the widespread deployment of IPv6 be made a high priority by all stakeholders."

When CIOs make firm decisions to deploy IPv6, the process is fairly straightforward. Staff will have to be trained, management tools will need to be enhanced, routers and operating systems will need to be updated, and IPv6-enabled versions of applications will need to be deployed. All these steps will take time.

The move to IPv6 will provide billions of further addresses through 128-bit addressing, which allows 50 billion, billion addresses for every person on the planet. Islands of IPv6 are already in use, but RIPE NCC argues that infrastructure support must be addressed in time for IPv6 to fulfill its predicted role as the catalyst for the next stage of Internet development.

Pawlik concludes: "We have well-established, open and widely supported mechanisms for Internet resource management and we're confident that our Policy Development Process meets and will continue to meet the needs of all Internet stakeholders through the period of IPv4 exhaustion and IPv6 deployment. The immediate challenge lies in making content available in IPV6 using the processes and mechanisms already available to ensure that service providers and content providers build adequate experience and expertise in good time."

Note that this warning was in 2008, and at that time 41 "/8" blocks remained. Today, no more /8 blocks are available at the IANA level.

Asia Pacific Network Information Center (APNIC): www.apnic.net

APNIC provides service to Internet users in

- *South Asia*: Afghanistan, Bangladesh, Bhutan, India, Nepal, Pakistan, and Sri Lanka

- *Eastern Asia*: China, North Korea, Hong Kong, Japan, Macau, Mongolia, South Korea, and Taiwan

- *Southeast Asia*: Cambodia, Indonesia, Laos, Malaysia, Myanmar, Philippines, Singapore, Thailand, and Vietnam

- Australia and New Zealand

- *Oceania*: Various islands in Polynesia, Melanesia, and Micronesia

APNIC is currently running a program called "Kickstart IPv6," in which anyone that owns or obtains an IPv4 address allocation from APNIC can get a free block of IPv6 addresses. If their IPv4 block is less than a "/22," then the IPv6 block is a "/48." For IPv4 blocks from "/22" and up, the free IPv6 block is a "/32" (this is 4 billion times 4 billion times the size of the entire IPv4 address space). You could also look at this as 65,536 "/48" blocks. These addresses are not tied to any ISP and can be routed from anywhere. There is no demonstration of need required for obtaining the IPv6 address block.

APNIC also runs an IPv6 resource site at *http://icons.apnic.net/display/ipv6/ Home*. The name "icons" stands for *Internet Community of Online Networking Specialists*.

Latin American and Caribbean Network Information Center (LACNIC): www.lacnic.net

LACNIC was started in 2002. It provides services to Internet users in

- *North America*: Mexico

- *Central America*: Costa Rica, El Salvador, Guatemala, Honduras, Nicaragua, and Panama

- *South America*: Argentina, Belize, Bolivia, Brazil, Chile, Columbia, Ecuador, French Guiana, Paraguay, Peru, Uruguay, and Venezuela

- *Caribbean Islands*: Aruba, Barbados, Cayman Islands, Cuba, Dominica, Dominican Republic, Grenada, Haiti, Jamaica, and various smaller islands

LACNIC runs an IPv6 resource site at *portalipv6.lacnic.net/en/portal-IPv6-2*.

Africa Region (AfriNIC): www.afrinic.net

AfriNIC provides services for Internet users in the entire African continent. It began in April 2005.

They run an IPv6 resource center at www.afrinic.net/IPv6 and an IPv6 virtual lab at www.afrinic.net/projects/cvl.htm. This is a test network with public access, with primarily Cisco equipment.

The Number Resource Organization (NRO): www.nro.net

NRO was formed in October 2003 by the four Regional Internet Registries that existed at the time, to formalize their cooperative efforts. Its goal is to protect the unallocated number resource pool, to promote and protect the bottom-up policy development process, and to act as a focal point for Internet community input into the RIR system. They run an IPv6 site at www.nro.net/ipv6.

Recently NRO issued the following statement, when the remaining IPv4 address pool dropped below 10%:

"This is a key milestone in the growth and development of the global Internet," noted Axel Pawlik, Chairman of the NRO. "With less than 10 percent of the entire IPv4 address range still available for allocation to RIRs, it is vital that the Internet community take considered and determined action to ensure the global adoption of IPv6. The limited IPv4 addresses will not allow us enough resources to achieve the ambitions we all hold for global Internet access. The deployment of IPv6 is a key infrastructure development that will enable the network to support the billions of people and devices that will connect in the coming years," added Pawlik.

Internet Architecture Board (IAB): www.iab.org

The IAB is chartered both as a committee of the Internet Engineering Task Force (IETF) and as an advisory body of the Internet Society (ISOC). Its responsibilities include architectural oversight of IETF activities, Internet Standards Process oversight and appeal, and the appointment of the RFC Editor. The IAB is also responsible for the management of the IETF protocol parameter registries.

Internet Engineering Task Force (IETF): www.ietf.org

The mission of the IETF is to make the Internet work better by producing high quality, relevant technical documents that influence the way people design, use, and manage the Internet.

The IETF

- Runs numerous working groups on technical topics relevant to the Internet that are the main source of RFCs

- Oversees the standards process

- Maintains the Internet Drafts and the RFC Pages

- Holds periodic meetings (fall, spring, and summer, each year)

- Runs various mailing lists, which anyone can subscribe to

Internet Research Task Force (IRTF): www.irtf.org

To promote research of importance to the evolution of the future Internet by creating focused, long-term and small Research Groups working on topics related to Internet protocols, applications, architecture and technology.

Internet Society (ISOC): www.isoc.org

The Internet Society (ISOC) is a nonprofit organisation founded in 1992 to provide leadership in Internet related standards, education, and policy. With offices in Washington D.C., USA, and Geneva, Switzerland, it is dedicated to ensuring the open development, evolution and use of the Internet for the benefit of people throughout the world.

The Internet Society provides leadership in addressing issues that confront the future of the Internet and is the organisational home for the groups responsible for Internet infrastructure standards, including the Internet Engineering Task Force (IETF) and the Internet Architecture Board (IAB).

The Internet Society acts not only as a global clearinghouse for Internet information and education but also as a facilitator and coordinator of Internet-related initiatives around the world. For over 15 years ISOC has run international network training programs for developing countries and these have played a vital role in setting up the Internet connections and networks in virtually every country connecting to the Internet during this time.

The Internet Society has more than 100 organisational and more than 28,000 individual members in over 80 chapters around the world.

IPv6 Forum Groups

There are many groups organized specifically to advocate for the adoption of IPv6, given the importance of the issue. There is an international umbrella group, called the IPv6 Forum, chaired by Latif Ladid, who wrote the foreword to this book. Their website is at *www.ipv6forum.org*.

Local IPv6 Forum Chapters

There are local chapters of the IPv6 Forum in many countries. Some of these national groups use the term *Forum* (e.g., IPv6 Forum Downunder, at *www.ipv6forum.org.au*). Some use the term *Task Force* (e.g., North American IPv6 Task Force, at *www.nav6tf.org*). Some use the term *Council* (e.g., the German IPv6 Council, at *www.ipv6council.de*). Altogether there are currently 58 national or regional groups under the international

IPv6 Forum. These groups advocate within their own country or region for the adoption of IPv6 and put on conferences usually called *IPv6 summits*. There are links to all of the chapters on the IPv6 Forum international site (`www.ipv6forum.org`), as well as announcements about coming summits and other IPv6-related events.

IPv6 Ready Logo Program

Affiliated with the IPv6 Forum is a group whose goal is to do testing of IPv6 equipment and applications, ISPs who offer IPv6, and websites that are available over IPv6. This testing and issuing of certifications is done under the IPv6 Ready Logo Program. Their website is at `www.ipv6ready.org`. There are three main parts to the IPv6 Ready Logo Program: Products, ISP, and website.

IPv6-Ready Product Testing and Certification

Product testing uses test suites developed by TAHI (part of the Japan WIDE project) and IPv6-ready test labs. This is overseen by the IPv6 Ready Logo Committee (v6LC). There are both phase 1 ("Silver") tests, which verify behavior of the MUST clauses of all relevant RFCs, and phase 2 ("Gold") tests, which verify behavior of both the MUST and the SHOULD clauses of all relevant RFCs. The hundreds of products that have passed these tests are published on the IPv6 Ready site at

- `http://cf.v6pc.jp/logo_db/approved_list_ph1.php`

- `http://cf.v6pc.jp/logo_db/approved_list_ph2.php`

There are several categories of test suites currently. The IPv6 Ready Logo can be obtained for passing the *Core Protocols* tests, which include both *Conformance* and *Interoperability* tests. There are advanced tests in the following areas:

- *IPsec*: End-Node and Security Gateway

- *IKEv2*: End-Node and Security Gateway

- *Mobile IPv6*: Correspondent Node, Home Agent, and Mobile Node

- *NEMO*: Home Agent and Mobile Router

- *DHCPv6*: Client, Server, and Relay Agent

- *SIP*: UA, Endpoint, B2BUA, Proxy, Registrar

- *SNMP*: Management (SNMP-MIBs) – Agent and Manager
- *MLDv2*: Multicast Listener Discovery protocol (version 2)

The current test sites include

- BII: Beijing Internet Institute (People's Republic of China)
- CableLabs (United States)
- CHT-TL: ChungHwa Telecom Labs (Taiwan)
- CNLabs (India)
- IRISA: Institut de Recherché en Informatique et Systemes Aleatories (European Union)
- TEC (India)
- TTA: Telecommunication Technology Association (Korea)
- UNH-IOL: University of New Hampshire InterOp Lab (United States)

IPv6-Enabled ISP and Website Certification

Information on how an ISP or a website can be certified as delivering IPv6-compliant service is available at

- www.ipv6forum.com/ipv6_enabled

The ISP certification process was created by the Beijing Internet Institute (BII). There is currently a basic level. The advanced level will be introduced shortly. The list of certified ISPs is available at

- www.ipv6forum.com/ipv6_enabled/isp/approval_list.php

Notably, Malaysia has taken this even further and has three levels of ISP certification, which has been mandated by the Malaysian government:

- *Phase 1*: Basic network connectivity tests
- *Phase 2*: Interconnectivity tests
- *Phase 3*: Commercial and advanced network services

In 2010, 12 ISPs had already passed the first two levels and were working on passing the third.

For websites, again there is a basic level currently available and an advanced level coming soon. The list of certified websites is available at

- *www.ipv6forum.com/ipv6_enabled/approval_list.php*

Informal IPv6 Network Administration Certification

Hurricane Electric, in addition to providing free tunneled service via 6in4, 6to4, and Teredo, has put together an informal, self-administered certification program for IPv6 network administration. This covers aspects of IPv6 technology and implementation on various platforms (Linux, Windows, Cisco routers, etc.). There are several levels. At each succeeding level, you must answer harder questions. Several levels involve accomplishing actual network administration tasks, such as deploying an IPv6-compliant email server. To obtain that level, you must exchange an email with their IPv6 email server. The site is available at *ipv6.he.net/certification/*. There are multiple levels, including Newbie, Explorer, Enthusiast, Administrator, Professional, Guru, and Sage.

This is not a formal program, like Microsoft or Cisco certification, but it is free and very educational. I have already qualified at the top level, as an IPv6 Sage. Here is my very cool certification badge:

WIDE Project, Japan

Japan was an early leader in IPv6, and a consortium of Japanese IT companies headed by Professor Jun Murai has done some very important work that has significantly advanced the state of IPv6. These include reference implementations of the IPv6 stack for BSD (Kame) and Linux (USAGI), the test suites for the IPv6 Ready testing program (TAHI). They also provide IPv6 service to many developing countries in the region, such as the Philippines. Their website is available at *www.wide.ad.jp*.

The Kame project completed its task and has been discontinued, but its website (`www.kame.net`) has been left up, complete with its famous turtle logo (kame is Japanese for turtle). If you connect over IPv4, the turtle just sits there. If you connect over IPv6, the turtle dances. In the early days of IPv6, it was a rite of passage to verify you had accomplished IPv6 connectivity by watching the kame dance.

Summary

This chapter covered some of the major organizations involved in the management of IPv6 addresses and the rollout of IPv6 globally.

IANA is the top-level organization that allocates IPv6 addresses (they ran out of public IPv4 addresses in 2011, so they no longer allocate those).

Below IANA are the five Regional Internet Registries (RIRs): APNIC, ARIN, RIPE, LACNIC, and AfriNIC. All these have reached end of normal allocation for IPv4 public addresses. They still provide blocks of IPv6 addresses to telcos, ISPs, and other organizations.

There are several other organizations listed (NRO, IAB, IETF, IRTF, etc).

We also covered the IPv6 Forum, which was chartered by the IETF to oversee rollout of IPv6 globally and provide testing and training for it. I have been heavily involved with them since about 2004.

I finally mentioned a great program for free self-directed IPv6 training provided by Hurricane Electric, who also provides free tunneled IPv6 via 6in4 tunneling. Even if your ISP is not yet supporting IPv6, you can tunnel it into your network from them. See `https://tunnelbroker.net`.

CHAPTER 13

IPv6 Projects

There are various projects you can do for free, given the information in this book and open source components (or evaluation versions of Microsoft products) readily available on the Internet.

It is possible to do the open source implementations based on FreeBSD, NetBSD, OpenBSD, or various Linux flavors. Use the platform you are most familiar with. The BSD variants have a powerful dual-stack packet filtering component called *pf*. This can be used to add a host-based firewall to any project (to block access via anything but the desired protocols) or even to build a multi-NIC router or firewall with complex rules. In Linux, the equivalent component is called Netfilter/IP Tables. The BSD and Linux packet filtering components have roughly the same functionality, but totally different deployment and configuration schemes. Both have one part that lives in the Kernel Space and one part that lives in user space. The configuration of the IPv4 and IPv6 stacks is done in different ways, but the functionality is almost the same. Both the BSD and Linux IPv6 implementations have passed IPv6 Ready Gold testing (at least one release, possibly not the most recent). For the most part, other open source components (Apache, Postfix, Dovecot, etc.) are pretty much the same regardless of what underlying platform is used.

Microsoft Windows since version 7, Windows Server since 2008, and Exchange Server each have excellent support for IPv6 and dual-stack operation. You can put together a viable testbed network with just Microsoft products if you like (except for the gateway router/firewall) or all with just open source or mix and match. It all depends on your expertise and requirements.

Some open source components (e.g., SMTP MTA, POP3/IMAP mail retrieval agents) are available in a variety of popular implementations (Postfix, QMail, EXIM, Dovecot, Cyrus IMAP, etc.). Pretty much all these have support for IPv6, but in some cases, the specifics to actually deploy these in dual-stack mode may be difficult to locate. I will

© Lawrence E. Hughes 2022
L. E. Hughes, *Third Generation Internet Revealed*, https://doi.org/10.1007/978-1-4842-8603-6_13

recommend components that I have actually deployed and where I have verified dual-stack operation, but if you happen to prefer a different component, chances are the necessary configuration information is available online somewhere.

Each project has a basic level of functionality described and various extensions that can add more functionality (e.g., a basic router can be enhanced by adding packet filtering and/or proxies).

Accompanying Website

Rather than include these projects in this book, I have put these on the corresponding website, at `https://thirdinternet.com`. You can download the installation guides in PDF, and I can update them easily as new operating systems and releases of open source projects come out.

These include

- How to deploy a dual-stack firewall with pfSense, including 6in4 tunneling (I use a version of this in my home network)

- How to deploy Windows Server with dual-stack operation in AWS

- How to deploy FreeBSD with dual-stack operation, both standalone and in AWS

- Exploring IPv6 on your phone

Hurricane Electric IPv6 Certification

I also strongly recommend that you do the projects in the Hurricane Electric IPv6 Certification sequence. See `https://ipv6.he.net/certification/`. Among other projects, you will do the following:

- Configure IPv6 on your node.

- Connect to the IPv6 Internet.

- Deploy a working website available over IPv6.

- Deploy a working email server that accepts messages over IPv6.

- Deploy a working DNS server that supports IPv6.

- Configure a reverse DNS record for your IPv6 email server.

- Do network troubleshooting with ping and traceroute with IPv6 addresses.

They have automated tools to verify that the projects you deploy actually work.

SixConf

On the preceding website (and on `https://ipv6forum.com`), you can find a very useful free application (for Windows) that allows you to see (and completely control) the internal details of IPv6 addresses and configuration called SixConf. Here is a screenshot of the main window to give you an idea. A user guide is also available with full details. This will help you understand the information in this book. I have used this app when teaching IPv6 certification courses and find it really helps the students to understand what is going on. It is also useful when deploying IPv6 even in complex networks.

For whatever reason, Microsoft chose to provide configuration tools for IPv6 on their operating systems that look and act a lot like the ones for IPv4. This is kind of like providing a 747 with controls based on those in a family car. IPv6 is far richer and more complex, and this tool provides visibility into and control of these aspects.

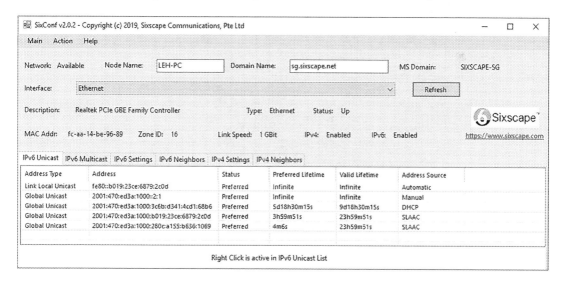

Note   We need to provide a way to download this from an Apress web page for this book.

Conclusion

If you have done all the exercises on the website, you now have a fairly complete dual-stack testbed network and are familiar with many of the things that you will need to do as a network administrator. Between the labs and the book, hopefully you now understand the following things:

- It is not particularly difficult to obtain free tunneled IPv6 service, even using free components. You do not need to wait for your ISP to provide IPv6 service to go fully operational. Simple transition mechanisms simplify the migration to full dual-stack operation. The only problem is you need at least one public IPv4 address to use 6in4. Failing that, 6rd is a reasonable substitute, but your ISP must provide it to you.

- Most operating systems and many existing network applications (BIND, Apache, Postfix, Dovecot, ssh/sshd, etc.) are already fully capable of supporting full dual-stack operation. Network configuration is not that different from IPv4.

- Most web applications (Apache or IIS based) get a "free ride," once the underlying web server has been migrated to dual stack. In addition (although not covered in these labs), most Microsoft ".Net" applications get a free ride.

- IPv4 NAT really doesn't provide *any* useful function other than extending the life of the IPv4 address space, and only then at a very high price (in terms of lost capabilities and additional complexity). It adds *no* security in firewall architectures. NAT is a crutch you no longer need. IPv6 without NAT actually provides a simpler, better firewall architecture (no need for BINAT, proxy ARP, NAT traversal, etc.). We are really just returning to the pre-NAT "classical" firewall architectures, not something new and untested.

- There are only a few really new concepts in IPv6 that current network administrators need to master, such as tunneling, Application Layer gateways, hexadecimal address representation, address scopes (e.g., link-local addresses), working without NAT, needing to provide Router Advertisement messages (for SLAAC to work), multicast and IPsec that actually work, etc. Everything else is remarkably similar to working with IPv4.

- The supply of IPv4 public addresses is *really* almost gone, and there is no alternative to this other than migration to IPv6. The timeline on this is sooner than most people realize. The four main RIRs have ended normal allocation of IPv4 to telcos, ISPs, and cloud providers, and the fifth one will soon. You have better be ready to support IPv6 if you want to keep your job (or have your organization continue operation) past that point.

Congratulations, and welcome to the Third Internet as its newest netizen!

Index

A

B

C

© Lawrence E. Hughes 2022
L. E. Hughes, *Third Generation Internet Revealed*, https://doi.org/10.1007/978-1-4842-8603-6

I, J

Printed in the United States
by Baker & Taylor Publisher Services